Corrosion in Concrete Structures

Scrivener Publishing
100 Cummings Center, Suite 541J
Beverly, MA 01915-6106

Publishers at Scrivener
Martin Scrivener (martin@scrivenerpublishing.com)
Phillip Carmical (pcarmical@scrivenerpublishing.com)

Corrosion in Concrete Structures

Integrating Advanced Technologies and Sustainable Practices

Edited by

Ruby Aslam
School of Civil and Hydraulic Engineering, Chongqing University of Science and Technology, Chongqing, China

Qihui Wang
School of Civil and Hydraulic Engineering, Chongqing University of Science and Technology, Chongqing, China

Zhitao Yan
School of Civil and Hydraulic Engineering, Chongqing University of Science and Technology, Chongqing, China

and

Afroz Aslam
Department of Chemistry, Constituent Government College, Thakurdwara, Moradabad, MJP, Rohilkhand University, Bareilly, U.P., India

Scrivener Publishing

WILEY

This edition first published 2026 by John Wiley & Sons, Inc., 111 River Street, Hoboken, NJ 07030, USA and Scrivener Publishing LLC, 100 Cummings Center, Suite 541J, Beverly, MA 01915, USA

For more information about Scrivener publications please visit www.scrivenerpublishing.com.

Wiley Global Headquarters
111 River Street, Hoboken, NJ 07030, USA

For details of our global editorial offices, customer services, and more information about Wiley products visit us at www.wiley.com.

The manufacturer's authorized representative according to the EU General Product Safety Regulation is Wiley-VCH GmbH, Boschstr. 12, 69469 Weinheim, Germany, e-mail: Product_Safety@wiley.com.

Library of Congress Cataloging-in-Publication Data

ISBN 978-1-394-34700-1

Cover image: Generated with AI using Adobe Firefly
Cover design by Russell Richardson

Set in size of 11pt and Minion Pro by Manila Typesetting Company, Makati, Philippines

Printed in the USA

10 9 8 7 6 5 4 3 2 1

Contents

Preface

Concrete is the backbone of modern infrastructure, forming the foundation of bridges, highways, buildings, and countless other structures worldwide. Yet despite its strength and versatility, concrete is highly susceptible to corrosion, leading to structural degradation, safety risks, and costly repairs. Addressing this challenge requires a multidisciplinary approach that integrates fundamental corrosion science with cutting-edge technologies and sustainable practices.

This book provides a comprehensive exploration of concrete corrosion, beginning with "Corrosion: Principles, Mechanism, and Key Insights" to establish a solid theoretical foundation. Then, the second chapter, "Concrete Material Properties and Corrosion," offers insights into the intrinsic factors that influence corrosion resistance. A detailed look at "Corrosion-Prone Environments in Concrete Corrosion," presented in the third chapter, is essential for predicting and mitigating damage under various exposure conditions.

To support proper maintenance and durability, the next chapter, "Corrosion Assessment and Inspection in Concrete Corrosion," presents state-of-the-art evaluation techniques. Then, "Impact of Concrete Corrosion on Environmental and Sustainability" highlights the broader implications of corrosion on environmental health and infrastructure longevity. The book also covers practical mitigation strategies, including "Role of Anti-Corrosion Coatings and Sealants in Preventing Concrete Corrosion," "Role of Cathodic Protection Systems in Preventing Concrete Corrosion," and "Role of Corrosion Inhibitors in Preventing Concrete Corrosion," offering effective approaches for extending the service life of concrete structures.

Real-world applications are featured in the chapter, *"Case Studies on Concrete Corrosion,"* where successful corrosion control efforts are analyzed. Next, *"Innovations in Concrete Corrosion Control"* explores emerging technologies such as nanomaterials, self-healing concrete, and AI-driven corrosion monitoring—offering a forward-looking perspective on sustainable infrastructure protection. Finally, *"Role of Nanotechnology in Concrete*

Corrosion Prevention" outlines the advantages of nanotechnology in safeguarding concrete from corrosion.

Designed for engineers, researchers, industry professionals, and academics, this book bridges the gap between theory and practice, offering valuable insights into both foundational principles and advanced applications. By fostering a deeper understanding of corrosion control strategies, we aim to contribute to the development of longer-lasting, safer, and more sustainable infrastructure.

We extend our sincere gratitude to the experts, researchers, and professionals whose work has informed and enriched this book. Special thanks to Martin Scrivener and the team at Scrivener Publishing for their guidance, support, and commitment to advancing the field. We also acknowledge the continued efforts of engineers and scientists dedicated to innovating corrosion prevention methods. It is our hope that this book serves as both a useful reference and a source of inspiration in the global effort to combat concrete corrosion.

The Editors
November 2025

1

Corrosion Fundamentals: Understanding the Science Behind the Damage

Kaoutar Zaidi[1], Walid Daoudi[2*], Selma Lamghafri[3], Omar Dagdag[4], Abdelmalik El Aatiaoui[2] and Abdelouahad Aouinti[1]

[1]University Mohammed Premier, Faculty of Sciences, Laboratory: Applied Chemistry and Environment (LCAE), Department of Chemistry, Oujda, Morocco
[2]Laboratory of Molecular Chemistry, Materials and Environment (LCM2E), Department of chemistry, Multidisciplinary Faculty of Nador, University Mohamed I, Nador, Morocco
[3]Laboratory of Applied Sciences, National School of Applied Sciences Al-Hoceima, Abdelmalek Essaadi University, Tetouan, Morocco
[4]Department of Mechanical Engineering, Gachon University, Seongnam, Republic of Korea

Abstract
This chapter explores the basics of corrosion, an electrochemical process that degrades materials, especially metals, on contact with the environment. It explains that corrosion involves redox reactions and describes the different types of corrosion, such as uniform, galvanic, and pitting. Accelerating factors such as humidity, chloride ions, and acidity are also discussed etc. The concepts of anode, cathode, electrochemical potential, and corrosion cell are explained in detail, providing a better understanding of the mechanisms involved. The role of electrolytes in corrosion cell formation is highlighted.

***Keywords*:** Corrosion, electrochemical process, redox reactions, corrosion types, environmental factors

**Corresponding author*: walid.daoudi@ump.ac.ma

Ruby Aslam, Qihui Wang, Zhitao Yan and Afroz Aslam (eds.) Corrosion in Concrete Structures: Integrating Advanced Technologies and Sustainable Practices, (1–12) © 2026 Scrivener Publishing LLC

1.1 Introduction

Metals and alloys are often used in service conditions that expose them to aggressive environments, making them vulnerable to corrosion in a variety of forms [1]. Knowledge on the fundamentals of corrosion is relevant to many industries from engineering to materials science, where it is well recognized that corrosion can have an unfavorable effect on structures and equipment. Corrosion can be defined broadly as the deterioration of materials caused by chemical and/or electrochemical reactions with the surrounding environment, which can generate safety and economic issues worldwide. There are many types of corrosive conditions that can develop based on the environmental conditions and the properties of the materials that are involved [2]. Corrosion can be in the form of uniform, galvanic, localized, and others, which can be affected by environmental conditions, materials, and surface morphology. The electrochemical reactions that are involved in the deterioration of metals underpin the understanding of corrosion, where metals degrade due to oxidation and reduction reactions taking place in the presence of moisture, oxygen, and other elements [3].

It is important to include the mechanical, physical, and chemical properties of metals. Mechanical and physical properties can be defined through constants; however, the chemical properties of a particular or specific metal are entirely dependent on the exact environmental conditions that the metal encounters while in use. The importance of mechanical, physical, and chemical properties is determined by the use of the metal. For example, railway tracks place importance on mechanical properties such as elasticity, tensile strength, hardness, and wear resistance, while electrical transmission relies primarily on electrical conductivity. Although a metal or alloy might often be chosen based on its mechanical or physical characteristics, in most applications, the interaction between the metal and its environment cannot be entirely disregarded [4]. However, the impact of this interaction differs in importance depending on the specific conditions. The impact of the interaction between a metal and its environment on the environment itself can be more critical than the metal's own degradation. For example, lead pipes are unsuitable for transporting plumbo-solvent waters, as lead concentrations above 0.1 ppm are toxic. Similarly, galvanized steel is not appropriate for certain food products due to the toxicity of zinc salts. In numerous chemical processes, the choice of metal is often driven by the need to prevent environmental contamination from metallic impurities that could alter the color or taste of products or trigger unwanted reactions. For instance, copper and its alloys are unsuitable for soap production because trace amounts of copper ions can cause discoloration and rancidity in the soap. In such cases, it becomes necessary to use unreactive and relatively costly metals, even if the environment would not cause significant deterioration of less expensive options like mild steel [5].

1.2 Basic Chemistry of Corrosion

1.2.1 Electrochemical Nature of Corrosion

Aqueous corrosion is fundamentally electrochemical, rooted in Michael Faraday's nineteenth-century principles of electrochemistry. Understanding corrosion and its prevention hinges on these principles. Each electrochemical corrosion cell necessitates four components: an anode (the corroding metal), a cathode (another conductor where environmental reactions occur), an electrolyte (aqueous environment facilitating ionic conduction between the anode and cathode), and an electrical connection (enabling electron flow between the anode and cathode). Typically, anodes and cathodes are close together, often on the same metal piece. Without any of these components, electrochemical corrosion cannot occur, so analyzing the corrosion cell is crucial for mitigation. Corrosion reactions can be divided into anode and cathode half-cell reactions for clearer understanding; the anode reaction involves the metal corroding and releasing ions into the electrolyte [6].

1.2.2 Chemical Reaction of Corrosion

The principles of corrosion as a heterogeneous chemical reaction refer to understanding how metals react with nonmetal environments (the reactants) at their interface, leading to the degradation of the metal. This approach simplifies the complex details of metal structures by focusing on the overall chemical reaction where the metal (A) and the nonmetal (B) reactants produce new compounds (C and D). The reaction can be represented by a simple chemical equation, highlighting that while many environmental factors may be present, only a few significantly influence the corrosion process. The nonmetallic reactants are often called the environment, but it is important to note that in a complex environment, the primary constituents might play a minor role in the reaction. For instance, in the atmospheric corrosion of steel, nitrogen makes up 75% of the atmosphere, yet its impact is negligible compared to that of moisture, oxygen, sulfur dioxide, solid particles, etc. However, in the high-temperature reaction of titanium with the atmosphere, nitrogen becomes a significant factor [7].

Electrochemical corrosion reaction equations contain symbols for electrons, reacting elements (ions and/or molecules), and ions or molecules produced by the corrosion reaction [8]. For example, the corrosion of iron is represented by the anodic electrochemical equation

$$Fe^{\circ} \rightarrow Fe^{+2}+2e^{-} \quad (1.1)$$

where Fe° represents iron atoms at the metal surface, Fe^{+2} represents iron ions, and $2e^{-}$ represents the two electrons produced by the anodic reaction. Equation 1.1 is referred to as an anodic half reaction because free electrons are produced.

The electrode at which chemical reduction occurs (or + current enters the electrode from the electrolyte) is called the cathode. Examples of cathodic reactions are:

$$H^{+} \rightarrow 1/2\ H_2 - e^{-} \quad (1.2)$$

$$Cu^{+2} \rightarrow Cu - 2e^{-} \quad (1.3)$$

$$Fe^{+3} \rightarrow Fe^{+2} - e^{-} \quad (1.4)$$

All of them represent reduction in the chemical sense.

The electrode at which chemical oxidation occurs (or + electricity leaves the electrode and enters the electrolyte) is called the anode. Examples of anodic reactions are:

$$Zn \rightarrow Zn^{+2}+2e^{-} \quad (1.5)$$

$$Al \rightarrow Al^{+3}+3e^{-} \quad (1.6)$$

$$Fe^{2+} \rightarrow Fe^{+3}+e^{-} \quad (1.7)$$

These equations represent oxidation in the chemical sense. Corrosion of metals usually occurs at the anode. Nevertheless, alkaline reaction products forming at the cathode can sometimes cause secondary corrosion of amphoteric metals, such as Al, Zn, Pb, and Sn, which corrode rapidly on exposure to either acids or alkalines [9].

1.3 Types of Corrosion

Corrosion can affect the metal in a variety of ways, which depend on its nature and the precise environmental conditions prevailing, and a broad classification of the various forms of corrosion. However, corrosion damage occurs in other ways as well, resulting, for example, in failure by cracking or in loss of strength or ductility. The five primary categories of corrosion are categorized based on their visual characteristics or changes in physical properties.

1.3.1 Uniform Corrosion

This is the most prevalent form of corrosion, characterized by uniform deterioration across the metal surface. It generally leads to a consistent reduction in material thickness [10]. When iron corrodes in acidic environments or aluminum in alkaline conditions, the entire surface can be compromised. The formation of protective oxide films on specific areas is not feasible, as these films cannot maintain stability in such liquids. Examples include oxidation and tarnishing, active dissolution in acids, anodic oxidation and passivation, as well as chemical and electrochemical polishing. Additionally, corrosion can occur both in atmospheric conditions and when immersed in certain environments. One approach to prevent this type of corrosion is to choose a material that forms a passive layer in the specific environment. Alternatively, conducting exposure tests before using the heat exchanger can help determine the corrosion rate (e.g., in mm/year), enabling the design of the heat exchanger with a thicker wall [11].

Figure 1.1 illustrates a cross-section of various tubes, revealing signs of corrosion on both the internal and external surfaces, along with noticeable

Figure 1.1 Cross-section of the tube exhibiting significant uniform corrosion and reduced wall thickness. Reproduced with permission from NACE International, from Xie *et al.* [13].

differences in wall thickness. The authors found that corrosion was most pronounced just below the top tube sheet. This was attributed to the build-up of non-condensable gases in the upper shell area, which led to an increased concentration of corrosive elements, such as chlorides, in the water at the vapor–liquid interface on the heated tube surfaces [12].

1.3.2 Pitting Corrosion

Highly localized corrosion occurs at specific sites, creating small pits that can penetrate the metal and potentially cause perforation. This is commonly seen in passive metals like stainless steels and aluminum alloys when exposed to certain ions, such as chloride ions (Cl^-) [10]. Metals that have a protective film either from spontaneous reactions with the environment or from surface treatments like coatings can still be prone to pitting corrosion [12]. At mechanical defects in the protective layer, pits can begin to form, and if they grow large enough, they may perforate the heat transfer surface. The initiation of pitting is affected by various factors, including metallurgical and structural characteristics, environmental conditions, polarization effects, and the presence of corrosion products. For stainless steel, factors such as high chloride concentration, elevated temperatures, and low pH levels enhance the risk of pitting corrosion. Other passivating metals commonly used in heat exchanger construction, such as aluminum, nickel, titanium, and their alloys, can also experience localized breakdown of their protective films, leading to pitting [14]. Figure 1.2 depicts a cross-section of a section where the lower and upper tubes were predominantly filled with brine and steam, respectively, while the middle tubes contained a mixture of condensing steam and brine.

Figure 1.2 Cross-section of a tube affected by pitting corrosion. Reproduced from Julian *et al.*, with permission from the International Geothermal Association [15].

1.3.3 Crevice Corrosion

Crevice corrosion is a severe and hazardous type of localized corrosion that occurs in environments that differ from the surrounding area. In crevices or shielded regions with stagnant liquid, oxygen depletion can take place. This differential oxygen concentration influences the initiation of crevice corrosion, and the presence of chlorides can further exacerbate the issue. As the crevice narrows, the rate of corrosion increases due to an expanded cathode area [16]. In plate heat exchangers, such crevices commonly occur at gaskets (where metal meets non-metal), at points of plate contact, and beneath deposits. An example of significant crevice corrosion in the gasket area of a plate made from alloy 31 (UNS N08031) is illustrated in Figure 1.3.

1.3.4 Intergranular Corrosion

The grain boundaries of a metal can be susceptible to corrosion, leading to damage along these boundaries instead of throughout the bulk material. This often results from improper heat treatment or impurities in the metal. When intergranular zones are more prone to attack than the grain interiors, intergranular trenches form. If corrosion affects most grain boundaries at the surface, the overall penetration and loss of strength may not be severe, though they will significantly exceed uniform corrosion rates. Typically, intergranular corrosion becomes pronounced when the liquid has a selective corrosive nature that targets grain boundaries while leaving the interiors relatively untouched. The electrochemical characteristics of intergranular attack emphasize the need for grain interiors to act as cathodes compared to the boundaries. Figure 1.4 from Ifezue and Tobins illustrates intergranular corrosion in an aluminum tube, commonly caused by inadequate heat treatment during production [18].

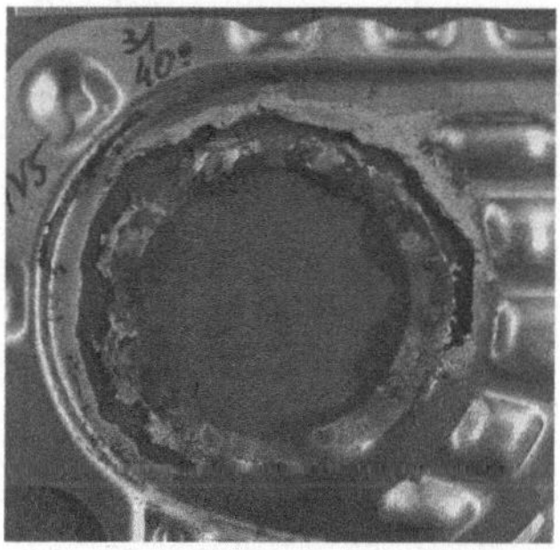

Figure 1.3 Crevice corrosion observed beneath the gasket of a plate heat exchanger. Reproduced with permission from NACE International, from Cassagne *et al.* [17].

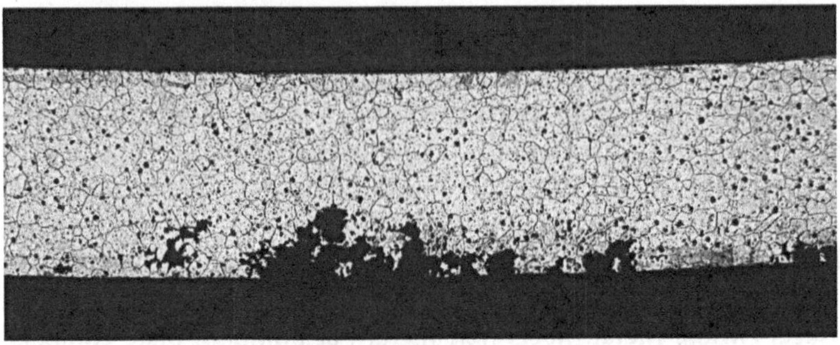

Figure 1.4 Intergranular corrosion in an aluminum tube. Reproduced with permission from Springer Nature: Journal of Failure Analysis and Prevention, Ifezue and Tobins, Copyright 2015 [18].

1.3.5 Stress Corrosion

When a corrosive environment is coupled with mechanical stresses, the corrosion process is intensified. One significant phenomenon that arises from this combination is stress corrosion cracking (SCC). This mechanism occurs when a static tensile stress is present in a susceptible metal within a corrosive environment, leading to failure even when the same structure would withstand the stress in a non-corrosive setting. These stresses can be external or arise from residual stresses, such as those from welding or cold working [16]. In shell-and-tube heat exchangers, residual stresses are commonly found at tube-to-tube sheet expanded joints and U-bends [14]. Additionally, residual stresses frequently contribute to SCC in cold-rolled plate heat exchangers [19]. When intergranular corrosion affects all or most of the intergranular paths that intersect with the metal's surface, both the penetration rate and the loss of strength will be relatively low for a given amount of metal loss.

Tawancy studied the failure of a shell-and-tube heat exchanger in an oil refinery, noting leaks in some bends just 48 h after the unit was restarted following downtime. The author concluded that polythionic acid formed from sulfur-bearing deposits and condensed water vapor during the shutdown. This acid, combined with internal stresses in the bends, led to cracking in the type 321 steel, as depicted in Figure 1.5 [20].

1.4 Electrochemical Corrosion Mechanisms

The electrochemical mechanisms of corrosion are based on redox (reduction–oxidation) reactions occurring on the surface of metals in contact with an

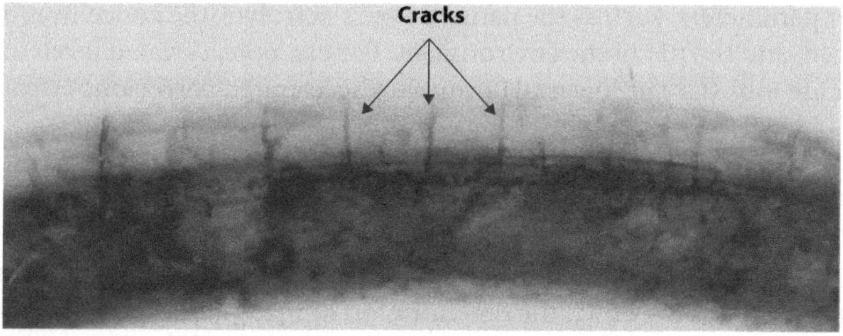

Figure 1.5 Stress corrosion cracks in a bend caused by polythionic acid. Reproduced from Tawancy with permission from Elsevier [20].

aggressive environment. This process begins with metal oxidation at the anode, where metal atoms lose electrons and are transformed into metal ions. For example, in the case of iron, the anodic reaction is as follows:

$$Fe \rightarrow Fe^{2+} + 2e^{-}$$

At the same time, at the cathode, the electrons released migrate through the metal and take part in reduction reactions, such as the reduction of oxygen in the presence of protons or water, depending on the medium. For example, in an acidic environment, the reduction reaction may be:

$$O_2 + 4H^{+} + 4e^{-} \rightarrow 2H_2O$$

These redox reactions create an electrochemical cell, where the anode corrodes while the cathode remains protected. An electrolyte, often an aqueous solution containing ions (such as seawater), is required to enable the transfer of ions between the anode and cathode. The electrolyte thus promotes the dissolution of metal ions and the transport of charges.

The corrosion cell is characterized by the progressive degradation of the metal at the anode. In some cases, such as galvanic corrosion, two different metals in contact accelerate this process: the more reactive metal (the one with the lower electrochemical potential) becomes the anode and corrodes more rapidly. In other forms of corrosion, such as pitting, small anodic zones form locally, creating holes or crevices on the metal surface. The electrochemical potential between anode and cathode determines the rate of corrosion—that is, the higher the potential difference, the greater the rate of corrosion. However, this process is also highly dependent on other parameters, such as the nature of the electrolyte, the concentration of oxygen, and the pH of the environment. For example, elevated levels of the chloride ions (Cl^-) or lower pH promotes accelerated corrosion.

1.5 Corrosion-Inducing Factors

Corrosion as a complicated process has many factors that can act together to speed up or slow down the deterioration of materials, especially metals. The first influencing factor is environment (nature of the environment). Humidity and/or water act as electrolytes transferring ions and contributing to corrosion. Furthermore, the presence of oxygen contributes to the reduction reaction at the cathode, significantly speeding up corrosion rates, particularly in environments with enough oxygen and ions, like chlorides (Cl^-), that promote pitting. The pH of the environment is also important, and often when the environment is acidic, corrosion occurs more quickly because the H^- ions facilitate the reactions related to degradation. The second major category affecting corrosion is the nature of the material. Pure metals will all respond differently to corrosion mathematically, depending on their electrochemi cal potential. For instance, zinc and magnesium corrode, particularly bioreactive or unprotected, much faster than copper. Alloys, such as stainless steel, can have a higher resistance due to the formation of a protective oxide surface layer. However, defects or impurities in the crystalline structure may incite local anodic (or cathodic) and thus increase the rate of corrosion. Temperature has an obvious, direct influence, simply because higher temperatures accelerate chemical reactions. In some instances, heat creates more localized forms of corrosion, such as high-temperature corrosion. Mechanical stresses also figure strongly; especially in the case of stress corrosion, the area that experiences a mechanical tension becomes more susceptible to corrosion. Additionally, friction/abrasion can act to remove layers on the surface of the metal, either naturally or artificially formed, rendering it available and exposed to an accelerated form of degradation. Corrosive exposure including strong acids (HCl, H_2SO_4) where an industry environment provides additional exposure to additional acidic gases (SO_2, CO_2) increases the corrosion rate. Saltwater (high concentration of chloride ions) is particularly aggressive toward metals, as metals in saltwater cause corrosion in metals in a galvanic corrosion process. Lastly, airflow or liquid circulation can affect the corrosion rate by increasing the availability of oxygen and reactants in the surrounding environment. Poorly aerated environments cause corrosion rates to slow down due to the lack of available oxygen for the corrosive reactions.

1.6 Conclusion

In conclusion, knowledge of the basic concepts of corrosion is important to help anticipate and prevent the destruction it can cause to materials, especially metals. Corrosion is a complicated electrochemical process influenced by chemical, environmental, and physical factors. If we can identify what kinds of corrosion exist, the underlying mechanisms involved, and what causes it to speed up or slow down, we can create strategies to effectively protect infrastructures and maximize the life of materials. Learning and understanding these concepts is important to minimize maintenance costs and provide safe operations of industrial systems.

AI disclosure statement: The author acknowledges the use of AI-based tools for minor language editing or grammatical correction. However, all intellectual contributions, analyses, and conclusions presented in this chapter are entirely the author's own.

References

1. Corrosion-fatigue: a review of damage tolerance models.
2. Aljourani, J., Raeissi, K., Golozar, M.A., Benzimidazole and its derivatives as corrosion inhibitors for mild steel in 1M HCl solution. *Corros. Sci.*, 51, 8, 1836–1843, 2009.
3. Shreir, L.L. *Basic Concepts of Corrosion*. Elsevier, Oxford, UK, 2010.
4. McCafferty, E., Societal Aspects of Corrosion, in: *Introduction to Corrosion Science*, pp. 1–11, Springer, New York, NY, 2010.
5. McCafferty, E., *Introduction to Corrosion Science*. Springer Science & Business Media, p. 302, 2010.
6. Bradford, S.A., Basic Corrosion Theory, in: *Corrosion Control*, Springer, Boston, MA, 1993, https://doi.org/10.1007/978-1-4684-8845-6_2.
7. Electrochemical Corrosion Basics. William S. Tait Pair O Docss Professionals L.L.C., Madison, WI, United States.
8. Faes, W., Lecompte, S., Ahmed, Z.Y., Van Bael, J., Salenbien, R., Verbeken, K., De Paepe, M., Corrosion and corrosion prevention in heat exchangers. *Corros. Rev.*, 37, 2, 131–155, 2019, https://doi.org/10.1515/corrrev-2018-0054.

9. Winston Revie, R., Corrosion and Corrosion Control: *An Introduction to Corrosion Science and Engineering*. 4th ed., John Wiley & Sons, Hoboken, NJ, p. 512, 2008.
10. Shreir, L.L., Basic Concepts of Corrosion, *Shreir's Corrosion*, vol. 1, pp. 1:3–1:15, Elsevier B.V., Oxford, UK, 2010.
11. Revie, R.W. and Uhlig, H.H., *Corrosion and corrosion control: an introduction to corrosion science and engineering*, John Wiley & Sons, Inc., New York, 2008.
12. Faes, W., Lecompte, S., Ahmed, Z.Y., Van Bael, J., Salenbien, R., Verbeken, K., De Paepe, M., Corrosion and corrosion prevention in heat exchangers, https://doi.org/10.1515/corrrev-2018-0054 Received June 22, 2018; accepted December 13, 2018.
13. Xie, J., Yazdanfar, K., Anning, G., Rockwell, T., Ahmed, A., Corrosion of a vertical shell and tube heat exchanger, in: *CORROSION 2015*, NACE International, Houston, TX, 2015.
14. Kuppan, T., *Heat exchanger design handbook*, Marcel Dekker, Inc., New York, 2000.
15. Julian, R., Lichti, K., Mroczek, E., Mountain, B., Heavy metal scaling and corrosion in a geothermal heat exchanger, in: *Proceedings World Geothermal Congress*, vol. 2015, 2015.
16. Roberge, P.R., *Handbook of corrosion engineering*, McGraw-Hill, New York, 2000.
17. Cassagne, T., Houlle, P., Zuili, D., Bluzat, P., Corrieu, J.M., Replacing titanium in sea water plate heat exchangers, in: *CORROSION 2010*, NACE International, Houston, TX, 2010.
18. Ifezue, D. and Tobins, F., Corrosion failure of aluminum heat exchanger tubes. *J. Fail. Anal. Prev.*, 15, 541–547, 2015.
19. Wang, L., Sundén, B., Manglik, R.M., *Plate heat exchangers: design, applications and performance, vol. 11*, Wit Press, Southampton, UK, 2007.
20. Tawancy, H., Failure of hydrocracker heat exchanger tubes in an oil refinery by polythionic acid-stress corrosion cracking. *Eng. Fail. Anal.*, 16, 2091–2097, 2009.

2

Concrete Material Properties and Corrosion

Rajae Salim[1]*, Jamila Lazrak[2], Elhachmia Ech-chihbi[1], Walid Ettahiri[2], Charafeddine Jama[3], Belkheir Hammouti[1] and Mustapha Taleb[2]

[1]*Euromed University of Fes, UEMF, Fez, Morocco*
[2]*Laboratory of Engineering, Electrochemistry, Modeling and Environment, Faculty of Sciences, Sidi Mohamed Ben Abdellah University, Fez, Morocco*
[3]*Univ. Lille, CNRS, INRAE, Centrale Lille, UMR 8207, - UMET - Unité Matériaux et Transformations, Lille, France*

Abstract
Concrete plays a critical role in modern construction due to its strength, durability, versatility, and relative ease of production and placement, making it one of the most commonly used construction materials worldwide. On the other hand, concrete corrosion refers to the deterioration or degradation of concrete structures due to chemical reactions involving its components, particularly the embedded reinforcing steel or other metallic elements. Therefore, reinforcement corrosion is one of the main durability problems that reduce the service life of reinforced concrete. Due to the increasing demand for longer-life infrastructure and the high cost of construction and maintenance, the repair of concrete struc tures has become extremely important. This chapter focuses on understanding the properties of concrete, its corrosion mechanism, and its prevention methods. Understanding these properties allows engineers and architects to design concrete structures that meet specific performance requirements while considering environmental conditions, loading conditions, and durability expectations.

Keywords: Concrete properties, reinforced concrete, prevention methods, corrosion mechanism

**Corresponding author*: salimrajae@gmail.com

Ruby Aslam, Qihui Wang, Zhitao Yan and Afroz Aslam (eds.) Corrosion in Concrete Structures: Integrating Advanced Technologies and Sustainable Practices, (13–58) © 2026 Scrivener Publishing LLC

2.1 Introduction

Corrosion is the deterioration of a metal surface due to chemical reactions, transforming a pure metal into a more chemically stable form, including oxides, sulfides, or hydroxides, in a corrosive environment [1]. Concrete is one of the most common construction materials due to its versatility, durability, and cost-effectiveness. It is a composite material primarily composed of cement, aggregates (such as sand and gravel), water, and various admixtures, influencing its properties [2].

The performance of concrete largely depends on its material properties, including strength, durability, workability, and resistance to environmental factors. One of the major concerns in concrete technology is corrosion, especially in reinforced concrete structures. Corrosion significantly impacts the performance of metals and alloys, particularly when exposed to aggressive environments. Corrosion poses a substantial and recurring challenge to industrial productivity and safety. As a result, corrosion represents a considerable economic burden worldwide [3].

On the other hand, corrosion primarily affects the steel reinforcement within the concrete, leading to significant structural degradation over time. The structural integrity and lifespan of concrete structures are compromised by steel reinforcement corrosion, which results in increased maintenance costs and safety risks [4].

The primary cause of concrete reinforcement corrosion is the ingress of aggressive agents like chlorides, carbon dioxide, and moisture. When these agents penetrate the concrete cover and reach the steel reinforcement, they initiate the corrosion process, which expands the steel and causes cracking, spalling, and further deterioration of the concrete [5].

Reinforced concrete is often exposed to the combined effects of multiple corrosion factors rather than just one. These factors typically reinforce one another, leading to more severe damage to the concrete. Therefore, understanding the material properties of concrete, including permeability, porosity, and alkalinity, is essential in addressing the corrosion issue. Advances in concrete technology, such as supplementary cementitious materials (SCMs), corrosion inhibitors, and surface treatments, are being developed to enhance the longevity and durability of concrete structures by mitigating the effects of corrosion [6]. Studying concrete material properties and corrosion is crucial for designing durable and sustainable concrete structures that can withstand the challenges posed by various environmental conditions. Addressing these challenges ensures concrete infrastructure's safety, functionality, and cost-effectiveness worldwide.

This chapter summarizes the current understanding of concrete properties and the corrosion mechanism of concrete infrastructure. In addition, this paper will discuss the prevention methods that can be used to enhance the durability performance of the concrete. Therefore, this study's conclusion can provide a solid basis for future research into the durability performance of concrete environments and relevant durability-enhancing techniques.

1. The Application of Concrete
Although concrete is primarily known for its critical role in building foundations, its versatility extends beyond this basic function. This section explores the wide range of concrete applications, highlighting their adaptability and essential role in various construction projects. Steel reinforcement in concrete significantly enhances the material's performance under tensile, compressive, and shear stresses, making it a necessary element in modern construction, particularly for large, high-stress, and long-lasting structures. The application of concrete spans various fields, showcasing its versatility and importance in contemporary engineering and environmental practices.

2.1.1 Urban Infrastructures

Steel-reinforced concrete can be used for a variety of projects, including the construction of roads, tunnels, bridges, and dams. Its strength and ability to support heavy loads make it an ideal choice for these projects. For example, in roads, reinforced concrete provides a long-lasting surface capable of resisting traffic loads, weathering, and temperature changes. Andrews Phaedonos *et al.* examine approaches for achieving high-quality concrete construction in significant infrastructure projects. The article highlights how quality control, proper material selection, and advanced techniques can significantly improve the durability and lifespan of structures. Critical factors like curing methods, mixing practices, and environmental considerations are crucial to achieving long-term performance in bridges, roads, and other vital infrastructure. These measures help mitigate corrosion and cracking, ensuring safety and cost-effectiveness [7].

Also, steel reinforcement in concrete bridges supports the massive weight of vehicles and resists tensile stresses, providing stability against dynamic forces like traffic, wind, and seismic activity. In another study, Zhou and their collaborators focus on advancements in concrete materials and their application in modern construction. It discusses innovative techniques for enhancing concrete's durability, strength, and sustainability, particularly in large-scale infrastructure projects. Key topics include using supplementary cementitious materials, high-performance concrete mixtures, and improved reinforcement strategies to mitigate corrosion and environmental degradation. These innovations aim to extend the lifespan of concrete structures while reducing environmental impact [8].

Furthermore, reinforced concrete provides structural integrity to bear underground pressure and prevent collapse in subway or road tunnels. Its ability to resist cracking under strain ensures durability in these confined spaces, even in challenging conditions like water or soil pressure. Indeed several studies were performed in this field. For example, Song *et al.* explore cutting-edge developments in advanced concrete materials' durability and mechanical properties, particularly for high-performance applications. It highlights research on optimizing concrete's resistance to environmental factors such as corrosion and mechanical wear while improving sustainability through eco-friendly materials. These innovations aim to enhance the longevity and resilience of concrete in demanding construction environments [9].

In addition, steel-reinforced concrete is essential in dam construction. It provides the structural strength to handle immense water pressure and potential seismic loads. It also helps prevent cracking or leakage, ensuring the dam remains secure. In each of these projects, the combination of steel and concrete ensures that the structures can bear compressive (from the weight of materials and loads) and tensile (from stress and forces) forces, offering the durability and resilience needed for long-term infrastructure in urban environments.

2.1.2 Residential Sector

Concrete is commonly used for foundations, walls, and floors. Concrete homes offer excellent thermal insulation and enhanced resistance to natural disasters like earthquakes and hurricanes. For example, Parab *et al.* discuss innovations in concrete technology and its applications in modern construction. It emphasizes the importance of quality control, sustainability, and enhanced durability in concrete structures, particularly in large infrastructure projects. The paper also addresses how advancements in materials and construction techniques can improve the performance and longevity of concrete, reducing maintenance costs and environmental impact [10].

Erin Ashley *et al.* show that concrete's ability to absorb and store heat through its high thermal mass creates energy-efficient homes with stable indoor temperatures. Although concrete has many advantages, its environmental impact must be considered, particularly concerning carbon emissions during production. However, advances in sustainable practices make concrete a greener option in the construction sector [11].

2.1.3 Commercial and Industrial Buildings

Concrete also benefits commercial and industrial buildings. Warehouses, factories, and shopping centers use concrete for its load-bearing capacity and durability. Concrete parking lots are famous for their longevity and low maintenance requirements.

Moreover, heavy-weight concrete is essential in structural and safety applications in nuclear power plants. It provides radiation shielding and fire resistance, with specific formulations demonstrating excellent thermal resistance. Physical and mechanical properties can also be preserved at high temperatures. In parallel, heavy concrete and aggregates are analyzed and validated *via* research for their essential role in creating neutron-protection and fire-resistant materials.

Its widespread local availability, cost-effectiveness relative to alternatives, underground construction and support technologies, and remarkable adaptability across production, transportation, and application processes collectively enhance operational efficiency. These advantages make it ideal for projects requiring sustainable and economical solutions [12].

2.1.4 Agricultural Application

Concrete storage crates are increasingly used in the agricultural sector to preserve the quality of fruit and vegetables. Indeed their resistance to temperature variations and impermeability make them a wise choice for keeping fruit and vegetables fresh and maintaining their nutritional value.

For example, He *et al.* [13] reported in their paper that agricultural wastes can be used as sustainable alternative pozzolans in concrete, enhancing performance and benefiting the environment. While rice husk ash (RHA) is currently the most promising option, other residues are also being studied. Despite their potential, few cost analyses have been conducted, and a comprehensive economic assessment is still needed. Challenges, such as increased susceptibility to scaling with high levels of agro-waste, also exist. In addition, they report that future research should focus on modifying agro-cement with superplasticizers, conducting large-scale applications, and developing effective strategies for using different agricultural residues in construction [13].

2.1.5 Sustainable Construction

With the growing emphasis on sustainable development in sustainable construction, concrete plays a crucial role in green building. Green concrete using recycled materials and production processes that reduce CO_2 emissions is becoming increasingly popular. This approach makes it possible to minimize construction projects' environmental impact while maintaining traditional concrete's advantages. Indeed concrete constructions can also incorporate sustainable technologies such as rainwater harvesting systems and solar panels. These innovations increase buildings' energy efficiency and reduce their dependence on natural resources.

Construction projects using concrete can obtain environmental certifications, including Leadership in Energy and Environmental Design (LEED), attesting to their commitment to sustainability. This adds value to property projects and attracts environmentally conscious investors. Due to their innovations and multiple advantages, concrete constructions play a vital role in the construction industry. With its varied applications and promising prospects, concrete is indispensable for building solid, durable, and environmentally friendly structures.

Vijayan *et al.* [14] review in their paper the applications of fibers steel in civil engineering, emphasizing their role in enhancing the performance of reinforced concrete (RC) structures. They discuss the benefits of steel fiber-reinforced concrete (SFRC), including durability and improved fire resistance. Various experimental tests, like Cl^- ion penetration, compression, and shear strength evaluations, were analyzed to demonstrate SFRC's superior properties. The study also investigates the utilization of steel fibers in various forms of concrete, such as geopolymer and self-compacting concrete, and their application in main structural elements like columns, beams, walls, and slabs. Moreover, SFRC significantly increased the ductility and dynamic stability of RC structures, limited crack development, and enhanced shear resistance, making it suitable for flat slab structures, military applications, and temporary structures like drain slabs. Although SFRC construction is more expensive due to its enhanced properties, its long-term maintenance costs are comparable to conventional concrete structures, supporting its use for improved structural performance and durability [14].

2.2 Importance of Steel in Concrete

Concrete is one of the most commonly used building materials in modern society. Its widespread acceptance can be attributed to its abundant availability, malleability in various configurations, economic viability, and exceptional compressive strength. However, despite these advantages, concrete is characterized by insufficient tensile strength and sub-optimal performance under adverse conditions, a significant drawback for any construction material. To remedy these shortcomings, it is common practice to incorporate steel reinforcement into concrete. Therefore, steel plays a crucial role in enhancing the properties of concrete, making it a vital component in modern construction. Integrating steel, particularly fibers and slag forms, significantly improves the mechanical performance and sustainability of concrete structures.

Moreover, reinforcing steel is the solution we use to counterbalance the weakness of concrete in tension. The addition of steel creates a composite material where the concrete continues to excel in compression, and the steel takes on the forces of tension. It's a perfect collaboration between two materials that compensate for each other's weaknesses.

2.2.1 Strength Enhancement

Compressive strength is generally considered to be the main characteristic property of concrete. Since the discovery of reinforced concrete, research has focused on increasing this strength. Adding steel fibers to concrete mixtures has increased compressive and tensile strength. Studies indicate that incorporating 0.5% to 3% steel fibers can substantially improve tensile strength, addressing concrete's inherent weakness in this area [15]. Therefore, this study encouraged the adoption of fiber reinforcement as a substitute for conventional steel reinforcement in structural engineering applications that are economically feasible.

Another study by Shewalul [16] investigated the impact of waste steel scrap utilized as a reinforcing agent. The compressive strength of concrete is enhanced by incorporating 0.5% and 0.75% steel scrap, resulting in increases of 26.8% and 30.7%, respectively; however, a reduction of 5.3% in compressive strength is observed with the addition of 1.5% steel scrap. The findings indicate that steel scrap significantly elevates the compressive strength when juxtaposed with traditional concrete. Moreover, the optimal proportion is determined to be 0.75% by volume of concrete [16].

2.2.2 Environmental Considerations

Utilizing steel slag in concrete reduces reliance on natural aggregates, addressing environmental concerns associated with resource depletion. This practice promotes recycling and minimizes waste from steel production. Combining concrete and reinforcing steel creates robust and resilient structures. Concrete, which is not very tense, gains capacity thanks to the reinforcing steel, which allows the structures to withstand significant dynamic forces. This synergy, essential for the safety and durability of buildings, is based on complex engineering principles, which require special attention for each project. Indeed recent research highlights innovative approaches to improve concrete's mechanical properties and environmental impact, making it suitable for various engineering applications. For example, as Abden *et al.* indicate in their paper, the development of energy-efficient concrete using recycled aggregates and phase change materials (PCM) has shown a 38.3% increase in compressive strength and a 61.3% reduction in water permeability, making it ideal for flood-prone areas [17].

Moreover, incorporating biomaterials into concrete has gained traction, with a bibliometric analysis revealing a 20.81% annual increase in related research, indicating a growing interest in sustainable construction practices [18]. In addition, self-compacting Concrete (SCC) has been evaluated for repair applications, demonstrating superior mechanical properties compared to traditional vibrated concrete, thus enhancing the longevity of structures [19].

2.3 Material Properties of Concrete

Concrete material properties are critical to understanding how concrete performs under various conditions and what influences its durability, strength, and overall performance. Understanding these properties is essential for evaluating and mitigating corrosion risks in concrete structures. Here's an in-depth look at the key properties of concrete:

2.3.1 Compressive Strength

Compressive strength is substantial in reinforced concrete structures' durability and corrosion resistance. Higher compressive strength generally indicates a denser concrete matrix, which reduces permeability. This helps limit the penetration of moisture, chlorides, and other corrosive agents that can reach the steel reinforcement. Consequently, higher compressive strength enhances the concrete's ability to protect the embedded steel, reducing the risk of corrosion and prolonging the structure's lifespan. Lower compressive strength, on the other hand, may lead to higher permeability and increased sed corrosion susceptibility.

Compressive strength measures a material's ability to withstand axial loads without failure, and several factors influence it, particularly in concrete and masonry. The composition of the material plays a crucial role; different types of cement, such as Portland or fly ash, exhibit varying strengths. Aggregates' type, size, and grading significantly affect the overall strength, with well-graded aggregates enhancing it. The water-cement ratio is also critical, as lower ratios typically result in higher compressive strength.

The mix design, which refers to the proportions of cement materials, aggregates, water, and additives, also significantly impacts compressive strength. Proper curing conditions are vital for increasing strength, as they allow the concrete to hydrate fully; factors such as temperature, humidity, and the duration of curing must be considered. The age of the material is another important factor, with compressive strength generally increasing over time as hydration continues.

Environmental conditions, including temperature and humidity during curing, can also affect strength, with extreme conditions potentially leading to poor hydration. Moreover, the rate at which loads are applied influences compressive strength, as rapid loading may result in lower strength due to inadequate time for stress redistribution. The presence of admixtures, both chemical (like plasticizers) and mineral (like silica fume), can improve workability and enhance long-term strength. Defects or imperfections such as cracks, voids, and poor workmanship can significantly reduce compressive strength.

As an investigation, Vu *et al.* examine how statistical size effects influ ence the compressive strength of concrete, drawing insights from experimental studies and evaluating existing data. The research highlights that size effects significantly influence compressive strength variability, which can be understood through finite-size effects and the interpretation of compressive failure as a critical transition. As the sample size increases, strength variability diminishes, making the asymptotic strength more accurately represent concrete's actual compressive strength properties. The study recommends using 4 sample sizes to assess these effects properly. It also finds that low-porosity concretes are less susceptible to side impacts than higher-porosity concretes. It suggests that the characteristic scale of the pore structure plays a crucial role in influencing these size effects. A new image analysis technique is proposed for measuring these pore structure parameters, although further research on diverse concrete types is needed to refine this method. The findings underscore the importance of considering size effects when estimating concrete's characteristic compressive strength and suggest revising current evaluation procedures [20].

In another paper, Lübeck *et al.* reported that electrical resistivity is a key property of concrete, as it helps evaluate the accessibility of aggressive agents and estimate corrosion propagation. Therefore, they employed Wenner's four-electrode method to examine the electrical resistivity of concrete mixes made with white Portland cement and those that did not contain blast furnace slag. The research compared 50% and 70% slag content mixtures to reference mixes of 100% white and grey Portland cement. Results indicated that higher slag content increased electrical resistivity and decreased pore solution conductivity. This suggests that blast-furnace slag can effectively replace some white Portland cement [21].

2.3.2 Durability

Durability denotes concrete's ability to resist chemical attack, weathering, abrasion, and other processes leading to material deterioration. Several factors influence the durability of concrete, including exposure conditions like freeze-thaw cycles, chemical exposure, and wet-dry cycles, all of which can significantly degrade the material.

The mix design is also essential. The right combination of materials, including supplementary cementitious materials (fly ash and slag), can enhance durability. The concrete cover is also necessary to protect reinforcement from exposure to aggressive agents. However, there are notable challenges to concrete durability, including sulphate attack, alkali-silica reaction (ASR), and chloride ingress, which can compromise the material's integrity over time. Indeed Shi *et al.* explained in their paper that the durability of steel-reinforced concrete in chloride environments is a significant concern for design engineers, infrastructure owners, maintainers, and researchers. They explained that this property is essential to enhance the concrete property and the advancements in understanding this durability, covering topics such as the impact of mineral admixtures and methods for measuring chloride ingress [22].

A novel method incorporating carbon and aramid fibers was developed to improve the brittleness and toughness of concrete. Zhang *et al.* examined concrete samples with varying fiber types and ratios, leading to several key conclusions. The study combined experimental and theoretical analyses to evaluate performance improvements in fiber-reinforced concrete and introduced a carbonation depth formula for practical application. First, the compressive strength increased with fiber addition before declining as the mixing ratio rose. Notably, aramid fiber-reinforced concrete was more substantial than ordinary concrete, while visible vertical cracks indicated improved toughness. Second, the elastic modulus and compressive strength of fiber-reinforced concrete followed similar trends, with minimal impact from fiber addition, allowing the elastic modulus of ordinary concrete to be used in calculations. Third, carbonation resistance was enhanced in fiber-reinforced concretes, particularly in carbon fiber-reinforced samples, with the anti-carbonation capacity inversely related to fiber mixing ratios. New carbonation depth calculation formulas, validated against test data, showed a good correlation with measured values [23].

2.3.3 Permeability and Porosity

Permeability refers to the ability of concrete to allow fluids to pass through its pores, while porosity denotes the total volume of pores within the concrete. The permeability of concrete significantly impacts its performance: high permeability increases the risk of aggressive agents and water penetrating the material, which can lead to corrosion. Conversely, low permeability enhances the concrete's resistance to chemical attacks and reduces the risk of corrosion. Measures such as lowering the water-cement ratio can effectively control permeability, as lower ratios typically decrease permeability. Additionally, using admixtures, like silica fume, can help reduce porosity by filling in voids within the concrete. Permeability can be assessed through various testing methods, including water absorption and chloride penetration tests, to examine the concrete's durability and resistance to harmful agents.

Permeability significantly impacts the durability of concrete in vari ous ways. High permeability permits water to infiltrate the concrete when water carries harmful substances. In cold climates, this infiltrated water can freeze and expand, causing cracking and spalling, known as freeze-thaw damage. Furthermore, it can transport aggressive agents like chlorides and sulphates, initiating corrosion of the steel reinforcement surrounding it [24]. This corrosion weakens the structural integrity of the concrete, potentially leading to failure.

Additionally, high permeability increases the risk of chemical attacks from external agents. For example, sulphates present in soil or groundwater can react with the hydration products of cement, resulting in expansion and cracking known as sulphate attack. Similarly, alkali-silica reactions (ASR) can occur when reactive aggregates are exposed to moisture.

Concrete with high permeability generally exhibits lower durability, which makes it more sensitive to various forms of deterioration. This can lead to increased maintenance costs and a reduced lifespan for concrete structures. Another issue related to permeability is efflorescence, which occurs when soluble salts dissolve in water and crystallize on the surface as the water evaporates, resulting in unsightly stains and surface deterioration.

The permeability of concrete also affects its thermal and acoustic properties; high permeability can result in greater heat and moisture transfer, which may impact the energy efficiency of buildings and structures. To mitigate the negative effects of permeability, it is essential to implement strategies like optimizing the water-cement ratio and ensuring proper compaction and curing. By reducing permeability, concrete becomes more durable and resistant to water ingress, corrosion, and chemical attacks, ultimately enhancing its longevity and performance [25].

2.3.4 Workability

Workability is the ease with which concrete can be mixed, placed, compacted and finished without encountering problems such as segregation or bleeding. Several factors influence workability, with water content being a primary consideration; while higher water content can enhance workability, it may also reduce the concrete's strength and durability. Additionally, the properties of aggregates play a significant role in workability, as rounded aggregates tend to improve workability more effectively than angular ones. Using admixtures, such as plasticizers and superplasticizers, can further enhance workability without the need for additional water. To assess workability on-site, the slump test is commonly employed, providing a practical means of measuring the consistency and performance of the concrete mix.

Workability is a crucial characteristic of concrete that significantly affects its performance and quality across various aspects. High workability allows for easier placement in complex or congested areas, reducing labor effort and enhancing construction efficiency. It also ensures effective compaction, which is vital for minimizing air voids and ensuring that the concrete mix thoroughly fills all spaces in the formwork, thus increasing density and contributing to strength and durability. Moreover, concrete with optimal workability can achieve a smooth surface finish, reducing the risk of defects such as cracks and uneven surfaces. This is particularly important for aesthetic applications like exposed aggregate finishes.

Additionally, workability influences the likelihood of segregation and bleeding; higher workability helps minimize these issues, preserving the concrete's integrity. However, increased workability can facilitate placement, but it may also impact the overall strength and durability if achieved through excessive water content, leading to lower compressive strength and increased permeability. Using admixtures like plasticizers can enhance workability without significantly raising water content, allowing for high-performance mixes that maintain strength. Furthermore, good workability aids moisture retention during curing, which is essential for optimal hydration and strength development, minimizing cracking and ensuring long-term durability. Therefore, while high workability can improve construction processes, balancing it with other properties is essential to ensure the final concrete product meets the required strength and durability standards [26].

2.3.5 Shrinkage and Creep

Shrinkage is the reduction in concrete volume caused by moisture loss, which can lead to cracking if the concrete is restrained. There are two main types of shrinkage: plastic shrinkage, which occurs when the concrete is still fresh, and drying shrinkage, which takes place as hardened concrete loses moisture over time. Strategies such as using shrinkage-reducing admixtures, ensuring proper curing, and incorporating control joints are essential to prevent shrinkage-related issues. Creep, conversely, is the gradual deformation of concrete when subjected to sustained loads, which can affect its long-term structural behavior. Factors such as high temperatures and elevated stress levels can exacerbate creep, leading to additional challenges in maintaining the integrity of concrete structures over time.

Shrinkage significantly impacts concrete, influencing its structural integrity, durability, and overall performance in various ways. One of the most direct effects of shrinkage is the potential for cracking. As concrete loses moisture and shrinks, it can develop tensile stresses if restrained by surrounding structures or reinforcements, leading to cracks. These cracks not only compromise the aesthetic appeal of the concrete surface but may also reduce its load-bearing capacity. Additionally, cracks resulting from shrinkage can allow water and aggressive chemicals to penetrate the concrete, leading to further deterioration. This may include deterioration of reinforcing steel, freeze-thaw damage, and chemical attacks, which can significantly reduce the structure's longevity. Shrinkage can also lead to dimensional changes in concrete elements, affecting their alignment and fit within a structural framework and creating challenges during construction [27].

The cracks can weaken the concrete's load-bearing capacity, posing safety concerns and potentially leading to structural failures. Moreover, the issues arising from shrinkage can increase maintenance and repair costs over time, placing an additional burden on property owners and contractors. Various strategies can be employed to mitigate the effects of shrinkage. These include using shrinkage-reducing admixtures to minimize volume change, ensuring proper curing to retain moisture, and incorporating control joints to accommodate movement without causing significant cracking. Proper mix design and selecting appropriate aggregates can also help reduce shrinkage [28].

2.3.6 Tensile Strength

Tensile strength refers to the capability of concrete to resist tension and is generally around 10-15% of its compressive strength. This property is crucial because concrete is inherently weak in tension, so steel reinforcement is commonly fused into concrete structures to enhance their tensile capacity. Various testing methods are employed to assess tensile strength, including direct tension and indirect methods such as split-cylinder and flexural tests. Concrete has a relatively low tensile strength compared to its compressive strength, so it is more susceptible to cracking under tensile forces. This limitation affects overall structural performance, particularly in elements subjected to bending, tension, or dynamic loads, leading to potential premature failure if not adequately addressed.

Concrete structures often require steel reinforcement to counteract their weakness in tension, which enhances the tensile capacity. Combining concrete and steel (reinforced concrete) allows for better load distribution and improved performance, enabling structures to withstand various forces without experiencing significant cracking or failure. If the tensile stress exceeds the concrete's tensile strength, cracks can form and propagate under continuous loading or environmental conditions. These cracks can compromise the durability and integrity of the structure, making it essential for engineers to understand tensile strength to predict potential cracking issues and design accordingly.

Additionally, tensile strength is critical in determining the overall flexural strength of concrete beams and slabs. Insufficient tensile strength can lead to bending failures, highlighting the importance of ensuring design specifications account for tensile stresses, particularly in load-bearing elements. Cracks resulting from insufficient tensile strength can provide a pathway for moisture and harmful chemicals to infiltrate the concrete, causing reinforcement corrosion and other types of degradation. By accounting for tensile strength during the design process, engineers can improve the durability and lifespan of concrete structures [29].

2.3.7 Modulus of Elasticity

The modulus of elasticity is a key property of concrete that indicates its stiffness and ability to deform elastically under load. This property significantly impacts structural performance, as a higher modulus of elasticity helps reduce deflections and cracking in structural members, enhancing overall stability and durability. Additionally, the modulus of elasticity is influenced by various factors, with higher compressive strength typically leading to a higher modulus of elasticity. This relationship underscores the importance of selecting appropriate concrete mixes to achieve desired structural characteristics.

The modulus of elasticity is a critical property of concrete that significantly influences its performance and behavior under load. It measures the stiffness of concrete, which is essential for structural applications. A higher modulus of elasticity indicates that concrete will deform less when subjected to applied loads, making it more resistant to bending and deformation. This stiffness is vital for maintaining the shape and function of structures under various load conditions [30].

One of the key impacts of a higher modulus of elasticity is its ability to control deflection in concrete structures, such as beams and slabs. Minimizing deflection helps maintain the serviceability of the structure, preventing excessive bending that could lead to aesthetic issues, misalignment of structural components, or discomfort for occupants in buildings. Furthermore, concrete with a higher modulus of elasticity tends to experience less cracking under load, which is crucial for durability and aesthetic appearance. Reduced cracking not only enhances the longevity of the concrete but also improves the overall quality of the structure [31].

The modulus of elasticity also affects how loads are distributed within a concrete element. A higher modulus means that loads are transferred more efficiently through the material, provides better load distribution and minimizes the risk of localized failures. This property is fundamental in structural applications where load-bearing capacity is critical. A relationship exists between compressive strength and the modulus of elasticity, where an increase in compressive strength typically results in a higher modulus. This relationship underscores the importance of selecting appropriate concrete mixes and materials to achieve structural designs' desired stiffness and load-bearing capacity. Additionally, in foundation applications, the modulus of elasticity is important in evaluating the interaction between the concrete foundation and the soil. A higher modulus indicates that the foundation will deform less under load, which is essential for maintaining stability in structures built on it.

2.3.8 Thermal Properties

Concrete experiences thermal expansion and contraction with temperature changes, which can lead to cracking. Additionally, its thermal conductivity influences the heat transfer rate through the material, making it an essential factor in fire resistance design.

Thermal properties are crucial in determining concrete's performance, durability, and overall behavior in various applications. Concrete expands when heated and contracts when cooled, leading to internal stresses and cracking if the material is restrained. It is essential to incorporate design features like expansion joints to accommodate temperature changes and minimize damage. Additionally, the thermal conductivity of concrete affects heat transfer through the material, which is significant in fire resistance design; understanding this property ensures that concrete can withstand high temperatures and maintain structural integrity [32]. Furthermore, concrete's thermal properties contribute to building energy efficiency; materials with appropriate thermal mass can absorb and release heat. Therefore, it helps regulate room temperatures and reduces energy used for heating and air conditioning. Concrete's low thermal conductivity and high thermal mass also enhance its fire-resistant qualities, protecting heat transfer during fire exposure. Moreover, temperature management during the curing and hydration processes is vital for achieving optimal strength and durability; improper temperatures can adversely affect hydration reactions, resulting in weak concrete. Overall, understanding and managing the thermal properties of concrete is essential for effective design and construction practices, enabling the development of structures that can withstand temperature variations while ensuring long-term integrity and safety.

2.4 Corrosion Properties of Steel in Concrete

Corrosion in reinforced concrete is an electrochemical process that occurs when the protective passive layer on steel reinforcement is disrupted. Normally, concrete's high pH (around 13) maintains a passive film on the steel surface, preventing significant corrosion. However, this layer can be broken by carbonation, which lowers the pH, or by chloride ion penetration, which directly attacks the passive film. Once disrupted, an electrochemical cell forms, with anodic zones where steel dissolves into ferrous ions and cathodic zones where oxygen reduces to form hydroxyl ions. These reactions produce expansive corrosion products like rust, which exert tensile stresses on the surrounding concrete, causing cracking and spalling (Figure 2.1).

Studying the properties of steel in concrete is essential for safeguarding the durability and longevity of reinforced concrete constructions. Corrosion is one of the primary causes of structural weakening, leading

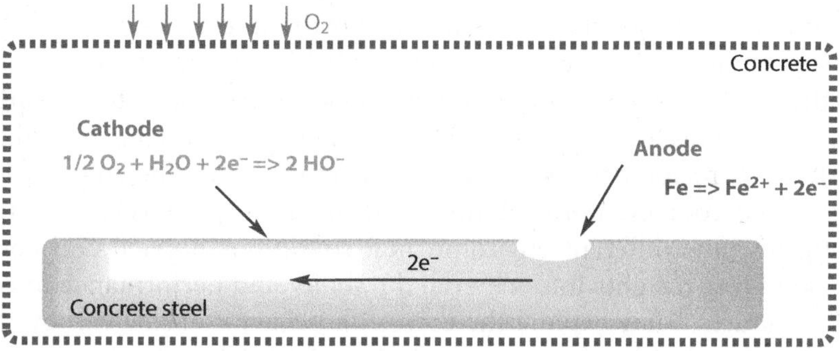

Figure 2.1 The anodic and cathodic reaction.

to reduced load-bearing capacity and, in severe cases, structural failure. Understanding how and why steel corrodes within concrete helps engineers design structures that resist environmental degradation, reducing the cost of repairs and replacement parts. This knowledge also supports the development of innovative materials and protective technologies, such as corrosion-resistant alloys, coatings, and cathodic protection systems. Additionally, by enhancing corrosion resistance, we can extend the lifespan of concrete structures, resulting in economic savings and contributing to sustainable construction practices. Improved corrosion resistance ensures compliance with building codes and standards, promoting safer infrastructure. Ultimately, a deep understanding of steel corrosion in concrete helps optimize construction methods, mitigate risks, and maintain the integrity and functionality of critical infrastructure over time.

2.4.1 Electrochemical Properties

The electrochemical properties of reinforced concrete play an essential duty in defining the behavior of steel embedded within it, especially with regard to corrosion. In a typical reinforced concrete environment, steel is protected by a passive film because of the concrete's high alkalinity. This passive film prevents the steel from corroding under normal conditions. However, changes in the electrochemical properties of the concrete can disrupt this film and initiate corrosion.

When corrosion begins, it follows an electrochemical process in which anodic and cathodic zones form on the steel surface. Iron (Fe) is oxidized at the anode to form ferrous ions (Fe^{2+}), releasing electrons. These electrons then travel to the cathodic regions, reducing oxygen and forming hydroxide ions (OH^-). This flow of electrons generates a corrosion current, and the reaction products (iron oxides and hydroxides) accumulate on the steel surface as rust. The expansion of rust, which occupies more volume than the original steel, creates internal stresses in the concrete, leading to cracking, spalling, and decreased structural integrity. Liu *et al.*, studied carbon steel's corrosion behavior when exposed to chloride and sulphate ions in simulated concrete pore solutions with different pH levels. The study investigates how different environmental conditions impact steel corrosion, providing insights into material durability and performance in concrete structures. They explore the corrosion behavior of low-carbon steel in alkaline simulated concrete pore solutions, focusing on how varying pH levels and the presence of chloride and sulphate ions, either separately or together, affect corrosion mechanisms. The findings indicate that higher pH reduces the adsorption of ions onto the passivation film, enhancing corrosion resistance. Chloride ions alone accelerate corrosion, while adding sulfate ions alters this process by competing for adsorption sites. This results in a two-stage corrosion mechanism, emphasizing the importance of managing ionic pH and composition of concrete to minimize corrosion risks and improve steel-reinforced durability [33].

Aspects like the resistivity of concrete, the availability of moisture and oxygen, and the presence of chlorides influence the corrosion process's electrochemical potential and current density. Low resistivity concrete, for example, allows higher corrosion currents, accelerating the rate of steel deterioration. Understanding these electrochemical properties is essential to developing strategies to minimize corrosion, such as using corrosion inhibitors, protective coatings, or designing concrete with lower permeability to restrict the ingress of harmful ions. Thus, the electrochemical parameters of reinforced concrete considerably impact the long-term durability and overall performance of steel-reinforced buildings.

2.4.2 Concrete Permeability

Concrete permeability significantly affects the durability of steel-reinforced concrete. It controls the rate at which aggressive elements (water, oxygen, chlorides, and carbon dioxide) infiltrate the concrete matrix. When concrete has high permeability, these substances can more easily infiltrate and reach the embedded steel reinforcement, increasing the risk of corrosion.

In highly permeable concrete, water and oxygen, two key elements required for the electrochemical corrosion process, can readily access the steel surface, making it more susceptible to rust formation. Also, harmful ions like chlorides (from deicing salts or seawater) can infiltrate the concrete and disrupt the passive oxide barrier naturally created on steel in alkaline conditions, initiating corrosion. The deterioration of the protective layer accelerates steel oxidation, resulting in rust formation, which occupies more volume than the original steel. This expansion creates internal stresses, causing cracking, spalling, and a reduction in the structural integrity of the concrete.

On the other hand, low-permeability concrete restricts the ingress of these aggressive agents, thereby maintaining the protective environment around the steel and prolonging the structure's service life. Concrete permeability can be reduced through proper mix design, cementitious materials, adequate curing, and a dense and well-compacted concrete matrix. By minimizing permeability, the risk of steel corrosion is significantly reduced, increasing the durability and longevity of reinforced concrete constructions.

2.4.3 Chemical Properties

The chemical properties of steel are crucial in determining its behavior in various environments, especially in relation to corrosion, reactivity, and overall durability. Steel primarily consists of iron, and its chemical properties are heavily influenced by the alloying elements added during production, such as carbon, chromium, nickel, manganese, and molybdenum [34]. Carbon is the most important element in steel, as it significantly affects its strength and hardness. However, higher carbon content can reduce steel's ductility, making it more prone to brittleness. The addition of chromium and nickel enhances steel's corrosion resistance, as seen in stainless steel, where chromium forms a protective oxide layer on the surface that prevents rusting. This makes stainless steel ideal for use in corrosive environments like marine and chemical applications.

Other alloying elements like manganese improve the strength and toughness of steel, while molybdenum increases its resistance to high temperatures and corrosion. Steel's oxidation resistance is a key chemical property, especially when exposed to oxygen, moisture, or aggressive chemicals. Without proper protection, steel can form iron oxides, commonly known as rust, which leads to material degradation over time. Additionally, steel reacts with acids, bases, and other chemicals, though the extent of these reactions depends on the composition of the steel. Steel's reactivity can be minimized by using coatings or by alloying it with elements that form more stable compounds, thus improving its durability in hostile environments. Ultimately, the chemical properties of steel, including its corrosion resistance and reactivity, play a significant role in determining the material's suitability for various industrial applications.

2.4.4 Mechanical Properties

The mechanical properties of steel are fundamental to its performance in various applications, determining how it reacts under different loads and conditions. One of the most important mechanical properties is strength, which refers to steel's ability to withstand applied forces without failure. Steel can be classified into different grades based on its tensile strength, with low-carbon steel being suitable for general applications, and high-strength steel being used for demanding structures like bridges and skyscrapers. Ductility is another key property, allowing steel to deform significantly before breaking. This ability to stretch or bend without fracturing is essential for shaping steel into complex forms and for its use in structures that need to absorb energy, such as in seismic regions.

Hardness is the resistance of steel to surface deformation or wear. Harder steels, achieved through processes like quenching and tempering, are used in tools and machinery where abrasion resistance is crucial. Toughness, which is the ability to absorb impact energy without breaking, is another vital mechanical property, especially in applications that are subjected to dynamic or fluctuating loads. Steel's elasticity allows it to return to its original shape after a stress is applied, which is particularly important in structural applications where steel must maintain its shape under load without permanent deformation. Together, these mechanical properties (strength, ductility, hardness, toughness, and elasticity) define the versatility and suitability of steel for a wide range of engineering and construction applications [34].

2.4.5 Physical Properties

The physical properties of steel are fundamental characteristics that define its behavior under various conditions and influence its suitability for different applications. One of the most important physical properties is density, which typically ranges around 7,850 kg/m^3. This relatively high density compared to other materials makes steel heavier, but it also contributes to its strength and structural integrity. Steel's thermal conductivity is moderate, typically ranging from 45 to 60 W/m.K, which means it can transfer heat but not as efficiently as metals like copper. This property is important in applications requiring heat resistance or insulation, such as in construction and industrial machinery.

Steel also has a high melting point, typically between 1370°C and 1510°C, depending on the alloy composition, making it suitable for use in high-temperature environments such as manufacturing and construction.

Its magnetic properties are another significant physical trait, as many types of steel are magnetic due to the iron content, which makes steel useful in electrical and magnetic applications, like motors and transformers. The electrical conductivity of steel is relatively low compared to metals like copper or aluminum, making it less ideal for applications that require efficient electrical conduction. Thermal expansion is another physical property of steel, which expands when heated and contracts when cooled. Considering this property in structures exposed to temperature fluctuations, such as bridges or large buildings, is essential as expansion or contraction can cause stress if not properly managed.

Lastly, the corrosion resistance of steel, which can be influenced by its alloying elements (like chromium in stainless steel), dictates how well it can withstand environmental factors such as moisture and air without degrading. Steel can form a protective oxide layer when exposed to oxygen, though in more aggressive environments, it can rust, especially if not adequately protected. Overall, the physical properties of steel, including its density, conductivity, magnetic behavior, and resistance to environmental factors, play a crucial role in determining its suitability for various engineering and industrial uses.

2.5 Causes of Concrete Corrosion

Corrosion of concrete, with the formation of rust on the reinforcement in the concrete, takes place in two phases. In the first phase (Figure 2.2), aggressive structures like carbon dioxide (CO_2) or chlorides (Cl^-) in the neighboring environment infiltrate the concrete. This is the incubation

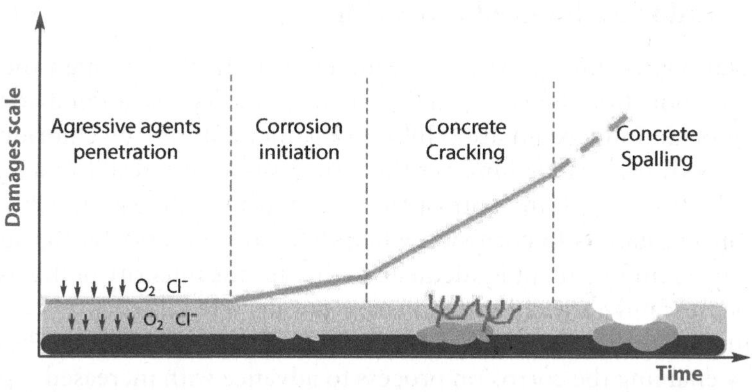

Figure 2.2 Conceptual diagram of concrete corrosion.

stage. The subsequent phase, propagation, starts when these aggressive substances reach relatively high concentrations in the reinforcement. This leads to rust formation, which can cause the concrete cover to crack and eventually burst.

2.5.1 Carbonation-Induced Corrosion

Carbonation occurs when atmospheric CO_2 reacts with the hydrates in the cement paste, causing a drop in pH to around 9. This drop in pH leads to the depassivation of the reinforcement when the carbonation front reaches the reinforcement, as described in the following equations. Under these conditions, the passive film is no longer stable [35].

$$CO_2 + H_2O \Rightarrow H_2CO_3 \tag{2.1}$$

$$Ca(OH_2) + H_2CO_3 \Rightarrow CaCO_3 + 2H_2O \tag{2.2}$$

Upon the complete carbonation of the concrete cover, the alkaline milieu that safeguards the passive film ceases to exist, thereby precipitating the initiation of corrosion on the steel bar's surface. In this circumstance, the steel surface can be conceptualized as comprising aggregates of localized microcells (i.e., a confluence of neighboring potentially anodic and cathodic sites). When the environmental conditions are sufficiently humid, electric currents will circulate, activating the local cells and heralding the commencement of corrosion. In this phenomenon, corrosion typically occurs relatively uniformly across the entirety of the steel surface.

2.5.2 Chloride-Induced Corrosion

The alkalinity within the pores of concrete is mitigated through the process of carbonation; however, the mechanism of depassivation associated with chloride-induced attack exhibits a distinct nature. The chloride ions interact with the passive film, yet this process does not result in a significant reduction in pH, in contrast to the effects of carbonation. Chlorides function as catalysts to corrosion when their concentration at the surface of the steel reinforcement is adequate to disrupt the integrity of the passive film. Significantly, the chloride ions are not depleted during this interaction; instead they assist in the degradation of the passive film on the steel, thereby enabling the corrosion process to advance with increased rapidity. The equation can be succinctly articulated as:

$$Fe^{2+} + 2Cl^{-} + 4H_2O \Rightarrow FeCl_2 \cdot 4H_2O \tag{2.3}$$

$$FeCl_2 \cdot 4H_2O \Rightarrow Fe(OH)_2 + 2Cl^{-} + 2H^{+}2H_2O \tag{2.4}$$

Moreover, depassivation happens when the chloride concentration in the reinforcement exceeds a critical threshold. This threshold is typically defined by a Cl-/OH- concentration ratio ranging from 0.6 to 1, corresponding to a chloride content of approximately 0.4% of the cement mass, as specified by the NF EN 206-1 standard for reinforced concrete. For prestressed concrete, this threshold is reduced to 0.1% [36].

2.6 Consequences of Concrete Corrosion

2.6.1 Degradation of the Embedding Concrete

The surrounding environment can alter the concrete that encases a reinforcement for several reasons: physical (frost can cause it to shatter), mechanical, or chemical, in particular, due to certain environmental substances. Generally, reinforced concrete structures are exposed to the atmosphere, water, or soil, environments that vary in pollution levels and contain substances capable of penetrating the concrete. These substances can alter the concrete's properties, particularly the chemical composition of the interstitial solution.

The most common aggressive agents affecting concrete are pure water, chlorides in water, and atmospheric carbon dioxide (CO_2). Pure water can leach concrete by dissolving cement components and increasing its porosity. Chloride salts, highly soluble in water, penetrate concrete through convection or diffusion, driven by a concentration gradient. While most chlorides remain dissolved in the concrete's interstitial solution, they can also react chemically with the material's components or be adsorbed.

Carbon dioxide (CO_2) exists as a gas in the atmosphere and can dissolve in the concrete's intersti tial solution, reacting with certain calcium compounds to form carbonates (carbonation). This reaction lowers the pH of the interstitial solution in concrete to about 9. The penetration of CO_2 into concrete occurs through diffusion, and it moves more quickly when the concrete is relatively dry. However, carbonation only happens if the concrete still contains an interstitial solution. Therefore, the most favorable conditions for CO_2 penetration are found at an average relative humidity of around 65% [37].

2.6.2 Initiation of Reinforcement Corrosion

Reinforcement corrosion begins when the protective layer on its surface becomes ineffective (depassivation) due to increased porosity. The first indicator of corrosion initiation is a change in these protective products. This process progresses through intermediate stages, forming more or less stable products (green rust). A more practical criterion is a noticeable change in the rate of metal dissolution, indicating increased corrosion activity. In practice, carbon dioxide (CO_2) initiates reinforcement corrosion when the concrete surrounding the steel undergoes carbonation and remains somewhat damp. Chlorides contribute to metallic corrosion and are dissolved ions in the chloride solution. Their critical concentration primarily depends on the concrete's pH and oxygen content.

2.6.3 Development of Corrosion

Protective products, such as ferrous hydroxide $Fe(OH)_2$, green rust, or magnetite, remain stable and are thin. Moreover, corrosion products that fail to halt the process accumulate on the reinforcement. As corrosion progresses, the steel undergoes localized dissolution and becomes coated with unstable products like reddish rust. In porous, damp concrete, these products infiltrate the concrete cover and may stain the surface. In drier concrete, the corrosion products expand, exerting pressure that deforms the surrounding layer, eventually causing cracks or spalling. This combination of reduced reinforcement cross-section and the swelling of rust weakens the bond between the steel and concrete, compromising structural integrity.

2.6.4 Structures Affected by Corrosion of Concrete

Various concrete structures and their susceptibility to damage depend on environmental and functional factors. Vertical elements like acroterias, balconies, and terraces in all climates or marine and industrial environments are particularly vulnerable due to their thin structure and the challenges of maintaining proper protective coverings. Similarly, industrial buildings, significantly posts, slabs, and beams, face increased risks from chemical exposure, with beams being highly sensitive when supporting floor slabs. Car parks in maritime or mountainous areas are affected by chlorides from seawater or de-icing salts, which attack beams and slabs.

Due to better quality control, no significant problems are reported for prefabricated structures, but specific elements like power line poles show signs of aging. Issues also arose in components built with calcium chloride-based accelerators, impacting elements like façade panels, acroteria, and planters. In bridges and engineering structures, decks, superstructure supports, and deck equipment are the most sensitive areas due to exposure to de-icing salts.

Reservoirs, whether underground, on-ground, or above ground, face leakage problems from cracks or defects in reinforcement coatings, particularly in chloride-rich environments. Silos storing granular materials experience severe stress during loading/unloading, leading to cracks and water penetration, which corrodes reinforcement.

Industrial air coolers are exposed to extreme conditions, leading to crack initiation due to external weather and internal mist. Similarly, industrial chimneys endure harsh acidic conditions (sulfuric and hydrochloric acid), causing deep concrete degradation [38]. Port structures near the sea are prone to chloride attack, with corrosion intensity linked to environmental aggressiveness. Mostly, buried, reinforced, and prestressed concrete pipes fail when protective layers are compromised by carbonation or chloride presence. Lastly, reinforced concrete historical monuments and listed buildings have unique repair challenges, while concrete pathologies due to inherent material issues (e.g., alkali-reaction, delayed ettringite formation) cause cracking and deformation in various structures.

2.7 Methods of Protecting Concrete against Corrosion

2.7.1 Using High-Quality Concrete Mix

Applying high-quality concrete mix is fundamental in minimizing the risk of corrosion in reinforced concrete structures. The properties of the concrete itself, such as its permeability, strength, and composition, play a critical role in determining its resistance to aggressive environmental factors that lead to corrosion, especially in chloride-rich environments like coastal areas. When the concrete mix is well proportioned with low water–cement ratios, proper compaction, and adequate curing, it produces a dense matrix that reduces permeability and limits corrosive elements' pathways to the embedded steel. This helps maintain the high alkalinity of the concrete, preserving the protective passivation layer on the steel surface and minimizing the risk of corrosion. Moreover, incorporating supplementary cementitious materials, such as fly ash or silica fume, can significantly improve concrete's resistance to chemical attacks while minimizing the formation of microcracks. A high-quality concrete mix enhances reinforced concrete's structural performance and durability and provides robust protection for the steel reinforcement, shielding it from early corrosion and degradation [39].

High-quality concrete is achieved by optimizing several key factors that ensure its strength, durability, and resistance to environmental damage. One of the most critical elements is maintaining a low water–cement ratio, which reduces porosity and permeability. Appropriate cement types and supplementary materials like fly ash or silica fume can enhance durability.

High-quality aggregates with proper gradation contribute to a dense mix with minimal voids. The mix design should balance strength and workability, and chemical admixtures can improve concrete properties, such as workability and freeze–thaw resistance. Proper mixing, placement, and compaction are essential to eliminate air voids and ensure a uniform distribution of ingredients. Adequate curing conditions and temperature control during mixing and setting support hydration prevent shrinkage and reduce cracking.

Finally, ensuring proper reinforcement placement and sufficient concrete cover helps protect the steel from corrosive elements. Addressing these factors holistically results in high-quality concrete that meets performance requirements and prolongs the lifespan of structures.

In literature, Garcia *et al.* investigated the performance of high-strength concrete in coastal environments. They found that a mix incorporating silica fume significantly reduced chloride penetration, extending the service life of marine structures by over 30% compared to standard concrete mixes. This study indicates that relying solely on volumetric parameters to assess mix quality may be misleading, as some mixes with balanced performance did not meet volumetric standards and vice versa. It suggests that a rational gradation selection can balance volumetric and mechanical properties more effectively rather than arbitrarily adjusting the asphalt content from the optimal asphalt content (OAC) determined in the mix design [40].

Nilimaa *et al.* explored the challenges of constructing with concrete in cold climates, focusing on issues like slower setting times, delayed curing and strength development, and damage from frost, early freezing, and freeze-thaw cycles. They proposed innovative approaches to address these problems, including adjusting concrete mix designs, using chemical additives, integrating supplementary cementitious materials, and employing advanced construction methods. Additionally, their study examined how cold weather affects workers, equipment, and machinery, highlighting the importance of careful planning, clear communication, and efficient management to mitigate these difficulties [41].

2.7.2 Coating Techniques

Coatings are extensively used to safeguard concrete and embedded steel reinforcement from corrosion by providing physical barriers that prevent the penetration of harmful agents like chloride ions, water, and oxygen. Different coatings are utilized depending on the environmental conditions and the level of protection required. For example, epoxy coatings provide excellent adhesion and chemical resistance, making them suitable for aggressive environments. Acrylic coatings offer good UV stability and moisture resistance, while polyurethane coatings are known for their durability and flexibility. Zinc-rich coatings, often used for cathodic protection, form a protective sheet that protects steel reinforcement from corrosion. The choice of coating type is crucial to ensuring adequate long-term protection and enhancing the structural lifespan.

2.7.2.1 *Electrocoating*

Electrocoating, also known as electrophoretic deposition (EPD), is a method that uses electrical current to deposit a layer of coating material onto a conductive surface. Usually, steel is embedded in concrete. This process ensures uniform coverage, even in complex geometries, which is crucial for protection against corrosion in hard areas to reach using traditional methods [42].

Indeed electrocoating is often used in the automotive industry to coat steel car parts. Still its application expands into construction materials like steel rebar in reinforced concrete to enhance durability in corrosive environments. In literature, Wang *et al.* [43] investigated the pattern size-controlled electroplating mechanism through flow field simulations and electrochemical measurements, revealing that larger patterns enhanced electrolyte flow. So, they incorporated 300 mg/L PEG and 6 mg/L SH110 as novel DC electroplating formulas that enabled ultra high rate copper electroplating. The key factors for achieving high-rate electroplating included enhanced cathodic polarization, minimized electrolyte convection, and the addition of suitable chloride ions. The results showed a faster electroplating rate at the edges of patterns, improving efficiency in electronic packaging production [43].

2.7.2.2 *Metallic Coatings*

Metallic coatings involve applying a corrosion-resistant metal layer, like zinc or aluminum, onto the steel surface. These coatings act as sacrificial anodes, protecting the underlying steel from corrosion. Metallic coatings can be applied using hot-dip galvanizing or thermal spraying methods. Additive manufacturing enables customizable structural design and performance for magnetoelectric sensors, but achieving effective design and fabrication requires precise alignment of size and modulus between magnetic and conductive components. Challenges arise from integrating magnetoelectric elements and the rigidity of coils, which limits sensor flexibility. To overcome these obstacles, Wu *et al.* propose a composite method combining selective laser sintering (SLS) with 3D transfer printing to develop flexible magnetoelectric sensors coated with liquid metal. This approach achieves outstanding sensing performance, with a sensitivity of 10.9 kPa^{-1} and a minimum detectable pressure of 0.1 kPa [44].

Indeed Randis *et al.* [45] highlight the advantages of using bio-based materials as nano-fillers in polymers. When paired with other nanomaterials and functional group modifications, these materials improve hydrophobicity and bonding with the polymer matrix, making them highly effective for barrier coatings against corrosive agents. The inherent properties of bio-based nanomaterials, including low density, high aspect ratio, superior mechanical strength, biocompatibility, and biodegradability, make them ideal for anti-corrosion nanocomposites. These qualities are promising for future high-performance metal coatings and other advanced applications [45].

2.7.2.3 *Organic Coatings*

Organic coatings, such as polyurethane, epoxy, and acrylic-based paints, are applied to protect concrete and steel from corrosion. These coatings create an impermeable barrier that blocks water, oxygen, and aggressive ions from reaching the surface of the concrete or steel. Organic coatings are also used to protect concrete surfaces from harsh environmental conditions, preventing cracking, spalling, and eventual degradation due to exposure to aggressive substances like chlorides and sulphates. In reinforced concrete structures exposed to de-icing salts, epoxy-coated rebars have been shown to reduce corrosion rate significantly. For example, Wang *et al.* [46] investigate the degradation behavior of polyurea and epoxy coatings, focusing on the correlation between bulk degradation, adhesion loss, and protective performance. The results indicate a gradual aging process in the coatings, characterized by increased water absorption and internal structural changes. While polyurea exhibits better weather resistance to UV aging than epoxy, it shows weaker adhesion strength and chloride ion resistance. The results underline that, in addition to overall degradation, adhesion between coatings and concrete is crucial to the effectiveness of protection [46].

Zhang *et al.* [47] offer robust experimental evidence for using epoxy resin and silane coupling agent-modified geopolymer in marine concrete coatings. However, they note the need for further investigation into practical performance aspects such as anti-aging properties and playability to assess the material's potential in real-world applications fully. The study synthesized new organic–inorganic composite coatings using waterborne epoxy and metakaolin-based geopolymer gel to protect marine concrete. Laboratory experiments showed that adding these organic modifiers significantly decreased drying shrinkage, enhanced adhesion strength, and improved abrasion resistance. Although the coatings demonstrated improved corrosion resistance and durability in simulated marine environments, carbonation resistance needs improvement. MG-WR composites with 30% waterborne resin and MG-SCA composites with 1% S823 showed the best performance against seawater corrosion [47].

2.7.3 Galvanization Protection

Cathodic protection is a method for controlling the corrosion of metal surfaces by turning them into cathodes in an electrochemical cell. This is achieved through various techniques, including sacrificial anodes, impressed current cathodic protection (ICCP), and hybrid systems. These methods are widely applied in protecting reinforced concrete structures, pipelines, and steel in marine environments [48].

2.7.3.1 Sacrificial Anodes

Sacrificial anode protection is a passive form of cathodic protection where a more reactive metal, such as zinc, magnesium, or aluminum, is used as the anode. The sacrificial metal corrodes in place of the steel, protecting it from corrosion. This method is often used in marine environments, buried pipelines, and reinforced concrete structures. For example, zinc anodes are commonly used in marine environments to protect steel components of offshore platforms and ship hulls. In reinforced concrete, sacrificial anodes can be embedded to protect steel rebars from chloride-induced corrosion. Indeed several studies have shown that this method significantly reduces the corrosion rate in marine environments. They explain that more reactive anode metal corrodes (oxidizes), releasing electrons that flow to the steel, preventing it from corroding. The anode continues to corrode until it is depleted, after which it must be replaced. Cheng *et al.* [49] developed a model to optimize sacrificial anode cathodic protection (SACP) for mitigating steel corrosion caused by chloride penetration in concrete. Their findings revealed that smaller anodes placed closer together were more effective in reducing steel corrosion currents than larger anodes spaced farther apart. During SACP, approximately 25%–30% of the zinc anodic current balanced the cathodic current, while most protected the steel bars. They concluded that increasing the zinc anodic current reduces corrosion but accelerates anode consumption, with an optimal ratio of 5.0–7.5 times the steel corrosion current for maximum efficiency [49].

2.7.3.2 *Impressed Current Cathodic Protection (ICCP)*

The ICCP technique is an active system with an external current to protect the metal. Unlike sacrificial anodes, ICCP systems use an external power source as a rectifier to provide a constant protective current to the steel. This method is highly effective in environments with high corrosion risks, such as bridges, piers, and industrial concrete structures. Research shows that ICCP significantly extends the lifespan of these structures by preventing corrosion, even in highly aggressive environments. The ICCP mechanism system supplies an external current to keep steel passive, effectively preventing corrosion-causing oxidation reactions. These systems can be precisely controlled to maintain the desired protective current, enhancing their efficiency and durability in protecting steel structures from corrosion over time. This level of control allows for tailored protection based on the specific conditions of the environment.

For example, a cable ICCP system was developed by Yao *et al.* to control corrosion proactively under simulated acid rain conditions. This study analyzed how ICCP protects bridge cables and identified its corrosion protection mechanisms. Key findings include that a more negative protection voltage enhances corrosion protection, but excessive negativity can cause hydrogen embrittlement in the steel wires. Additionally, corrosion products can delay cable deterioration, and the protection efficiency shifts from joint to dominated by the impressed current as corrosion progresses [50].

In another study, the acidification issue of ICCP systems was investigated by Fatemi *et al.*, developing novel acid-resistant mortar mixes using supplementary cementitious materials like GGBFS, fly ash, and silica fume. Laboratory tests showed that the HP2 mortar mixture exhibited superior acid resistance against HCl, with reduced mass loss and limited penetration. The study confirmed that these hybrid geopolymer mortars demonstrate enhanced durability and adhesion, making them suitable for long-term use in acidic environments. Successful application of the HP1 mortar on an actual bridge showed effective acid diffusion prevention and compliance with protection standards [51].

2.7.3.3 Hybrid Cathodic Protection

Hybrid cathodic protection combines aspects of both sacrificial anode and impressed current systems. This approach is used when the structure requires an initial high level of protection followed by a long-term passive system. A hybrid system may start with ICCP and later switch to a sacrificial anode system once the protective current is no longer needed. Moreover, hybrid systems have been shown to be cost-effective, providing long-term protection with minimal maintenance.

For example, Zhou *et al.* used ICCP technology on hybrid-reinforced concrete beams with CFRP bars and SFCBs as anodes to mitigate steel corrosion in high chloride environments. Experimental results show that ICCP effectively reduced corrosion, enhanced the crack and ultimate loads, and improved stiffness compared to untreated beams. The CFRP anode provided better protection than the SFCB anode, which showed degradation under ICCP. Overall, the system demonstrates promise for protecting steel in marine environments and promoting using sea sand and seawater in construction [52].

Recently, Nguyen *et al.* developed a dense sol-gel coating using TEOS, MAPTMS, and ZTP to protect HDG steel from corrosion in a 0.1-M NaCl solution. It examined the influence of Ce (III) concentrations, revealing that low concentrations (especially 650 ppm) enhanced corrosion resistance, while higher concentrations (2,000 and 5,000 ppm) significantly decreased protective performance over time. The rheology and FTIR analyses indicated that excess Ce (III) destabilized the Si–O–Si network, adversely affecting long-term coating efficacy [53].

2.7.4 Surface Treatments

Surface treatments protect concrete structures from corrosion, particularly in environments exposed to aggressive agents such as chlorides, sulfates, and moisture. These treatments include sealants, coatings, and waterproof membranes, which act as barriers, preventing water penetration and other corrosive elements into the concrete [54].

2.7.4.1 Sealants and Coatings

Sealants and coatings are functional directly to the concrete surface to block water, air, and harmful chemicals that can accelerate the reinforced steel corrosion. There are various types of sealants, such as silane, siloxane, epoxy, and acrylic-based coatings [55].

Silane and siloxane sealants are commonly used for their water-repellent properties. Silane penetrates deep into the concrete, forming a hydrophobic layer that reduces water absorption without altering the appearance of the concrete surface. Conversely, siloxane is often used on surfaces like brick and stone and helps prevent water ingress. Indeed several studies show that silane-based sealants significantly reduce the permeability of concrete, thus preventing the ingress of chloride ions and extending the structure's service life.

Epoxy and acrylic coatings are organic coatings that are highly effective in creating a durable, impermeable barrier. Epoxy coatings, in particular, offer excellent adhesion to concrete and steel and provide resistance to chemicals, making them suitable for industrial environments. Acrylic coatings are often preferred for their ease of application and UV resistance. For example, in coastal and industrial environments, epoxy coatings have been shown to improve the durability of reinforced concrete by blocking moisture and aggressive agents like chlorides.

2.7.4.2 Waterproof Membranes

Waterproof membranes provide additional protection by preventing water and moisture from penetrating the concrete. These membranes are often applied as sheets or liquid-applied systems and can be particularly useful for structures exposed to constant moisture, such as basements, tunnels, and marine structures.

- Sheet membranes: These are typically made from bitumen, polyvinyl chloride (PVC), or rubberized asphalt. They are applied in layers, providing a continuous waterproof barrier. For example, bituminous sheet membranes are widely used for waterproofing foundations and underground structures. Studies indicate that these membranes effectively prevent water infiltration, reducing the risk of concrete deterioration and steel corrosion [56].
- Liquid-applied membranes: These are applied in liquid form and cure to form a seamless, rubber-like membrane. Liquid-applied systems are ideal for irregularly shaped surfaces and can be used in horizontal and vertical applications. For example, liquid-applied polyurethane membranes are used to waterproof bridge decks and roofs. Research has shown that these membranes provide excellent durability and flexibility, preventing cracking and water ingress even under dynamic loads [57].

2.7.5 Use of Corrosion Inhibitors

Corrosion inhibitors are substances added to concrete to slow down or prevent corrosion of the embedded steel. They work by either passivating the metal surface, reducing its reactivity with corrosive agents, or creating a protective film around the steel. These inhibitors are beneficial in environments where chloride exposure or carbonation is prevalent, such as marine structures, bridges, and industrial buildings. Green corrosion inhibitors have gained attention as sustainable alternatives in concrete research due to their high diversity and potential effectiveness. Several review papers discuss various green inhibitors, their advantages, and their limitations. It highlights how pH and ion type affect their protection mechanisms. These materials often contain active groups like carboxylates and amines, which form protective films on steel, providing elevated efficiencies. However, further research is required to evaluate their long-term behavior in concrete and their cost-effectiveness for commercial applications [58].

Corrosion inhibitors can be categorized into anodic, cathodic, and mixed types, each serving different protective functions. Anodic inhibitors, like calcium nitrite, form a protective oxide layer on steel surfaces, particularly effective in highly corrosive environments such as those with de-icing salts or high chloride levels. Cathodic inhibitors, such as sodium molybdate and phosphates, reduce corrosion by decreasing the rate of cathodic reactions, often by lowering oxygen availability. Mixed inhibitors, such as amines, combine both anodic and cathodic inhibition mechanisms, offering comprehensive corrosion protection by forming a protective film on steel surfaces that reduces both types of reactions.

Corrosion inhibitors offer several advantages, such as enhancing the durability of concrete structures by delaying corrosion onset and slowing its progression, which ultimately reduce long-term maintenance and repair costs. They are versatile and can be used in new constructions or repair work, making them suitable for various concrete structures exposed to harsh environments. However, their effectiveness depends on proper concentration, as insufficient dosage may not provide adequate protection, while excessive amounts could negatively impact concrete properties. Additionally, specific inhibitors pose environmental risks, prompting the development of more sustainable, green alternatives.

2.7.6 Electrochemical Chloride Extraction (ECE)

ECE is a non-destructive technique used to remove chloride ions from reinforced concrete structures, especially when these ions accumulate and cause the corrosion of steel reinforcement. Chloride ions are the main reason of corrosion in concrete exposed to marine environments, de-icing salts, or industrial chemicals. ECE helps prevent further corrosion by reducing chloride concentrations around the steel reinforcement to safe levels, thus extending the structure's lifespan [59]. This technique effectively removes chlorides from the steel-concrete interface, delaying further corrosion and extending the lifespan of structures. However, ECE comes with some limitations. It is relatively expensive compared to other protection methods like corrosion inhibitors or surface coatings. While it temporarily removes chlorides, it does not prevent future chloride ingress, necessitating periodic maintenance. Additionally, the efficiency of ECE depends on the porosity and quality of the concrete, with low-quality concrete allowing faster chloride re-entry.

Electrochemical chloride extraction (ECE) involves temporarily applying an electrical current to the concrete structure. A direct current flows between the steel reinforcement, which serves as the cathode, and an external electrode, typically made of a conductive material like titanium mesh, positioned on the concrete surface. An alkaline solution is applied to the surface to aid the migration of chloride ions. The electrical current drives the chloride ions away from the steel rebar towards the external electrode, where they are either neutralized or collected.

2.7.7 Design and Construction Practices

Proper design and construction practices prevent corrosion in reinforced concrete structures. By integrating adequate concrete cover and effective drainage design, engineers can significantly reduce the risk of corrosion, prolong the structure's lifespan, and lower maintenance costs. Practical applications of concrete protection techniques vary based on environmental exposure and structural requirements. Ensuring adequate cover thickness and efficient drainage systems is crucial for bridges and highways to protect against de-icing salts, rainfall, and water spray from vehicles. In marine structures, thicker concrete covers and well-designed drainage systems help shield reinforcement from seawater and salt spray. Similarly, basements and tunnels benefit from waterproof membranes combined with adequate drainage to prevent water ingress, ensuring long-term durability and resistance to seepage.

2.7.7.1 Adequate Concrete Cover

The concrete cover is the layer between the structure's outer surface and the embedded steel reinforcement. Its primary function is to protect the steel by acting as a physical barrier against environmental factors such as moisture, chlorides, and carbon dioxide. An adequate concrete cover is essential for preventing the ingress of these aggressive agents [60].

- The thickness of the cover: The thickness depends on the exposure conditions. A thicker cover (typically 50–75 mm) is recommended in marine environments where chloride penetration is a significant concern. Thicker covers provide more protection and delay the time for corrosive agents to obtain the steel.
- Quality of concrete: Along with the thickness, the quality of the concrete (low permeability and high strength) is important in ensuring that the concrete cover effectively protects the steel. Using high-quality, low-permeability concrete reduces the rate at which chlorides and carbon dioxide penetrate the cover.
- Standards and guidelines: Eurocode 2 and ACI 318 set the guidelines on the minimum cover thickness based on the structure's environmental exposure.

2.7.7.2 Good Drainage Design

Proper drainage design prevents water accumulation on and around concrete structures. Standing water can lead to the ingress of moisture and chlorides, accelerating the corrosion of steel reinforcement. Well designed drainage systems ensure that water is directed away from vulnerable areas, reducing the risk of corrosion [61].

- Effective slope design: Sloping surfaces, such as bridge decks or parking structures, should be designed with adequate gradients to allow water to drain off efficiently. Improper drainage can lead to pooling, which increases the risk of water infiltration into the concrete.
- Use of drains and gutters: Incorporating drains and gutters in the design can help prevent water from accumulating on the surface of the concrete. These systems should be designed to handle the expected volume of water, especially in areas exposed to heavy rainfall or snow.
- Waterproofing and membranes: To further protect the structure, waterproofing membranes or coatings can be applied to surfaces that are exposed to standing water. This is particularly useful for flat roofs, balconies, and bridge decks.

2.8 Techniques for Assessing Corrosion Concrete

Inspectors perform an essential and intricate task assessing concrete corrosion during evaluations of concrete structures. The process involves multiple steps and uses advanced tools to account for various external factors such as humidity, temperature, light exposure, and chlorides and carbonates. It begins by estimating the probability of corrosion using the half-cell potential method with devices like the corrosion profometer. Inspectors then detect and map the concrete cover with tools such as the Profometer PM8000 Pro, as inadequate cover increases the vulnerability of reinforcing bars to environmental damage and corrosion.

Thereafter, the resistivity of the concrete is measured using devices like the Proceq Resipod, which helps identify areas with low resistivity; these zones are more prone to corrosion due to higher permeability and deeper penetration of harmful substances like chlorides. Finally, deterioration maps are generated using tools such as the Proceq GP8x00 and GPR Insights, providing a detailed visual representation of areas with a high probability of damage, including zones with corrosion, weaker materials, lower density, and higher permeability.

By integrating data from these tools, inspectors can enhance the quality of half-cell potential inspections, which environmental factors can otherwise affect. This comprehensive, data-driven approach offers a complete understanding of corrosion risks, enabling informed decisions on necessary maintenance and repairs.

There are several techniques for assessing corrosion in concrete, each providing valuable insights into the condition of the concrete structure. These methods can be broadly categorized into non-destructive or direct inspection (Figure 2.3), which require probes or sensors to be tapped into the structure to make physical contact.

Firstly, group 1 regrouped the non-destructive techniques that were used for assessing the corrosion of concrete. For example, radiography is a non-destructive testing technique used to evaluate reinforced concrete structures by employing short-wave electromagnetic beams like X-rays or gamma rays. These beams penetrate the material, and their attenuation or deviation helps detect internal defects such as voids, cracks, or density variations. This technique is valuable for locating reinforcement bars and identifying issues like pitting corrosion or thickness loss. While radiography provides detailed imaging, it has limitations, including difficulty in determining the depth of defects and detecting small cracks oriented non-parallel to the beam. It is also costly, requires skilled operators, and involves

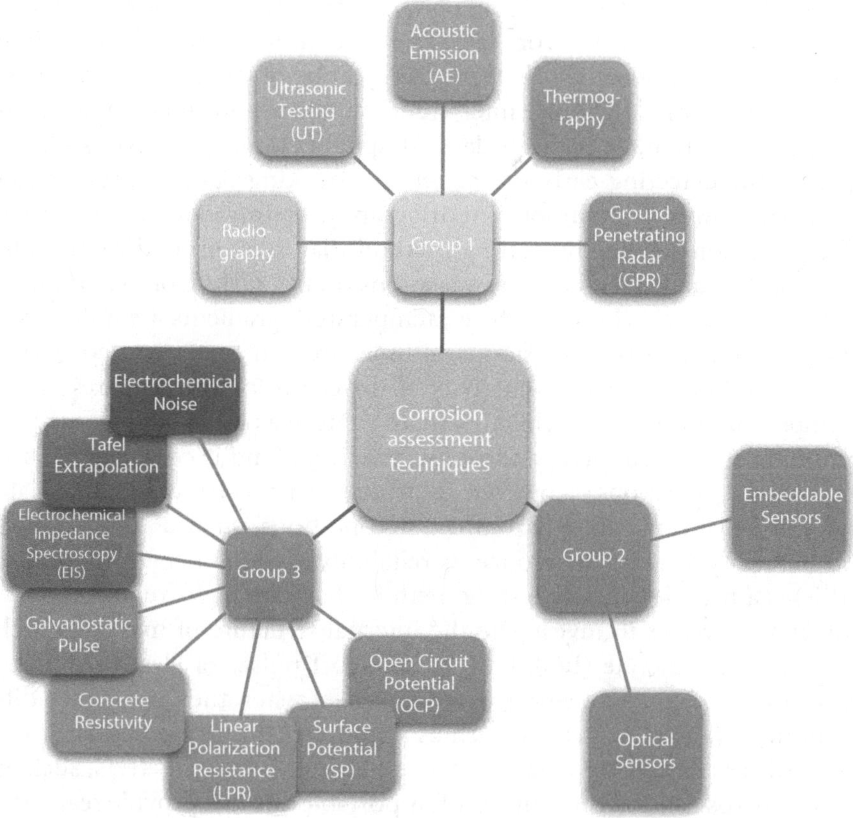

Figure 2.3 Corrosion assessment techniques for reinforced steel concrete.

radiation hazards, necessitating safety precautions. Moreover, ultrasonic testing (UT) uses high-frequency sound waves (1–10 MHz) to assess reinforced concrete (RC) structures. Mechanical vibrations are introduced into the material, and the waves reflect at boundaries or flaws, providing insights into internal defects like voids, cracks, and inhomogeneities.

The time waves travel through the material helps estimate thickness and detect damage. UT is portable, easy to operate, and effective for field and laboratory inspections. However, its limitations include difficulty evaluating crack depth if filled with water and difficulty interpreting data, as results depend on multiple factors. Furthermore, acoustic emission (AE) testing is a non-destructive method used to monitor stress waves generated by the rapid release of energy due to cracking, deformation, or other damage in materials. AE sensors on the structure detect these high-frequency waves, allowing for early detection of micro-cracks and corrosion-induced deterioration in reinforced concrete (RC) structures. AE provides real-time monitoring, high sensitivity, and precise location of defects. However, it is prone to environmental noise interference, requiring advanced signal processing to differentiate relevant data. Despite its limitations, AE is highly effective for detecting early corrosion and tracking damage progression in RC structures. In addition, thermography, or infrared thermography (IRT), is a non-destructive testing method that uses infrared cameras to detect and visualize temperature variations on the surface of materials. In reinforced concrete (RC) structures, temperature gradients are influenced by defects like cracks, corrosion, and moisture content. When corrosion occurs, the low thermal conductivity of the corrosion products results in a temperature difference, which IRT can capture as infrared images. This technique is fast, portable, and non-contact, making it cost-effective for large-scale inspections. However, it primarily provides qualitative data and can be affected by environmental conditions, such as temperature and humidity, which may reduce its reliability. Ground penetrating radar (GPR) is a non-destructive testing method that uses high-frequency electromagnetic waves to investigate the internal structure of materials such as reinforced concrete (RC). GPR sends short pulses of electromagnetic waves into the concrete, and the waves reflect when they encounter different materials or structures, such as reinforcement bars or voids. These reflected signals are analyzed to detect and locate internal defects, such as cracks, corrosion, and moisture. GPR is portable and can provide real-time results. However, it requires skilled interpretation, as the data analysis can be complex. It is most effective in detecting corrosion early and locating reinforcement, but challenges remain in interpreting results due to signal distortion from environmental factors like moisture content.

The second group represents the techniques that treat physical contact, including embeddable and optical sensors. Indeed embedded sensors are devices integrated into reinforced concrete (RC) structures to monitor various parameters related to corrosion and structural health over time. These sensors measure factors such as corrosion potential, corrosion rate, concrete resistivity, chloride content, temperature, and moisture levels, providing continuous, real-time data. This allows for long-term monitoring without needing to access the structure externally. Embeddable sensors are beneficial for assessing RC structures' condition, helping predict future deterioration and guide maintenance decisions. However, they can be expensive, and their installation requires careful planning, as they are often permanent and difficult to replace once embedded in the concrete. In addition, optical sensors are used to monitor the health of reinforced concrete (RC) structures by measuring changes in light or optical properties, often through fiber optics. These sensors, such as Fiber Bragg Grating (FBG) sensors, detect changes in strain, temperature, or displacement due to corrosion-induced expansion or cracking of the reinforcement bars. Optical sensors are immune to electromagnetic interference, making them ideal for harsh environments, and they can provide high-accuracy real-time monitoring. They are small, lightweight, and flexible, which allows for easy installation in large structures. However, the high cost of equipment and the need for specialized installation and analysis limit their widespread use. Despite these challenges, optical sensors are promising for long-term monitoring and early detection of corrosion in concrete structures.

The third group focuses on electrochemical methods that provide valuable insights into the corrosion process by measuring various electrical properties. These techniques are essential for detecting corrosion early and assessing the condition of steel reinforcement. One of the primary methods in this group is open circuit potential (OCP) measurement, which evaluates the potential difference between the steel reinforcement and the surrounding concrete. This method helps identify areas of active corrosion by detecting shifts in the electrochemical potential of the reinforcement. A similar technique, surface potential (SP) measurement, assesses the surface potential of the reinforcement relative to the concrete to map the condition of the steel and identify areas with corrosion activity. Moreover, linear polarization resistance (LPR) measurement is another key technique that measures the polarization resistance of the steel surface. Since this resistance is inversely related to the corrosion rate, LPR provides a quantitative estimate of the structure's corrosion rate.

Additionally, concrete resistivity measurement evaluates the electrical resistivity of concrete, which is an important indicator of the likelihood of corrosion. Low resistivity suggests that corrosive agents, such as chloride ions, can easily penetrate the concrete, increasing the risk of corrosion. Other techniques include the galvanostatic pulse transient method, which applies a current pulse to the reinforcement and monitors its response to estimate the corrosion rate, and electrochemical impedance spectroscopy (EIS), which measures the impedance at the concrete-steel interface across a range of frequencies. EIS provides detailed information about corrosion, such as localized corrosion and the overall corrosion resistance of the steel. Tafel extrapolation involves measuring the steel's electrochemical polarization curve and extrapolating it to determine the corrosion current density, which correlates with the corrosion rate. Lastly, electrochemical noise analysis analyzes natural fluctuations in the steel reinforcement's electrical signals, helping detect corrosion initiation and progression without applying external current.

2.9 Future Perspectives and Research Directions

Advancements in corrosion-resistant materials have significantly evolved, focusing on innovative approaches to mitigate corrosion in various applications. Recent research highlights the integration of nanotechnology, advanced coatings, and alloy design strategies as key developments in this field.

- ✓ Nanomaterials as corrosion inhibitors: Nanomaterials lock active metal sites, preventing corrosion and enhancing properties like hardness and durability at low costs. Their application as corrosion inhibitors is gaining traction, showcasing potential in various industrial settings.
- ✓ Advanced coating technologies: Corrosion-resistant coatings form barriers protecting metals from moisture and oxygen, which is crucial for infrastructure longevity. Emerging technologies include smart coatings that release inhibitors in response to environmental stimuli, significantly improving corrosion resistance.
- ✓ Alloy design strategies: High-temperature alloys, particularly those incorporating reactive elements like Ti, Al, and Cr, form protective oxide layers that enhance stability and resistance to environmental degradation. Combining inhibitors with coatings, such as plasma electrolytic oxidation (PEO) on magnesium alloys, has shown promising results in extending service life and providing self-healing properties.
- ✓ Ecological materials in construction and renovation

In a world faced with environmental challenges, the use of eco-friendly materials in construction has become a priority for building professionals and homeowners concerned about the future of our planet. Traditional building materials can harm the environment, whereas eco-friendly materials offer a sustainable, nature-friendly alternative. This article explores the many benefits of eco-friendly materials for sustainable construction.

2.10 Conclusion

Steel corrosion in concrete is a multifaceted process governed by electrochemical reactions, the material characteristics of the concrete, and the mechanical performance of reinforced concrete structures. It is influenced by environmental conditions, chloride penetration, carbonation, and the materials' quality. Over the years, researchers from various disciplines, including corrosion chemistry, physics, materials science, and civil engineering, have contributed to developing advanced detection methods and protective strategies to address this issue. This paper provides a comprehensive review of the properties of corrosion-prone concrete, the underlying causes of steel reinforcement corrosion, and the techniques employed to mitigate it in infrastructure construction, emphasizing both traditional approaches and emerging innovations.

AI disclosure statement: The author acknowledges the use of AI-based tools for minor language editing or grammatical correction. However, all intellectual contributions, analyses, and conclusions presented in this chapter are entirely the author's own.

References

1. Salim, R., Ech-Chihbi, E., Oudda, H., El Hajjaji, F., Taleb, M., Jodeh, S., A review on the assessment of imidazo [1, 2-a] pyridines as corrosion inhibitor of metals. *J. Bio- Tribo-Corros.*, 5, 14, 1, 2019.
2. Safiuddin, M., Gonzalez, M., Cao, J., Tighe, S.L., State-of-the-art report on the use of nano-materials in concrete. *Int. J. Pavement Eng.*, 15, 10, 940, 2014.
3. Habib, S., Shakoor, R.A., Kahraman, R., A focused review on smart carriers tailored for corrosion protection: Developments, applications, and challenges. *Prog. Org. Coat.*, 154, 106218, 2021.
4. Khalaf, A.H., Xiao, Y., Xu, N., Wu, B., Li, H., Lin, B., Nie, Z., Tang, J., Emerging AI technologies for corrosion monitoring in oil and gas industry: A comprehensive review. *Eng. Fail. Anal.*, 155, 107735, 2023.
5. Green, W.K., Steel reinforcement corrosion in concrete–an overview of some fundamentals. *Corros. Eng. Sci. Technol.*, 55, 4, 289, 2020.
6. Raghav, M., Park, T., Yang, H.M., Lee, S.Y., Karthick, S., Lee, H.S., Review of the effects of supplementary cementitious materials and chemical additives on the physical, mechanical and durability properties of hydraulic concrete. *Materials*, 14, 23, 7270, 2021.
7. Andrews-Phaedonos, F., Ensuring quality and durability in concrete construction for major infrastructure, in: *Austroads Bridge Conference, 10th, 2017, Melbourne*, 2017 Apr.
8. Zhou, M., Lu, W., Song, J., Lee, G.C., Application of ultra-high performance concrete in bridge engineering. *Constr. Build. Mater.*, 186, 1256, 2018.
9. Song, Y., Wang, C., Huang, M., Li, R., Wang, Q., Mechanical characteristics and stability analysis of an innovative type of steel–concrete composite support in shallow-buried excavation tunnels. *Tunnelling Underground Space Technol.*, 145, 105581, 2024.

10. Parab, R.R., Jariwala, H.I., Karale, K.N., Subhan, K.M., Sameen, R., Innovations in Disaster-Resistant Construction and Practices. *Int. J. Res. Eng. Sci. Manag.*, 5, 1, 178, 2022.
11. Ashley, E. and Lionel, L., Concrete's contribution to sustainable development. *J. Green Build.*, 3, 4, 37, 2008.
12. Fabian, E., Hansson, M.V., Krutrök, B,., Concrete: an enabler of large-scale block and sublevel cave mining projects globally. *Aust. Cent. Geomech.*, 2022.
13. He, J., Kawasaki, S., Achal, V., The utilization of agricultural waste as agro-cement in concrete: A review. *Sustainability*, 12, 17, 6971, 2020.
14. Vijayan, D.S., Sivasuriyan, A., Parthiban, D., Jakimiuk, A., Bayat, H., Podlasek, A., Vaverková, M.D., Koda, E., A comprehensive analysis of the use of SFRC in structures and its current state of development in the construction industry. *Materials*, 15, 19, 7012, 2022.
15. Lage, B., Suwal, P., Baidhya, B., Sulu, A., Rajthala, B., Structural Performance of Concrete: Exploring the Limits of Steel Fiber Reinforcement. *Int. J. Multi. Res.*, 6, 3, 1, 2024.
16. Shewalul, Y.W., Experimental study of the effect of waste steel scrap as reinforcing material on the mechanical properties of concrete. *Case Stud. Constr. Mater.*, 2021.
17. Abden, M.J., Tam, V.W., Afroze, J.D., Le, K.N., Energy efficient sustainable concrete for multifunctional applications. *Constr. Build. Mater.*, 418, 135213, 2024.
18. Jaramillo, H.Y., Vasco-Echeverri, O.H., Moreno-Pacheco, L.A., García-León, R.A., Biomaterials in Concrete for Engineering Applications: A Bibliometric Review. *Infrastructures*, 8, 11, 161, 2023.
19. Benaddi, F.Z., Boukhattem, L., Benhamou, B., Tabares-Velasco, P.C., Ait Nouh, F., Study on the performance of thermal energy of a classroom built with natural materials. *Proc. Inst. Civ. Engineers-Energy*, 176, 2, 89, 2023.
20. Vu, C.C., Plé, O., Weiss, J., Amitrano, D., Revisiting the concept of characteristic compressive strength of concrete. *Constr. Build. Mater.*, 263, 120126, 2020.
21. Lübeck, A., Gastaldini, A.L., Barin, D.S., Siqueira, H.C., Compressive strength and electrical properties of concrete with white Portland cement and blast-furnace slag. *Cem. Concr. Compos.*, 34, 3, 392, 2012.
22. Shi, X., Xie, N., Fortune, K., Gong, J., Durability of steel reinforced concrete in chloride environments: An overview. *Constr. Build. Mater.*, 30, 125, 2012.
23. Zhang, J., Cheng, J., Dou, Y., Xin, Q., Mechanical Properties and Durability of Fiber-reinforced Concrete. *J. Eng. Sci. Technol. Rev.*, 10, 5, 68, 2017.
24. Singh, A., Sampath, P.V., Biligiri, K.P., A review of sustainable pervious concrete systems: Emphasis on clogging, material characterization, and environmental aspects. *Constr. Build. Mater.*, 261, 120491, 2020.
25. Luhar, S., Luhar, I., Shaikh, F.U., A review on the performance evaluation of autonomous self-healing bacterial concrete: mechanisms, strength, durability, and microstructural properties. *J. Compos. Sci.*, 6, 1, 23, 2022.

26. Han, H., Han, K.R., Cheong, I.W., Optimizing latex-modified concrete for enhanced workability and durability in pavement construction. *Constr. Build. Mater.*, 444, 137722, 2024.
27. Zhang, B., Zhu, H., Cheng, Y., Huseien, G.F., Shah, K.W., Shrinkage mechanisms and shrinkage-mitigating strategies of alkali-activated slag composites: A critical review. *Constr. Build. Mater.*, 318, 125993, 2022.
28. Tran, N.P., Gunasekara, C., Law, D.W., Houshyar, S., Setunge, S., Cwirzen, A., A critical review on drying shrinkage mitigation strategies in cement-based materials. *J. Build. Eng.*, 38, 102210, 2021.
29. Chemrouk, M., The deteriorations of reinforced concrete and the option of high performances reinforced concrete. *Procedia Eng.*, 125, 713, 2015.
30. Korolev, A.S., Kopp, A., Odnoburcev, D., Loskov, V., Shimanovsky, P., Koroleva, Y., Vatin, N.I., Compressive and tensile elastic properties of concrete: Empirical factors in span reinforced structures design. *Materials*, 14, 24, 7578, 2021.
31. Rajadesingu, S., Mendonce, K.C., Palani, N., Monisha, P., Vijayakumar, P., Ayyadurai, S., Exploring the potential of bacterial concrete: A sustainable solution for remediation of crack and durability enhancement–A critical review. *Constr. Build. Mater.*, 439, 137238, 2024.
32. Sargam, Y., Wang, K., Alleman, J.E., Effects of modern concrete materials on thermal conductivity. *J. Mater. Civ. Eng.*, 32, 4, 04020058, 2020.
33. Liu, G., Li, J., Zhang, Y., Corrosion of carbon steels subjected to chloride and sulfate in simulated concrete pore solutions with different pH. *Constr. Build. Mater.*, 440, 137445, 2024.
34. Schlegel, J., Alloying Elements and Steel Properties, in: *The World of Steel: On the History, Production and Use of a Basic Material*, pp. 63–103, Springer Fachmedien Wiesbaden, Wiesbaden, 2023.
35. Asmara, Y.P., Corrosion of Steel Reinforcement, in: *Concrete Reinforcement Degradation and Rehabilitation: Damages, Corrosion and Prevention*, pp. 59–70, Springer Nature Singapore, Singapore, 2023.
36. Mahima, S., Moorthi, P.V., Bahurudeen, A., Gopinath, A., Influence of chloride threshold value in service life prediction of reinforced concrete structures. *Sādhanā*, 43, 1, 2018.
37. Pochat-Bohatier, C., Sanchez, J., Gontard, N., Influence of relative humidity on carbon dioxide sorption in wheat gluten films. *J. Food Eng.*, 77, 4, 983, 2006.
38. Schorr, M. and Valdez, B., The phosphoric acid industry equipment materials and corrosion. *Corros. Rev.*, 34, 1-2, 85, 2016.
39. Yang, H., Xiong, C., Liu, X., Liu, A., Li, T., Ding, R., Shah, S.P., Li, W., Application of layered double hydroxides (LDHs) in corrosion resistance of reinforced concrete-state of the art. *Constr. Build. Mater.*, 307, 124991, 2021.
40. Garcia, V.M., Barros, L., Garibay, J., Abdallah, I., Nazarian, S., Effect of aggregate gradation on performance of asphalt concrete mixtures. *J. Mater. Civ. Eng.*, 32, 5, 04020102, 2020.

41. Nilimaa, J. and Zhaka, V., An overview of smart materials and technologies for concrete construction in cold weather. *Eng.*, 4, 2, 1550, 2023.
42. Han, J., Wang, Z., Zhi, A., Li, Y., Zhao, S., Yan, H., Han, Q., A smart electroplating approach to fabricate mechanically robust and fluorine-free Ni-W alloys based superhydrophobic coating on Al alloy. *Vacuum*, 217, 112501, 2023.
43. Wang, Q., Su, P., Lei, Z., Chen, M., Luo, X., Toward ultra-high-rate copper pattern electroplating with simultaneously improved coating properties *via* simulations and experiments. *J. Manuf. Processes*, 131, 369, 2024.
44. Wu, H., Luo, R., Li, Z., Tian, Y., Yuan, J., Su, B., Zhou, K., Yan, C., Shi, Y., Additively Manufactured Flexible Liquid Metal–Coated Self-Powered Magnetoelectric Sensors with High Design Freedom. *Adv. Mater.*, 36, 34, 2307546, 2024.
45. Randis, R., Darmadi, D.B., Gapsari, F., Sonief, A.A., Akpan, E.D., Ebenso, E.E., The potential of nanocomposite-based coatings for corrosion protection of metals: a review. *J. Mol. Liq.*, 390, 123067, 2023.
46. Wang, H., Feng, P., Lv, Y., Geng, Z., Liu, Q., Liu, X., A comparative study on UV degradation of organic coatings for concrete: Structure, adhesion, and protection performance. *Prog. Org. Coat.*, 149, 105892, 2020.
47. Zhang, M., Xu, H., Zeze, A.L., Liu, X., Tao, M., Coating performance, durability and anti-corrosion mechanism of organic modified geopolymer composite for marine concrete protection. *Cem. Concr. Compos.*, 129, 104495, 2022.
48. Vaira Vignesh, R. and Sathiya, P., Sacrificial anode materials to protect marine grade steel structures: a review. *Corros. Rev.*, 42, 3, 303, 2024.
49. Cheng, X., Xia, J., Wu, R.J., Jin, W.L., Pan, C.G., Optimisation of sacrificial anode cathodic protection system in chloride-contaminated reinforced concrete structure. *J. Build. Eng.*, 45, 103515, 2022.
50. Yao, G., He, X., Liu, J., Guo, Z., Chen, P., Test study of the bridge cable corrosion protection mechanism based on impressed current cathodic protection. *Lubricants*, 11, 1, 30, 2023.
51. Fatemi, H., Hadigheh, S.A., Tao, Y., Adam, G., Development of a novel and specialised cementitious matrix overlay for anode embedment in impressed current cathodic protection (ICCP) systems for reinforced concrete bridges. *Case Stud. Constr. Mater.*, 20, e02908, 2024.
52. Zhou, Y., Zheng, Y., Sui, L., Hu, B., Huang, X., Study on the flexural performance of hybrid-reinforced concrete beams with a new cathodic protection system subjected to corrosion. *Materials*, 13, 1, 234, 2020.
53. Nguyen, T.T., Thai, T.T., Paint, Y., Trinh, A.T., Olivier, M.G., Effect of cerium nitrate concentration on corrosion protection of hybrid organic/inorganic Si/Zr sol-gel coating applied on hot-dip galvanized steel. *Surf. Coat. Technol.*, 480, 130562, 2024.
54. Pelz, U. and Karlovšek, J., Spray-applied waterproofing membranes in tunneling: A construction perspective. *Tunnelling Underground Space Technol.*, 142, 105409, 2023.

55. Pandiyarajan, N. and Nunthavarawon, P., Recent Advancements in Sealants Solutions for Surface Coatings: A Comprehensive Review. *J. Bio- Tribo-Corros.*, 10, 3, 61, 2024.
56. Ren, G., Li, Z., Tian, L., Lu, D., Jin, Y., Zhang, Y., Li, B., Yu, H., He, J., Sun, D., Environmentally friendly waterproof and breathable electrospun nanofiber membranes *via* post-heat treatment. *Colloids Surf., A*, 658, 130643, 2023.
57. Nair, P.S. and Gettu, R., Commercially available waterproofing agents in India: A review. *Indian Concr. J.*, 90, 36, 2016.
58. Casanova, L., Ceriani, F., Messinese, E., Paterlini, L., Beretta, S., Bolzoni, F.M., Brenna, A., Diamanti, M.V., Ormellese, M., Pedeferri, M., Recent Advances in the Use of Green Corrosion Inhibitors to Prevent Chloride-Induced Corrosion in Reinforced Concrete. *Materials*, 16, 23, 7462, 2023.
59. Jiang, C., Song, C., Gu, X.L., Zhang, Q., Zhang, W.P., Modeling electrochemical chloride extraction process in cement-based materials considering coupled multi-ion transports and thermodynamic equilibriums. *J. Clean. Prod.*, 401, 136778, 2023.
60. Arya, C., *Design of Structural Elements: Concrete, Steelwork, Masonry and Timber Designs to Eurocodes*, CRC Press, p. 520, 2022.
61. Lou, Y., Wang, P., Li, Y., Chen, C., Li, J., Hu, T., Management of the designed risk level of urban drainage system in the future: Evidence from haining city, China. *J. Environ. Manage.*, 351, 119846, 2024.

3

Fundamentals and Understanding of Corrosion-Prone Environments in Concrete Corrosion

Demian I. Njoku[1,2*], Annuncieta C. Njoku[3], Paul C. Uzoma[3,4], Ini-Ibehe Nabuk Etim[5,6] and Inime I. Udoh[7]

[1]Department of Applied Science, School of Science and Engineering, Hong Kong Metropolitan University, HK, SAR China
[2]Department of Industrial Chemistry, Madonna University Elele, Rivers State, Nigeria
[3]Department of Polymer Engineering, Federal University of Technology Owerri, Imo State, Nigeria
[4]Zju-UIUC Institute, International Campus, Zhejiang University, Haining, China
[5]State Key Laboratory of Advanced Marine Materials, Key Laboratory of Marine Environmental Corrosion and Bio-Fouling, Institute of Oceanology, Chinese Academy of Sciences, Qingdao, China
[6]Marine Chemistry and Corrosion Research Group, Department of Marine Science, Akwa Ibom State University, Akwa Ibom State, Nigeria
[7]School of Environmental and Chemical Engineering, Shenyang Ligong University, Shenyang, China

Abstract

This book chapter provides a detailed exploration of the different environments that can lead to corrosion in concrete structures, a significant concern in construction and civil engineering. The chapter delves into marine, industrial, deicing salt, underground, agricultural, and coastal environments, highlighting the unique factors contributing to the corrosion process. It discusses the corrosive impact of elements such as salts, pollutants, and certain chemicals on the deterioration of concrete structures. The chapter also explores underground concrete structures, focusing on the roles of moisture and lack of oxygen in the corrosion process and the current practices and techniques that have been implemented to

**Corresponding author*: demianifeanyi@mail.com

Ruby Aslam, Qihui Wang, Zhitao Yan and Afroz Aslam (eds.) Corrosion in Concrete Structures: Integrating Advanced Technologies and Sustainable Practices, (59–84) © 2026 Scrivener Publishing LLC

mitigate corrosion in these environments. Furthermore, some cases of failures resulting from concrete corrosion in various settings and the mitigation strategies were discussed. The information presented in this chapter would be pivotal to construction and civil engineering professionals, researchers, and students interested in concrete corrosion.

Keywords: Concrete, inhibition, chloride ions, steel rebar, structural collapse, acidic gases, and pollution

3.1 Introduction

The major structural materials used to construct civil and industrial structures, including bridges, dams, and nuclear power plants, are primarily concrete and cement-based materials. This is attributed to their high mechanical strength and low porosity. Concrete structures are often designed to last for decades, if not centuries [1]. However, unpredicted corrosion can drastically reduce the lifespan of concrete structures, leading to premature failures. Steel rebar is often employed to improve the strength and performance of concrete structures without altering the macroscopic cementitious matrix properties [2]. Steel plays an important role in providing the tensile strength to the reinforced concrete to withstand the tension and compressive strength of the concrete in columns and walls as well as extra shear strength beneath and above the concrete in beams [3, 4]. Interestingly, concrete pH is alkaline, and steel corrosion is protected in the alkaline pH often encountered in concrete [5]. This concrete protection in alkaline pH conditions is related to forming a stable passive film that can impede chloride access to the steel, provided the threshold level has not been exceeded. Regrettably, this protection can be bridged in specific environments due to the continued percolation of salty water or environmental waters containing corrosive entities [6]. The access of corrosive environmental waters to the concrete can overcome the chloride threshold, while the diffusion of CO_2 and/or H_2S from some polluted environments can lower the concrete alkaline pH [6]. These can readily compromise the concrete structures' passive film layers, thereby triggering corrosion. Corrosion forms products with larger volumes, leading to cracking, spalling, delamination, or the separation of steel from the concrete, making the steel lose its reinforcing function. The precipitated corrosion product generates tensile stress that provokes the appearance of cracks which compromise the service life of the concrete structures [7]. In practice, the appearance of cracks in the structure is a strong indication that a heightened need for maintenance is required. The corrosion-induced failure of concrete structures has been viewed as the most damaging and scary type of material failure caused by corrosion.

Many efforts have been introduced to minimize the corrosion of rebar in reinforced concretes, ranging from metallic coating (galvanized, stainless steel coating) to organic coatings (epoxy coatings) of the rebar before deploying it into the steel [8, 9]. Though organic coatings have been reported as cheaper surface protection that can be manipulated to offer multifunctionality in steel protection in marine environments [10–12], stainless steel and galvanized steel are believed to exhibit better performance than organic coatings. However, the application of stainless steel is not sustainable due to cost. It is prone to other types of corrosion, such as chloride-induced pitting corrosion, which may occur in coastal environments. Secondly, applying galvanized steel is still confronted with comparably higher costs, and the possible reactions with the concrete materials during the initial curing time, forming hydroxyzincates, remain a bottleneck to its application. Furthermore, organic coatings such as the fusion bonded epoxy coating, which seem to be the most feasible alternative to afford the protection of reinforced concrete rebar in specific applications, are prone to defects, voids, cracks, pits (holiday defeat), etc., which are inevitable during the handling, application, transportation, and installation of steel rebar in reinforcement concrete.

Overall, the aggravation of concrete corrosion largely depends on the ingress of environmental waters with corrosive entities. The corrosion of steel rebars is an electrochemical process containing the dissolution of iron at the anodic sites and the oxygen reduction at the cathodic sites, and the pores in the concrete containing the solution act as the electrolyte. Thus, this chapter discusses the corrosive impact of different environments with elements such as salts, pollutants, and certain chemicals on concrete structures, focusing on how these factors accelerate deterioration and reduce the lifespan of concrete structures. It further evaluates the current materials and construction techniques used to prevent or minimize corrosion in these environments, identifying potential areas for further research or improvement in concrete corrosion prevention and mitigation.

3.2 Factors That Contribute to Concrete Corrosion

Concrete corrosion occurs due to several chemical and physical factors. Several factors intensify concrete corrosion, which can be categorized into physical- and chemical-based factors like salts, pollutants, corrosive chemicals, and moisture.

Physical properties: Several physical factors contribute to the degradation of concrete structures, including erosion, salt scaling, abrasions, and frosting effects. These phenomena can worsen the deterioration of concrete over time [13]. Salty water can penetrate concrete to varying depths under pressure in marine environments. This can trigger reactions between the hydrated cement products and the reinforcing steel, leading to defects that allow harmful ions to infiltrate [13]. Additionally, freezing and thawing cycles and scaling can physically break down concretes. Erosion caused by scaling often results from the combined effects of interfacial salt ions and cold temperatures [14]. In colder regions, the freeze–thaw cycles within the pores and cracks in concretes can lead to disintegration due to expansion.

Salt: salt, or sodium chloride, is a significant factor in concrete corrosion, especially in marine environments, coastal areas, and regions where de-icing salts are used. Salt dissolves inside the water and releases chloride ions, which infuse the concrete, interact with the rebar, and trigger degradation by breaking down the alkalinity and reducing the compressive strength of the concrete structure. Salt-induced corrosion can be affected by the salt concentration [15], the quality of the concrete, exposure time, and environmental conditions. Chloride can interact with tricalcium aluminate (C_3A) and appear as calcium chloroaluminate components of cement, resulting in the swelling of the concrete. The swelling induced by chloride is 2.5 times greater than would have expanded in H_2O permeation. This exhibits deleterious effects on the concrete strength, highlighting the importance of controlling chloride access on concrete [16].

Pollutants: Pollutants contribute to concrete corrosion, particularly in urban and industrial areas. The types of contaminants that can cause concrete to deteriorate include gases like carbon dioxide (CO_2), sulfur dioxide (SO_2), nitrogen oxides, acidic rain [17], dust, etc. The pollutants react with concrete's alkaline compounds and other constituents, reducing compressive strength and causing structural damage. Ivaskova [18] reported air pollution as an essential factor in the deterioration of construction materials. They stated that industrial activities could pollute the atmosphere and hydrosphere particularly. These pollutants aggravate the corrosion of building materials such as concrete, plaster of buildings, art, sculptures, etc. In this report, they collated data on air pollution impact on stone materials and carbon steel in the Slovak Republic [18]. The result was used to generate a corrosion map for the different pollutants, which predicted risk in the analysis, design, and maintenance of steel and concrete structures. The $Ca(OH)_2$ cement component can react with CO_2 and moisture to produce $CaCO_3$. This carbonation has been reported to lower the alkaline pH of concretes, and the process causes a volume gradient in the concrete, ultimately leading to their breakdown [2, 19].

Chemicals: Chemicals like hydrochloric acid, sodium hydroxide, agricultural chemicals, and industrial chemicals can attack concrete structures *via* chemical reactions leading to the expansion and chemical breakdown of the concrete structures [20]. Huang [21] investigated the effect of HCl-induced corrosion on a concrete's mechanical properties. They studied corrosion damage on concrete C25, C45, and C55 as a result of HCl, and the test samples were cured for 360 days in different percent (%) values of aqueous HCl (5%–20%) for 24 h. The dynamic modulus loss, mass loss, and compressive and flexural strength of the concrete samples were taken, and the results showed that the corrosion rate increased with an increase in HCl content.

Moisture and lack of oxygen: Concretes are a mixture of cement, water, sand, and aggregates [22]. The ratio of water to cement is crucial as it influences the workability of the concrete [23]. Moisture reacts with the cement paste, breaks down calcium silicate hydrate, and facilitates the penetration of sulfate ions, causing water infiltration through cracks, pores, and capillaries. Lack of oxygen (anoxic or hypoxic conditions) can also impact corrosion significantly, especially in aquatic and underground environments. Lack of oxygen can shift corrosion from an oxidative to a reductive process, increasing the susceptibility to microbiologically influenced corrosion. Anoxic conditions favor anaerobic microorganisms and increase the production of hydrogen sulfide (H_2S), which improves corrosion [24]. This type of corrosion can be seen in oil and gas, water treatment, etc. Li [25] studied the synergistic effects of moisture, chloride ions, and carbonation on rebar corrosion on mortar. They found that moisture content, which is also increased by chloride and carbonation, significantly alters the resistivity of the mortar, leading to binary corrosion behavior through ohmic control, anode-to-cathode ratio, and area effects. The research provided valuable insight into the improvement of reinforced concrete structures' lifetime. It also highlighted the importance of considering moisture content to mitigate rebar corrosion in concrete structures accurately.

3.3 Concrete Corrosion-Prone Environments

3.3.1 Moisture and Availability of Oxygen

The presence of moisture and oxygen supports concrete corrosion [26]. Oxygen can quickly diffuse through concrete pores, accelerating the steel bars' rusting and resulting in cracks and structural damage [27]. Thus, decreasing the thickness of the concrete cover can shorten the path needed for oxygen to travel to reach the embedded steel. The faster the oxygen can travel, the more oxygen is available for the cathodic reaction [28]. Moisture can weaken the protective environment surrounding concrete structures. Along with moisture, corrosive salts can penetrate the concrete, causing the concrete to corrode and break off. Moreover, excessive moisture creates a favorable environment for mold and mildew, which can cause harm to the concrete surface [27, 29].

3.3.2 Marine Environment

The real-life service time for concrete structures in marine environments is usually much less than the designed predictions because the seawater chemically interacts with the cementitious materials, resulting in concrete deterioration. Generally, chloride-triggered deterioration of concretes has been described as one of the prominent factors of degradation in reinforced concrete structures exposed to marine environments [30]. When the ingress of chloride ions is above the critical chloride content, the degradation of the passive oxide film on rebars results in corrosion initiation. This can critically degrade the concrete structures' capacity and reduce the service life. The presence of chloride in marine environments enhances the formation of Friedel's salt ($3CaO{\cdot}Al_2O_3{\cdot}CaCl_2{\cdot}10H_2O$) and/or Kuzel's salt ($3CaO{\cdot}Al_2O_3{\cdot}0.5CaCl_2{\cdot}0.5CaSO_4{\cdot}10H_2O$) in concrete *via* dissolution and precipitation and ionic exchanges [31, 32]. Besides chlorides and sodium, seawater has other potentially destructive ions to concrete structures. Chlorides alone are not harmful. However, these other ions influence the penetration of chloride, either by controlling the chloride-binding capacity of the paste and the porosity of the concrete cover or by initiating the deterioration of the concrete cover [33]. Figure 3.1(a) shows the relative concentration of the major ions in seawater from different marine areas [34]. Figure 3.1(b) presents a schematic overview of the phase changes in structural concrete uncovered in the marine vicinity and the elemental profiles [33]. Magnesium ions in seawater can lead to a reaction with calcium hydroxide in concrete and precipitate as brucite, decreasing the alkalinity of the pore solution and destabilizing the C-S-H. The decalcified C-S-H would separate or completely transform into the MSH (which is described as a non-cementitious magnesium silicate hydrate (M-S-H) [35].

The sulfate ion in seawater can cause the formation of $CaSO_4{\cdot}2H_2O$ (gypsum) *via* the interaction with $Ca(OH)_2$ or ettringite ($3CaO{\cdot}Al_2O_3{\cdot}3CaSO_4{\cdot}32H_2O$) by reacting aluminate phases [34]. The reaction with sulfates can result in

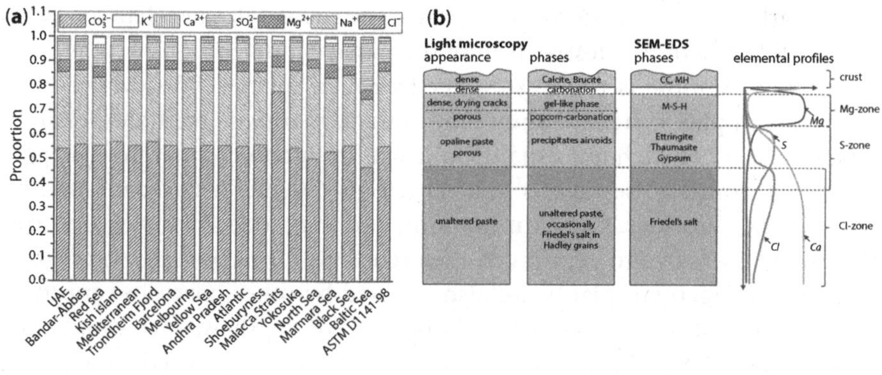

Figure 3.1 (a) Plot of the proportion of elements in seawater from different places. Reprinted with permission [33]. (b) Illustration of the different zones near the concrete surface and the observations with light microscopy and SEM-EDS. Reprinted with permission [34].

cracking of the concrete. Furthermore, the presence of sulfates decreases the chloride-binding capacity of the paste. At temperatures <15°C, when concretes containing calcium carbonate encounter sulfate from seawater, it can transform C–S–H into thaumasite. Thaumasite is a product of a non-cementing reaction that can cause concrete scaling [33].

Furthermore, the carbonates in seawater can interact with the Ca^{2+} in the pore medium (in the concretes), resulting in the development of calcium carbonates, which can seal the concrete surface. In addition, carbonates in the paste can decrease the chloride-binding ability of the cement [33].

3.3.3 Industrial Environments

Concrete structures that construct many industrial complexes are usually exposed to corrosive attacks. For instance, during the steel/iron fabrication process, vast amounts of CO_2 and SO_2 are discharged, which can provoke aggressive corrosion actions in concrete structures within the environment. CO_2 and SO_2 can diffuse into the concrete structure and react with the hydration products of the cement. CO_2 can convert $Ca(OH)_2$, calcium silicate hydrate (C–S–H), and calcium aluminate into $CaCO_3$ [36]. SO_2 interacts with all calcium compounds of hydration products, converting them into sulfur-bearing compounds. Moreover, both sulfuration and carbonation of concrete can decrease the pH of the solution in the concrete apertures (pores). Usually, the pH of the concrete aperture solution is between 12.5 and 13.8, where the thin iron oxide protective film around the rebar is stable [36]. The passive film destabilizes when the pH decreases due to carbonation and sulfuration. Once the pH is ~9, the rebar starts to corrode because of the destruction of the passive film [37].

Another example is the wastewater treatment plants. Wastewater systems are complex systems comprising sewers, maintenance holes, chambers, tanks, etc., which are usually built with reinforced concrete. The pH of normal domestic sewage is neutral (pH 6–8), and it has compounds like sulfates, chlorides, CO_2, ammonium, and magnesium ions, which are potentially corrosive to reinforced concrete [38]. The concrete structures in waste management plants are also affected by other corrosion mechanisms such as abrasion erosion, carbonation, freeze–thaw, cavitation, and microbiologically induced corrosion (MIC). MIC is regarded as a main factor in the corrosion of wastewater concrete structures [39].

Another example is the nuclear power plant. Many of the structural components of nuclear power plants are fabricated using reinforced concrete because it can prevent the release of radiation and attenuate gamma/neutron radiation [40]. Reinforcement corrosion is the leading cause of concrete degradation in nuclear power plants [41]. Some aspects of the corrosion can be linked to radiation from the reactor core, which can cause corrosion in the construction materials and water radiolysis. Impartation of neutron irradiation escalates the chances of stress corrosion cracking in the material because new corrosion mechanisms will be initiated *via* radiation damage and the high corrosion potential of the reactor coolant [41]. Besides reinforcement corrosion, liner plates, structural steel, and prestressing steel can be degraded. The corrosion activities in liner plates and structural steel are the same as the reinforcing steel's corrosion process, which can be pitting, crevicing, galvanic corrosion, and stray electrical currents [42].

3.3.4 De-Icing Salt Environments

In cold climates, de-icing salts is essential for constructing concrete roads to ensure they remain operational during winter [43]. However, this practice can lead to a significant issue known as salt scaling, which is particularly problematic in these regions. Salt scaling is a multifaceted problem that results in the spalling of hardened mortar and the accumulation of coarse material on the surface of concrete roads, ultimately diminishing their lifespan [44]. This surface spalling occurs when the stress imbalance between de-icing salt and water, driven by thermal expansion, surpasses the concrete's tensile strength, causing cracks to form and propagate. Although salt scaling alone may not severely damage the concrete, it often exposes coarse aggregates, allowing water and other corrosive substances to infiltrate it, thus further compromising its durability and longevity [45].

3.3.5 Underground Environments

In most cases, the underground parts of civil engineering structures are made with reinforced concrete due to their excellent anticorrosion properties in rock and soil environments compared to other building materials [45]. Notwithstanding, some conditions can still cause corrosion of steel in underground reinforced concrete. For example, when the steel is not completely covered by concrete because of delamination, cracks, honeycombs, or very low cover depths, the uncovered steel can easily undergo corrosion in most soil environments. The soil environment is generally characterized by near-neutral pH conditions, moisture, and corrosive species like carbonates or chloride ions, which can result in high rates of anodic dissolutions. This is made worse by the development of galvanic elements within the structure. This can occur when the uncovered steel, acting as the anode, is in an electrical connection with reinforcing steel working as the cathode [45]. Moreover, in connection with the underground, reinforced concrete, the soil offers an electrolytic pathway *via* which the galvanic current can flow [45].

3.3.6 Agricultural Environments

Concrete structures in agricultural environments are exposed to aggressive bio-waste and agricultural and agro-food effluents [46, 47]. Although these effluents have diverse compositions, they share some common characteristics. They comprise complex organic compounds in particulate or dissolved form and microbial communities. Through microbial actions, the chemical compounds in the effluents are broken down into simpler forms, and the end products are primarily organic acids, ammonium, and gases (e.g., CO_2, H_2S, CH_4, H2S, etc.) [48]. Both the end products and the microbes are aggressive for the cementitious materials.

Furthermore, mechanical activities like high-pressure cleaning, movement of heavy machinery, scraping, etc., can add to the physicochemical adjustment. This can lead to erosion of cover concrete, loss of alkalinity, reduced mechanical strength, and a loss of impenetrability of the concrete to liquids and gases [48]. The resultant effect is the degradation of the structural integrity of the concrete.

3.3.7 Coastal Environments

Coastal environments are known to be rich in chloride ions. Thus, concrete structures in coastal areas are exposed to various environmental conditions like splash, tidal, and atmospheric zones, which can accelerate chloride-induced corrosion [49]. The diffusion of chloride into the concrete is considered one of the main external factors that cause corrosion on reinforced concrete structures. The detrimental effects of reinforcement corrosion can manifest in several ways, including a reduction in the steel cross-section, a weakened bond between the concrete and steel, the formation of cracks, and the spalling or flaking off of the concrete cover [50].

3.4 Distinctive Analysis of the Process of Concrete Corrosion

The distinctive analysis of various processes of concrete corrosion is vital in providing solutions to concrete deterioration due to corrosion [51–53]. The corrosion processes are generally described using the corrosion rate and reduction in concrete lifespan. Thus, this section will elaborate on the mechanism of corrosion acceleration and concrete lifespan reduction.

3.4.1 Mechanism Associated with the Acceleration of Concrete Corrosion

Different mechanism approaches have been established to describe the acceleration of concrete corrosion [54, 55]. One of the studies described rebar concrete constructions that show the mechanisms related to advancements in electrical techniques for corrosion monitoring [4]. According to Chen *et al.* [56] effective accelerated corrosion assessment technique refers to the basis for the laboratory analysis of corrosion characteristics of rebar metal placed within concrete and mortar. Based on the studies of Chen *et al.* [56], the electrochemical features of standard electrolytic accelerated corrosion techniques frequently used by scientists are surface coating, embedded auxiliary electrodes, part soaking, and soaking techniques. A computational model for forecasting the development of corroded products surrounding rebar was developed according to the unique electrochemical process. The factors that affected electrolytic accelerated corrosion were examined, such as the electrode placement for the embedded electrode technique and the extent of electrolyte media with other techniques measured. It was determined that the dispersal patterns of the corrosion product changed depending on the

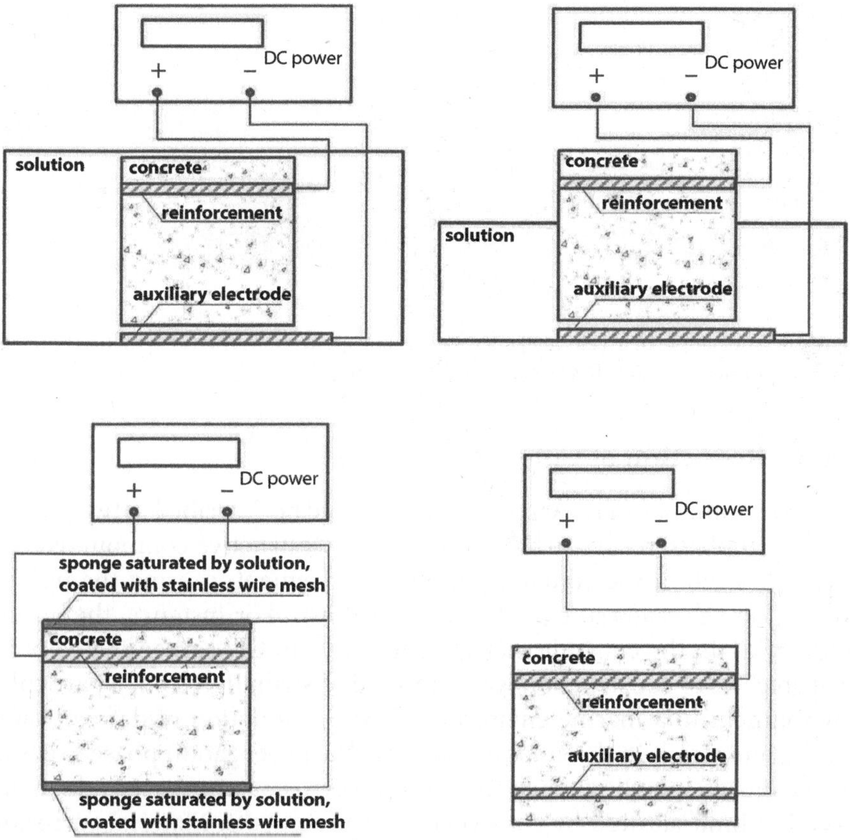

Figure 3.2 Different designs of electrolytic accelerated corrosion techniques. Reprinted with permission [56].

electrochemical technique used; an irregular–uniform distribution form of corroded particles within reinforced steel that resembles the natural corrosion may be quickly created by regulating and adjusting the important features of the techniques as shown in Figure 3.2 [56].

Jin *et al.* [57] demonstrated an example of accelerated corrosion in rebar. In this case, accelerated corrosion within a certain constant voltage of about 30 V was adopted for the rebars by linking it with the positive electrode of an electricity supply [57]. The humidity was kept at 60% ± 10% and the experimental–simulated temperature was regulated within 25°C ± 0.5°C. The activities on the concrete material surface were regularly seen through a photographic camera. The study revealed that Cl^- ions could migrate and induce rebar corrosion within concrete with cracks. The schematic figure of the characterization is shown in Figure 3.3.

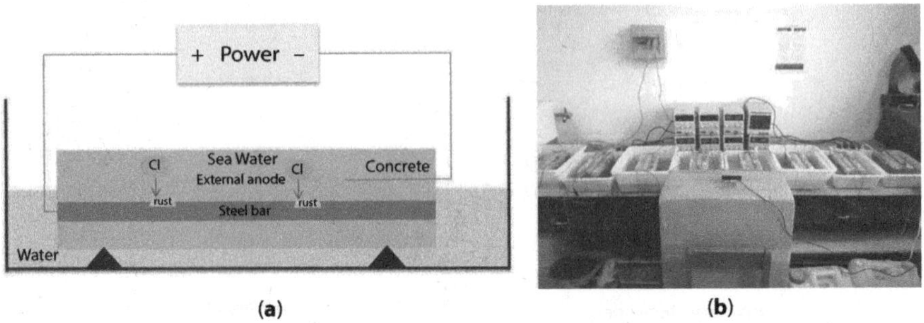

Figure 3.3 A sample of accelerated corrosion using applied current: (a) Schematic picture and (b) Laboratory image. Reprinted with permission [57].

3.4.2 Reduction of Concrete Lifespan

The service lifespan of rebar constructions could be described in two phases. The initial phase relates to the movement of destructive compounds—Cl^- and CO_2—within the concrete to the reinforcement, causing the passive coating on the material exterior to break down. For instance, the occurrence of Cl^- in the concrete matrix could be from Cl^--contaminated parts of aggregates of construction water, through distribution from a particular environment to a marine environment, or the adoption of defrost NaCl, $MgCl_2$, and $CaCl_2$ in the cold period [58]. The propagation phase exists in an active state pending when the level of corrosion approaches the deterioration limit allowed by construction requirements. Material corrosion is largely regarded as passive when the current density is reported to be around 0.1 $\mu A\ cm^{-2}$ and in the active stage when the rate exceeds 1 $\mu A\ cm^{-2}$ [59]. Regarding Tuutii's model, several studies measured the fluctuation in the corrosion process when the service lifespan of reinforced structure [60–62] or involved extra phases for distinguishing corrosion products development, enclosed fracture, and spalling/delamination within the description of concrete service lifespan are as shown in Figure 3.4.

A modern roadmap required for evaluating carbonation, lifespan, and embodied C of reused concrete in structures was reported by Dervishaj *et al.* [63]. The report showed that carbonation was recorded at about 19 kg CO_2/m^3 especially in the first service life, when the projection was nearly 5 kg CO_2/m^3. Furthermore, about 6% of the embodied C was absorbed as CO_2 during the service lifespan, including the reuse. The reuse far outperformed carbonation in terms of climatic advantages. Moreover, carbonation had no discernible impact on the outcomes of assessments of the embodied C of re-used resources in future life cycles that employed

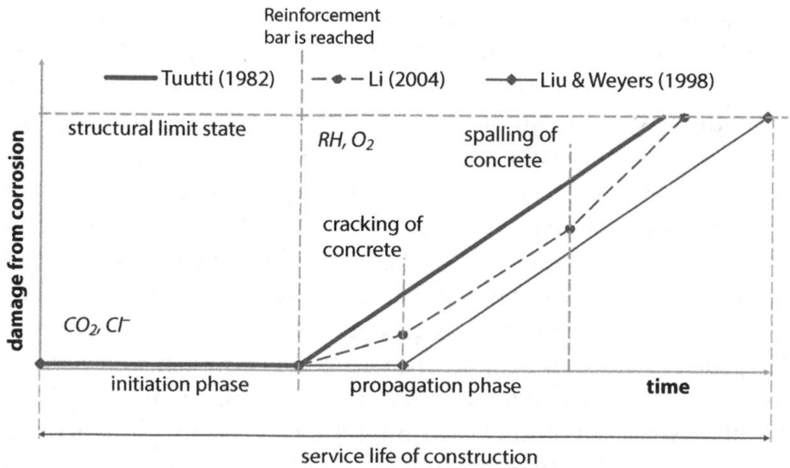

Figure 3.4 Diverse stages of acceleration of concrete corrosion. Reprinted with permission [63].

cutoff and scattered allocations. The digital method was useful for quickly assessing lifespan, embedded C, and carbonation of concrete material. This method could serve as a policymaking aid while preparing to recycle [14]. Another study by Ram *et al.* [64] gave some suggestions regarding the carbon footmark drop impairing machine-driven characteristics and the service lifespan of notable concrete materials [64]. The findings showed that the cement contents significantly impacted the durability of the concrete. Since all alternate mixes had a higher or similar sustainability score with less cement, the concrete mix design was deemed "over-designed".

3.5 Current Practices Used to Minimize Corrosion in Different Environments

3.5.1 Surface and Environmental Modification

The use of corrosion-resistant alloys, corrosion-inhibitive additives, and coatings (metallic and organic) dominate the materials often employed to control rebar corrosion in concretes. Among the materials used are inhibitors. The usage of inhibitors has been widespread to stop corrosion in reinforced concrete caused by chloride. Corrosion inhibitors are considered a valuable method for delaying the beginning of corrosion and/or lowering its incidence. They have been tested in steels/alloys in different environments [65–68]. According to ISO 8044, a corrosion inhibitor is any additive that, upon introduction at the proper (preferably low) concentration, impedes the corrosion rate without significantly altering the corrosive medium system. Numerous organic compounds, such as amines, alkanolamines, carboxylates, and esters, have been investigated as inhibitors since the 1990s [69]. For example, concrete could be efficiently treated using triethylenetetramine (TETA), a corrosion inhibitor. In a work by Xu *et al.* [70], the concentration of TETA surrounding the rebar was sufficient to prevent corrosion. Another study investigated the corrosion resistance of steel rebars by incorporating pozzolanic components into the concrete mixture, such as silica fume, polypropylene fibers, fly ash, and FerroGard 901 (industrial 2-dimethylaminoethanol) inhibitors. An analysis comparing the effectiveness of three electro-migrating corrosion inhibitors, guanidine, 3-aminopropyltriethoxysilane (APS), and 1,6-hexamethylenediamine (HMDA), in restoring deteriorated reinforced concrete and extending the lifespan of concrete has been enumerated [71]. APS and HMDA effectively increase carbonation and water resistance. Guandine has limited potential to improve durability. However, it is good in re-alkalization.

Ostnor and Justnes [72] studied steel rebar cylinders exposed in the sea tidal zone, rooftop, for 4 years at the ambient conditions of 38°C of air temperature and 90% of RH (relative humidity) and investigated the influence of some anodic inhibitors including $Ca(NO_3)_2$, $Ca(NO_2)_2$, and $NaNO_2$. Their analysis showed that negative half-cell potentials were obtained when anodic inhibitors were used, compared to what the visual examinations showed. The inhibitor dosage of 3%–4% of cement weight seems to be sufficient to prevent corrosion on the rebar, as demonstrated in Figure 3.5.

The surface of concrete is sometimes treated with inhibitors. These inhibiting agents should be able to migrate into the concrete's pores through capillary action and become absorbed into the metal reinforcement inside the structure. They have also been described as migration corrosion inhibitors. They can prevent and slow down the corrosive process of concrete structures and reinforcing [73–75]. Green corrosion inhibitors have also experienced tremendous interest in preventing concrete corrosion. A wide range of green corrosion inhibitors, including biopolymers, medicinal compounds, surface active agents, and other natural products, are becoming more popular. These inhibitors are used to prevent corrosion in reinforced concrete structures because they are readily available, environmentally friendly, biodegradable, low toxicity, and renewable. When the organic compounds found in these eco-based inhibitors are added to the mixing water in the reinforced concrete specimen, they are absorbed on the metal surface, creating a passivation layer that shields the active area from corrosion [75, 76].

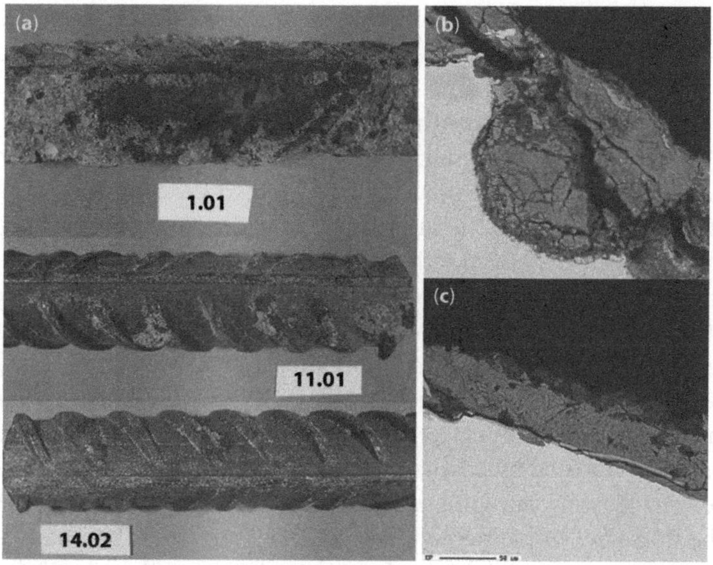

Figure 3.5 (a) Concrete cylinders that have had their rebars separated with chloride without inhibitor (mix 1), calcium nitrate (mix 11), and nitrite (mix 14) added have been kept for 4 years at 38°C and 90% relative humidity with air access. (b) Rebar implanted in concrete (mix 4) with additional chlorides and kept for 4 years at 38°C and 90% relative humidity developed porous corrosion products on its surface. The products of corrosion eat away at the steel body, causing pitting. (c) A dense iron oxide layer around the steel body of rebar inserted in concrete (mix 5) with chloride and nitrate inhibitors added and kept for 4 years at 38°C and 90% relative humidity. Reprinted with permission [72].

To enhance structural corrosion resistance, steel rebars have been treated with various primers and coatings, including hot-dip galvanization, zinc-rich epoxy primers, alkyd-based primers, polyamide epoxy top coatings, polyamide epoxy primers, polyurethane coatings, and combinations such as double layers of epoxy primer with alkyd top coatings and double layers of alkyd primer with top coatings. The optimal reinforced concrete mix design incorporated 25% fly ash, 10% silica fume, and 3% FerroGard 901 inhibitors by weight of cementitious materials and rebars coated with zinc-rich epoxy primer.

3.5.2 Trends and Best Design Practices for Improvement in the Field of Concrete Corrosion Prevention and Mitigation

In curbing the menace of concrete corrosion and prioritizing smart prevention protocols, attempts have been made to incorporate corrosion detection viz-a-viz protection techniques to proffer robust engineering solutions. Both direct detection techniques based on physical and electrochemical concepts associated with the corrosion process of steel and indirect techniques based on the measurement of corrosion-induced damages in reinforced concrete have been examined. According to the protecting mechanism, there are two types of corrosion protection methods: "prevention solutions" and "therapy solutions." The former include corrosion inhibitors, anti-corrosion coatings, and high-performance fiber-reinforced cementitious composite (HPFRCC) overlay, while the latter include cathodic protection (CP) and electrochemical chloride extraction (ECE). Although it is primarily used for strengthening due to its superior mechanical qualities, HPFRCC overlay is thought to be effective in preventing corrosion because of its high durability, whereas carbon-fiber-reinforced polymer (CFRP) can function as both an anode and a strengthening material in CP and ECE. These materials' twin purposes make them extremely promising for repairing RC structures that have been harmed by corrosion [77].

Moreover, it is encouraging when corrosion prevention and structural strengthening are implemented together. To restore the load-carrying capacity, the conventional strengthening technique often involves applying fiber-reinforced polymer (FRP) composites, steel plates, or concrete patches around the columns or beneath the bottom of beams or slabs. The repaired system may experience ongoing corrosion, which could be much more severe due to galvanic coupling of the repaired and unrepaired portions unless corrosion prevention measures are implemented concurrently to minimize or eliminate additional corrosion [78]. But until now, corrosion protection has not gotten nearly as much attention as strengthening and restoring corrosion-affected structures. The need to extend the service life of existing structures will become more pressing soon due to the growing number of aging infrastructures in many nations worldwide and the increasing scarcity of natural resources like water and sand used to make concrete.

3.6 Cases of Failures due to Corrosion of Concrete Structures in Different Environments and the Mitigations

Corrosion of concrete structures, particularly the corrosion of reinforcing steel within the concrete, can significantly harm the integrity, durability, and safety of concrete structures. The impact of corrosion can vary depending on the environment in which the concrete structure is located. In chloride-rich environments such as in the marine environment, chloride ingress to the steel region can lead to the breakdown of the passive oxide layer that protects the steel and initiates corrosion with deleterious effects on marine structures like (a) bridges and piers, (b) offshore platforms used for oil and gas exploration and production, (c) seawalls and breakwaters constructed to protect the coastal areas from erosion and wave action, (d) harbor and port structures, (e) concrete ships and floating structures, (f) foundation piles which support various marine structures, and (g) tunnels that are used for transportation and utility purposes [79]. Some notable accidents caused by the corrosion of marine structures attributed to chloride-induced corrosion in history include the Silver Bridge collapse (1967), the Mianus River Bridge collapse (1983), the or-Babeldaob Bridge collapse (1996), the -35W Mississippi River Bridge collapse (2007), etc. [79]. These cases illustrate the severe consequences of chloride-induced concrete corrosion on the structural integrity and safety of various infrastructure elements. They underscore the importance of implementing effective corrosion prevention and mitigation strategies. Going further, in industrial and urban environments, diffusion of chemical agents such as acids, alkalis, and sulfates may reach the steel of the concrete structures, triggering corrosion. This attack leads to decreased concrete strength and increased permeability, compromising structural integrity. In urban environments, carbon dioxide (CO_2) from vehicle emissions and industrial activities can diffuse and react with the calcium hydroxide in concrete to form calcium carbonate through carbonation. This lowers the pH of the concrete, causing rust formation, cracking, and spalling of the concrete cover. Many highway bridges, especially those built in the mid-20th century, have experienced significant carbonation-induced cracking. For example, the M4 highway bridge in the UK has undergone extensive repairs due to carbonation and subsequent reinforcement corrosion [80]. Several apartment buildings in Hong Kong have reported structural issues due to carbonation, necessitating costly repairs and maintenance [80]. Cold environments and hot climates have also been reported to influence concrete corrosion. Higher temperatures can provoke thermal cracking, especially if the structure is restrained and cannot freely expand or contract.

Repeated thermal cycles can cause fatigue, which can lead to structural failure over time. Thermal-induced expansion has been reported to cause highway cracking, leading to costly repairs and maintenance. Additionally, the use of de-icing salts can introduce chlorides into the concrete. In buried structures (subterranean environments), soil with high sulfate content or acidic groundwater has caused the cracking and deterioration of concrete. Acidic conditions can degrade the concrete matrix and promote the corrosion of the reinforcement. Underground concrete structures at various industrial sites, such as chemical plants, have suffered from acid attack due to the presence of acidic waste in the soil. This has led to significant structural damage and the need for frequent repairs. An instance can be related again to the Paris Catacombs that have experienced deterioration due to acidic groundwater, which has led to the weakening of the limestone and concrete structures within the subterranean network. There are reports on the reduction in porosity of permeable reactive barriers due to bio-geochemical clogging caused by acidic groundwater flow [81]. These have severe consequences, including loss of structural integrity, reduced load-carrying capacity, and increased maintenance and repair costs.

Preventive measures and mitigation: Chloride-induced concrete degradation can be mitigated by ensuring a high-quality concrete mix design with the capability to afford very low permeability to chloride penetration (low permeable concretes). Other recommended measures include the use of surface sealers or coatings to create a barrier against chloride penetration, the inclusion of corrosion-inhibiting additives into the concrete mix to impede corrosion progress of the reinforcing steel in the concrete, cathodic protection (either sacrificial or impressed current method), etc. In addition, using electrochemical techniques to extract chlorides from the concrete, optimized materials, inspection, and repairs are additional strategies to protect concrete structures from chloride-induced corrosion degradation. Reducing the risk of corrosion CO_2 ingress into concrete structures can be mitigated by designing low-permeability concrete modern construction practices. Secondly, appropriate protective coatings can significantly reduce carbonation rates, coupled with routine inspections to detect signs of carbonation and corrosion. Thermal cracking effects on concrete highways can be mitigated by incorporating expansion joints in concrete structures that accommodate movement due to thermal expansion and contraction, thereby reducing the risk of cracking. Secondly, applying reflective coatings to concrete surfaces can help reduce reflected sunrays, reducing the structures' heat absorption, which alleviates the thermal stresses.

3.7 Conclusions

Concrete corrosion is a multifaceted issue influenced by various physical and chemical factors. Physical factors such as erosion, salt scaling, abrasions, and freeze–thaw cycles significantly contribute to the degradation of concrete structures. These processes can lead to the penetration of harmful ions and the physical breakdown of the concrete, especially in marine environments where salty water and freezing temperatures are prevalent. Chemical agents, such as chlorides, play a crucial role in concrete corrosion, particularly in coastal areas and regions using de-icing salts. The chloride ions from dissolved salts penetrate the concrete, react with reinforcement steel, and cause concrete deterioration, highlighting the importance of controlling chloride access to maintain concrete integrity. The role of other chemical species present in distinct industrial or polluted environments, such as carbon dioxide, sulfur dioxide, nitrogen oxides, acidic rain, and dust, in concrete corrosion was also highlighted.

The environmental and surface modifications employed to control concrete corrosion include using corrosion-resistant alloys, corrosion-inhibitive additives, and various coatings. While inhibitors work by altering the electrochemical environment around the rebar, thus reducing the potential for corrosion, coatings provide an impervious barrier against corrosive entities from the environment. Coatings and primers on steel rebars also enhance the structural resistance to corrosion. Integrating these methods makes it possible to significantly reduce the incidence of concrete corrosion and extend the service life of concrete structures. Recent trends in controlling concrete corrosion emphasize integrating detection and protection techniques to provide comprehensive solutions. The chapter also highlights essential approaches for extending the service life of structures, especially as aging infrastructures increase and natural resources become scarcer. It underscores the need for comprehensive strategies that address both the prevention and repair of concrete corrosion to ensure the longevity and safety of concrete structures.

Acknowledgment

D. I. Njoku acknowledges the Hong Kong Metropolitan University for granting me a Research Associate Position in 2024.

AI disclosure statement: The author acknowledges the use of AI-based tools for minor language editing or grammatical correction. However, all intellectual contributions, analyses, and conclusions presented in this chapter are entirely the author's own.

References

1. Gagg, C.R., Cement and concrete as an engineering material: An historic appraisal and case study analysis. *Eng. Fail. Anal.*, 40, 114–140, 2014.
2. Qu, F., Li, W., Dong, W., Tam, V.W.Y., Yu, T., Durability deterioration of concrete under marine environment from material to structure: A critical review. *J. Build. Eng.*, 35, 102074, 2021.

3. Elliot, D.F., Wright, C.H., Yuers, K.L., Better reinforcement for concrete. *High Perform. Text.*, 00, MAR., 6–7, 1999.
4. Rodrigues, R., Gaboreau, S., Gance, J., Ignatiadis, I., Betelu, S., Reinforced concrete structures: A review of corrosion mechanisms and advances in electrical methods for corrosion monitoring. *Constr. Build. Mater.*, 269, 121240, 2021.
5. Yang, K.-H., Lim, H.-S., Kwon, S.-J., Kim, J.-H., Repair cost estimation techniques for reinforced concrete structures located at the seashore: Considering various probabilistic service life functions and actual mix proportions. *Constr. Build. Mater.*, 256, 119469, 2020.
6. James, A., Bazarchi, E., Chiniforush, A.A., Panjebashi Aghdam, P., Hosseini, M.R., Akbarnezhad, A., Martek, I., Ghodoosi, F., Rebar corrosion detection, protection, and rehabilitation of reinforced concrete structures in coastal environments: A review. *Constr. Build. Mater.*, 224, 1026–1039, 2019.
7. Taheri-Shakib, J. and Al-Mayah, A., 4D evolutions of cracks, voids, and corrosion products in reinforced concrete materials. *Sci. Rep.*, 13, 1, 22455, 2023.
8. Astuti, P., Argenta, A.V., Fahma, R.K., Permatasari, D., Corrosion prevention on reinforced concrete using steel coating of rebar. *AIP Conf. Proc.*, 2846, 1, 90002, 2023.
9. Afshar, A., Jahandari, S., Rasekh, H., Shariati, M., Afshar, A., Shokrgozar, A., Corrosion resistance evaluation of rebars with various primers and coatings in concrete modified with different additives. *Constr. Build. Mater.*, 262, 120034, 2020.
10. Njoku, D.I., Njoku, A.C., Udoh, I.I., Uzoma, P.C., Etim, I.I.N., Li, B., Li, Y., Understanding the multifunctional anticorrosion protective mechanism of epoxy-based coatings modified with hydrogel and benzotriazole conveying nanotubes for Q235 steel protection in 3.5 % NaCl. *Surf. Coat. Technol.*, 484, May, 130851, 2024.
11. Njoku, D.I., Li, B., Khan, M.S., Chinonso, U.P., Njoku, C.N., Onyeachu, I.B., Li, Y., Quadruple-action coatings provided by doping epoxy with inhibitor laden clay nanotubes functionalized with layer-by-layer of cross-bridged chitosan and anionic polyelectrolytes. *Prog. Org. Coat.*, 157, 106312, 2021.
12. Njoku, D.I., Cui, M., Xiao, H., Shang, B., Li, Y., Understanding the anticorrosive protective mechanisms of modified epoxy coatings with improved barrier, active and self-healing functionalities: EIS and spectroscopic techniques. *Sci. Rep.*, 7, 1, 1–15, 2017.
13. Uthaman, S. and Vishwakarma, V., Assessment of causes and consequences of concrete deterioration and its remediation. *J. Build. Eng.*, 79, 107790, 2023.
14. Müller, M., Wiedey, R., Hoheisel, W., Serno, P., Breitkreutz, J., Impact of co-administered stabilizers on the biopharmaceutical performance of regorafenib amorphous solid dispersions. *Eur. J. Pharm. Biopharm.*, 169, 189–199, 2021.
15. Fang, Y., Yao, Z., Li, X., Xu, Y., Huang, X., Durability of reactive powder concrete of drilling shaft and triaxial compression damage constitutive model under composite salt erosion. *J. Build. Eng.*, 62, 105395, 2022.

16. Watanabe, A., Furukawa, H., Miyamoto, S., Minagawa, H., Non-destructive chemical analysis of water and chlorine content in cement paste using near-infrared spectroscopy. *Constr. Build. Mater.*, 196, 95–104, 2019.
17. Yong, D., Effect of Acid Rain Pollution on Durability of Reinforced Concrete Structures. *IOP Conf. Ser. Earth Environ. Sci.*, 450, 1, 12115, 2020.
18. Ivaskova, M., Kotes, P., Brodnan, M., Air Pollution as an Important Factor in Construction Materials Deterioration in Slovak Republic. *Procedia Eng.*, 108, 131–138, 2015.
19. Possan, E., Felix, E.F., Thomaz, W.A., CO2 uptake by carbonation of concrete during life cycle of building structures. *J. Build. Pathol. Rehabil.*, 1, 1, 1–9, 2016.
20. Chakraborty, S., Mandal, R., Chakraborty, S., Guadagnini, M., Pilakoutas, K., Chemical attack and corrosion resistance of concrete prepared with electrolyzed water. *J. Mater. Res. Technol.*, 11, 1193–1205, 2021.
21. Huang, P., Bao, Y., Yao, Y., Influence of HCl corrosion on the mechanical properties of concrete. *Cem. Concr. Res.*, 35, 3, 584–589, 2005.
22. Zou, Y.-X., Zuo, X.-B., Zhang, H.-L., Wang, S.-Q., Influence of fly ash and chlorides on the behavior of sulfate attack in blended cement pastes. *Constr. Build. Mater.*, 394, 132231, 2023.
23. Shi, W., Najimi, M., Shafei, B., Reinforcement corrosion and transport of water and chloride ions in shrinkage-compensating cement concretes. *Cem. Concr. Res.*, 135, 106121, 2020.
24. Goethals, J., De Windt, L., Miron, G.D., Wittebroodt, C., Abdelouas, A., Anoxic corrosion of carbon steel in high pH cementitious media and high temperature conditions: New insights on the formation of (Fe,Al) Si-hydrogarnet corrosion product. *Corros. Sci.*, 237, 112318, 2024.
25. Li, G., Evitts, R., Boulfiza, M., Interactive effects of moisture, chloride, and carbonation on rebar corrosion in mortar. *Constr. Build. Mater.*, 440, 137440, 2024.
26. Ahmad, S., Reinforcement corrosion in concrete structures, its monitoring and service life prediction—-a review. *Cem. Concr. Compos.*, 25, 4, 459–471, 2003.
27. Li, S., Jin, Z., Chang, H., Gao, Y., Oxygen diffusion into unsaturated seawater and sea sand concrete and its effect on non-uniform corrosion of steel bars. *J. Build. Eng.*, 84, 108665, 2024.
28. Scott, A. and Alexander, M.G., The influence of binder type, cracking and cover on corrosion rates of steel in chloride-contaminated concrete. *Mag. Concr. Res.*, 59, 7, 495–505, 2007.
29. Jahandari, S., Tao, Z., Alim, M.A., Li, W., Integral waterproof concrete: A comprehensive review. *J. Build. Eng.*, 78, 107718, 2023.
30. Val, D.V. and Stewart, M.G., Life-cycle cost analysis of reinforced concrete structures in marine environments. *Struct. Saf.*, 25, 4, 343–362, 2003.
31. Glasser, F.P., Kindness, A., Stronach, S.A., Stability and solubility relationships in AFm phases. Part I. Chloride, sulfate and hydroxide. *Cem. Concr. Res.*, 29, 6, 861–866, 1999.

32. Birnin-Yauri, U.A. and Glasser, F.P., Friedel's salt, Ca2Al(OH)6(Cl,OH)·2H2O: Its solid solutions and their role in chloride binding. *Cem. Concr. Res.*, 28, 12, 1713–1723, 1998.
33. Jakobsen, U.H., De Weerdt, K., Geiker, M.R., Elemental zonation in marine concrete. *Cem. Concr. Res.*, 85, 12–27, 2016.
34. Yi, Y., Zhu, D., Guo, S., Zhang, Z., Shi, C., A review on the deterioration and approaches to enhance the durability of concrete in the marine environment. *Cem. Concr. Compos.*, 113, 103695, 2020.
35. De Weerdt, K. and Justnes, H., The effect of sea water on the phase assemblage of hydrated cement paste. *Cem. Concr. Compos.*, 55, 215–222, 2015.
36. Lv, Y., Niu, D., Liu, X., Lin, M., Li, Y., Corrosion damage and life prediction of concrete structure in the coking ammonium sulfate workshop of iron and steel industry. *Sci. Rep.*, 13, 1, 2826, 2023.
37. Neves, R., Branco, F., de Brito, J., Field assessment of the relationship between natural and accelerated concrete carbonation resistance. *Cem. Concr. Compos.*, 41, 9–15, 2013.
38. Von Sperling, M., *Wastewater Characteristics, Treatment and Disposal*, IWA Publishing, 2007.
39. Woyciechowski, P., Łukowski, P., Szmigiera, E., Adamczewski, G., Chilmon, K., Spodzieja, S., Concrete corrosion in a wastewater treatment plant – A comprehensive case study. *Constr. Build. Mater.*, 303, 124388, 2021.
40. Bruck, P.M., Esselman, T.C., Elaidi, B.M., Wall, J.J., Wong, E.L., Structural assessment of radiation damage in light water power reactor concrete biological shield walls. *Nucl. Eng. Des.*, 350, 9–20, 2019.
41. Rasheed, P.A., Nayar, S.K., AlFantazi, A., Concrete corrosion in nuclear power plants and other nuclear installations and its mitigation techniques: a review. *Corros. Rev.*, 42, 1, 57–73, 2024.
42. Kambayashi, D., Sasano, H., Sawada, S., Suzuki, K., Maruyama, I., Numerical Analysis of a Concrete Biological Shielding Wall under Neutron Irradiation by 3D RBSM. *J. Adv. Concr. Technol.*, 18, 10, 618–632, 2020.
43. Zhang, J., Wang, J., Li, Y., Yuan, J., Wu, Y., Research progresses on salt scaling and protective methods for concrete pavements. *Constr. Build. Mater.*, 342, 127993, 2022.
44. Wang, K., Nelsen, D.E., Nixon, W.A., Damaging effects of deicing chemicals on concrete materials. *Cem. Concr. Compos.*, 28, 2, 173–188, 2006.
45. Pfändler, P., Bircher, L., Angst, U., Inspecting the corrosion state of underground reinforced concrete structures. *J. Infrastruct. Preserv. Resil.*, 3, 1, 17, 2022.
46. Maraveas, C., Durability Issues and Corrosion of Structural Materials and Systems in Farm Environment. *Appl. Sci.*, 10, 3, 990, 2020.
47. Assaad Abdelmseeh, V., Jofriet, J., Hayward, G., Sulphate and sulphide corrosion in livestock buildings, Part I: Concrete deterioration. *Biosyst. Eng.*, 99, 3, 372–381, 2008.

48. Bertron, A., Peyre Lavigne, M., Patapy, C., Erable, B., Biodeterioration of concrete in agricultural, agro-food and biogas plants: state of the art and challenges. *RILEM Tech. Lett.*, 2, 83–89, 2017.
49. Quaranta, G., Giaccu, G.F., Briseghella, B., Nuti, C., Probabilistic corrosion hazard maps for reinforced concrete infrastructure in coastal regions. *Struct. Infrastruct. Eng.*, 1–20, 2024.
50. Rincon, L.F., Moscoso, Y.M., Hamami, A.E.A., Matos, J.C., Bastidas-Arteaga, E., Degradation Models and Maintenance Strategies for Reinforced Concrete Structures in Coastal Environments under Climate Change: A Review. *Buildings*, 14, 3, 562, 2024.
51. Etim, I.-I.N., Ikeuba, A.I., Njoku, C.N., Predictive Models of Corrosion Advances in Corrosion Modelling: A Case Study of Fuzzy Modelling in Pipeline Coating BT, in: *Advances in Corrosion Modelling*, R. Javaherdashti, (Ed.), pp. 167–188, Springer Nature Switzerland, Cham, 2024.
52. Etim, I.-I.N., Wei, J., Dong, J., Xu, D., Chen, N., Wei, X., Su, M., Ke, W., Mitigation of the corrosion-causing Desulfovibrio desulfuricans biofilm using an organic silicon quaternary ammonium salt in alkaline media simulated concrete pore solutions. *Biofouling*, 34, 10, 1121–1137, 2018.
53. Lv, X., Wang, C., Liu, J., Sand, W., Etim, I.N., Zhang, Y., Xu, A., The Microbiologically Influenced Corrosion and Protection of Pipelines : A Detailed Review, pp. 1–20, 2024.
54. Etim, I.-I.N., Dong, J., Wei, J., Nan, C., Felix Daniel, E., Babu Subedi, D., Xu, D., Prasad Yadav, A., Su, M., Ke, W., Mitigation of sulphate-reducing bacteria attack on the corrosion of 20SiMn steel rebar in sulphoaluminate concrete using organic silicon quaternary ammonium salt. *Constr. Build. Mater.*, 257, 119047, 2020.
55. Li, Z., Jin, Z., Wang, P., Zhao, T., Corrosion mechanism of reinforced bars inside concrete and relevant monitoring or detection apparatus: A review. *Constr. Build. Mater.*, 279, 122432, 2021.
56. Chen, J., Fu, C., Ye, H., Jin, X., Corrosion of steel embedded in mortar and concrete under different electrolytic accelerated corrosion methods. *Constr. Build. Mater.*, 241, 117971, 2020.
57. Jin, Z., Zhao, X., Zhao, T., Yang, L., Interaction between compressive load and corrosive-ion attack on reinforced concrete with accelerated potentiostatic corrosion. *Constr. Build. Mater.*, 113, 805–814, 2016.
58. Cements and Cement Paste, in: *Corrosion of Steel in Concrete*, pp. 1–20, 2013.
59. Andrade, C., Propagation of reinforcement corrosion: principles, testing and modelling. *Mater. Struct.*, 52, 1, 2, 2018.
60. Chun, Q.L., Reliability Based Service Life Prediction of Corrosion Affected Concrete Structures. *J. Struct. Eng.*, 130, 10, 1570–1577, 2004.
61. Raupach, M., Models for the propagation phase of reinforcement corrosion – an overview. *Mater. Corros.*, 57, 8, 605–613, 2006.

62. Otieno, M.B., Beushausen, H.D., Alexander, M.G., Modelling corrosion propagation in reinforced concrete structures – A critical review. *Cem. Concr. Compos.*, 33, 2, 240–245, 2011.
63. Dervishaj, A., Malmqvist, T., Silfwerbrand, J., Gudmundsson, K., A digital workflow for assessing lifespan, carbonation, and embodied carbon of reusing concrete in buildings. *J. Build. Eng.*, 96, August, 110536, 2024.
64. Ram, K., Serdar, M., Londono-Zuluaga, D., Scrivener, K., Does carbon footprint reduction impair mechanical properties and service life of concrete? *Mater. Struct.*, 56, 1, 6, 2022.
65. Njoku, D.I., Li, Y., Lgaz, H., Oguzie, E.E., Dispersive adsorption of Xylopia aethiopica constituents on carbon steel in acid-chloride medium: A combined experimental and theoretical approach. *J. Mol. Liq.*, 249, 371–388, 2018.
66. Eziuka, J.E., Onyeachu, I.B., Njoku, D.I., Nwanonenyi, S.C., Chidiebere, M.A., Oguzie, E.E., , Elucidating the inhibition behavior of Pterocarpus santalinoides leaves extract on mild steel corrosion in H2SO4 solution–GC-MS Experimental and computational approach. *Morrocan J. Chem.*, 11, 3, 579–593, 2023.
67. Cheng, C.-R., Emori, W., Wei, K., Louis, H., Unimuke, T.O., Okonkwo, P.C., Njoku, D.I., Okafor, P.C., Natural triterpenoids of Ganoderma lucidum as new, green, and effective corrosion inhibitor for steel in acidic medium: characterization, experimental and theoretical investigations. *J. Adhes. Sci. Technol.*, 36, 1–24, 2022.
68. Onyeachu, I.B., Solomon, M.M., Adama, K.K., Nnadozie, C.F., Ahanotu, C.C., Akanazu, C.E., Njoku, D.I., Exploration of the potentials of imidazole-based inhibitor package for heat exchanger-type stainless steel during acid cleaning operation. *Arab. J. Chem.*, 15, 103837, 2022.
69. Bolzoni, F., Brenna, A., Ormellese, M., Recent advances in the use of inhibitors to prevent chloride-induced corrosion in reinforced concrete. *Cem. Concr. Res.*, 154, 106719, 2022.
70. Xu, C., Jin, W.L., Wang, H.L., Wu, H.T., Huang, N., Li, Z.Y., Mao, J.H., Organic corrosion inhibitor of triethylenetetramine into chloride contamination concrete by eletro-injection method. *Constr. Build. Mater.*, 115, 602–617, 2016.
71. Shen, L., Jiang, H., Cao, J., Zhang, M., Zhang, H., A comparison study of the performance of three electro-migrating corrosion inhibitors in improving the concrete durability and rehabilitating decayed reinforced concrete. *Constr. Build. Mater.*, 238, 117673, 2020.
72. Østnor, T.A. and Justnes, H., Anodic corrosion inhibitors against chloride induced corrosion of concrete rebars. *Adv. Appl. Ceram.*, 110, 3, 131–136, 2011.
73. Wang, H., Jiang, M., Hang, M., Yang, Y., Zhou, X., Liu, X., Xu, G., Inhibition resistance and mechanism of migrating corrosion inhibitor on reinforced concrete under coupled carbonation and chloride attack. *J. Build. Eng.*, 76, 107398, 2023.

74. Sun, C., Sun, M., Liu, J., Dong, Z., Fan, L., Duan, J., Anti-Corrosion Performance of Migratory Corrosion Inhibitors on Reinforced Concrete Exposed to Varying Degrees of Chloride Erosion. *Mater. (Basel).*, 15, (15), 2022.
75. Tiwari, A., Goyal, S., Luxami, V., Chakraborty, M.K., Prabhakar, G., Assessment of corrosion inhibition efficiency of generic compounds having different functional groups in carbonated pore solution with chlorides and migration ability in concrete. *Constr. Build. Mater.*, 290, 123275, 2021.
76. Singh, J.K., Yang, H.-M., Lee, H.-S., Mandal, S., ASLAM, F., Alyousef, R., Role of L-arginine on the formation and breakdown of passive film onto the steel rebars surface in chloride contaminated concrete pore solution. *J. Mol. Liq.*, 337, 116454, 2021.
77. Hu, J.Y., Zhang, S.S., Chen, E., Li, W.G., A review on corrosion detection and protection of existing reinforced concrete (RC) structures. *Constr. Build. Mater.*, 325, 126718, 2022.
78. Bertolini, L., Gastaldi, M., Pedeferri, M.P., 6 - Risk of galvanic corrosion induced by CFRP strengthening in reinforced concrete, in: *European Federation of Corrosion (EFC) Series*, vol. 38, pp. 62–74, Woodhead Publishing, 2007.
79. ChatGPTHKMU Concretes", Response to "Cases of concrete failures caused by chloride-induced corrosion of.
80. ChatGPTHKMU Response to" cases of concrete failures caused by CO2-induced cracking, and spalling of the concrete structures".
81. Medawela, S., Indraratna, B., Rowe, R.K., The reduction in porosity of permeable reactive barriers due to bio-geochemical clogging caused by acidic groundwater flow. *Can. Geotech. J.*, 60, 2, 151–165, 2022.

4

Corrosion Assessment and Inspection in Concrete Corrosion

Ichraq Bouhouche[1], Khalid Bouiti[1], Nabil Lahrache[1], Najoua Labjar[1*], Hamid Nasrellah[2], Said Laasri[3], Ayoub Cherrat[1] and Souad El Hajjaji[1]

[1]*Laboratory of Spectroscopy, Molecular Modeling, Materials, Nanomaterials, Water and Environment, CERNE2D, ENSAM, Mohammed V University in Rabat, Rabat, Morocco*
[2]*Higher School of Education and Training, Chouaib Doukkali University, El Jadida, Morocco*
[3]*Laboratory of Engineering Sciences for Energy, ENSA- El Jadida, Chouaib Doukkali University, El Jadida, Morocco*

Abstract

This chapter examines essential processes for assessing and inspecting corrosion in concrete structures. It covers various forms of corrosion—carbonation, chloride, and sulfate-induced—and stresses the importance of early detection to maintain structural integrity. Key assessment methods, including visual inspection, electrochemical techniques (half-cell potential, linear polarization resistance), and non-destructive testing (ultrasonic testing, radiography, and GPR) are discussed for their ability to reveal corrosion and internal defects. Advanced technologies like remote sensing and digital imaging are highlighted for enhancing assessment accuracy. Case studies demonstrate these methods' practical effectiveness in real-world applications.

Keywords: Concrete durability, corrosion monitoring, corrosion detection, structural health monitoring, corrosion management

**Corresponding author*: najoua.labjar@ensam.um5.ac.ma

Ruby Aslam, Qihui Wang, Zhitao Yan and Afroz Aslam (eds.) Corrosion in Concrete Structures: Integrating Advanced Technologies and Sustainable Practices, (85–118) © 2026 Scrivener Publishing LLC

4.1 Introduction to Corrosion Assessment and Inspection

4.1.1 Overview of Corrosion in Concrete Structures

Concrete corrosion edifices represent a significant concern impacting global infrastructure, culminating in considerable economic detriment and safety implications. This phenomenon transpires when steel reinforcement embedded within concrete is subjected to harmful environmental conditions, thereby facilitating rust development, as shown in Figure 4.1. The resultant rust exhibits a more significant volumetric expansion than the original steel, engendering internal stresses that precipitate cracking, spalling, and, ultimately, the degradation of structural integrity [1, 2]. The predominant catalysts for corrosion in concrete involve the penetration of chloride ions, carbonation, and sulfate assault. Chloride-induced corrosion, frequently originating from de-icing agents or marine environments, undermines the passive oxide layer that safeguards the steel reinforcement. Carbonation reduces the pH of concrete, compromising this protective layer and permitting the progression of corrosion. Conversely, sulfate assaults instigate expansive reactions that can compromise the concrete

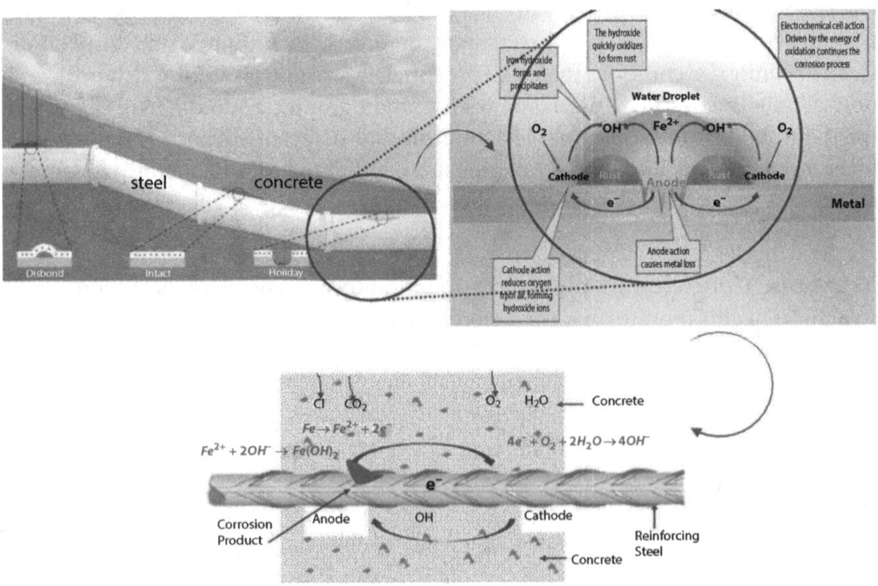

Figure 4.1 Steel corrosion reinforcement in concrete, illustrating the electrochemical process and its impact on structural integrity.

matrix and further deteriorate the reinforcement [3]. A thorough comprehension of the diverse forms of corrosion and the particular conditions that promote each variant is paramount for formulating effective assessment, prevention, and remediation methodologies. This holistic understanding further informs the selection of suitable materials and construction practices that diminish the likelihood of corrosion from the initial stages of the construction process [4].

4.1.2 Significance of Early Detection

Early corrosion detection is paramount in mitigating its adverse effects on concrete structures. As corrosion progresses, the cost and complexity of repairs increase exponentially. Detecting corrosion at an early stage extends the structure's service life and reduces the likelihood of catastrophic failures, which can have severe consequences for public safety and economic stability [5].

Various techniques are employed for early corrosion detection, each with strengths and limitations. Visual inspections, though straightforward, may not reveal subsurface damage or early-stage corrosion. Electrochemical methods, such as half-cell potential measurements, offer more detailed insights by assessing the corrosion potential of steel reinforcement. These methods can indicate areas of active corrosion before visible signs appear on the surface [6].

Non-destructive testing (NDT) techniques, such as ultrasonic pulse velocity and ground-penetrating radar (GPR), are increasingly used to detect internal defects and corrosion without causing damage to the structure. Integrating these techniques with digital tools and remote monitoring systems has significantly enhanced the accuracy and efficiency of early detection, enabling continuous monitoring and timely intervention [7].

4.1.3 Challenges in Corrosion Assessment

Corrosion assessment in concrete structures is fraught with challenges, primarily due to the material's complex nature and the degradation processes involved. Concrete is a heterogeneous material with composition, density, and porosity variations that can influence corrosion progression and complicate its detection [8].

One of the primary challenges is difficulty accessing and accurately assessing the condition of embedded steel reinforcement. Direct assessment methods, such as core sampling, are invasive and can further damage the structure. Non-destructive testing methods offer a non-invasive alternative, but they require specialized equipment and expertise, and their effectiveness can be limited by the reinforcement depth and surface anomalies [9].

Environmental factors pose additional challenges. Variations in moisture content, temperature, and the presence of aggressive agents like chlorides can influence both the progression of corrosion and the performance of assessment techniques. For instance, high humidity levels can obscure half-cell potential readings, while fluctuating temperatures may affect the precision of ultrasonic measurements [10].

Understanding the results of corrosion assessments is another significant challenge. The data obtained from various testing methods often require expert analysis to correlate findings with the actual condition of the structure. Misinterpretation of data can lead to either unnecessary repairs or, conversely, the overlooking of critical issues, which have profound implications for the safety and longevity of the structure [11].

Overcoming these challenges requires a multidisciplinary approach, combining advanced technologies with a deep understanding of corrosion mechanisms and material behavior. Innovations in sensors, data analytics, and predictive modeling are increasingly being integrated into corrosion assessment practices, offering new possibilities for more accurate and reliable evaluations.

4.2 Forms of Corrosion in Concrete

Though robust and durable, concrete structures are susceptible to various forms of corrosion that can severely compromise their integrity and lifespan [12]. The primary forms of corrosion in concrete include carbonation-induced corrosion, chloride-induced corrosion, and sulfate attack. These mechanisms involve distinct processes and environmental conditions, leading to different deterioration patterns. Understanding these forms of corrosion is essential for developing effective strategies for assessment, prevention, and remediation [13, 14].

4.2.1 Carbonation-Induced Corrosion

Carbonation-induced corrosion is a frequent cause of damage in reinforced concrete structures. This degradation happens as carbon dioxide (CO_2) from the air seeps into the concrete. It interacts with calcium hydroxide ($Ca(OH)_2$) present in the cement paste, leading to the formation of calcium carbonate ($CaCO_3$). Known as carbonation, this reaction slowly reduces the pH of the concrete from its initial high alkalinity (approximately 12.5–13) down to levels below 9.5 [15, 16].

The high alkalinity of concrete initially provides a passive layer on the surface of the embedded steel reinforcement, protecting it from corrosion. However, as the carbonation front advances and the pH drops, this protective passive layer breaks down, exposing the steel to corrosive elements such as moisture and oxygen. Once the steel is depassivated, corrosion begins, forming rust, which results in an expansion that leads to the cracking and peeling of the concrete surface [17, 18].

Carbonation is particularly prevalent in structures located in urban or industrial environments where CO_2 concentrations are elevated. It is also more common in older structures or those with poor-quality concrete, which is more porous and allows for faster CO_2 diffusion. The speed at which carbonation occurs is affected by variables like the quality of the concrete, moisture levels, environmental factors, and the thickness of the concrete cover. Regular monitoring and using protective coatings or corrosion inhibitors can help mitigate the risks associated with carbonation-induced corrosion [19, 20].

4.2.2 Chloride-Induced Corrosion

Chloride-induced corrosion is the most aggressive form of corrosion in concrete structures and is a significant concern, especially in marine environments, coastal areas, and regions where de-icing salts are used. Chloride ions (Cl^-) penetrate the concrete through cracks and pores or by diffusion, accumulating around the steel reinforcement. Unlike carbonation, which causes a general lowering of the concrete's pH, chloride ions can trigger corrosion, even when the environment has a high pH level [21, 22].

Once the concentration of chloride ions at the steel surface reaches a critical threshold, they disrupt the passive oxide layer, making the steel susceptible to corrosion. The anodic reaction of iron with chloride ions leads to the formation of iron chloride ($FeCl_2$); this then reacts with water and oxygen, leading to the formation of iron oxides and hydroxides (rust), a process that results in a notable increase in volume, which induces tensile stresses in the concrete, resulting in cracking, delamination, and spalling [23, 24].

Chloride-induced corrosion is particularly dangerous because it often progresses unnoticed until a significant damage has occurred. The presence of cracks in the concrete can worsen the infiltration of chloride ions and result in inadequate concrete cover or poor-quality concrete with high permeability. Preventive measures include using low-permeability concrete, sufficient concrete cover, and corrosion-resistant steel and applying surface sealants or coatings to limit chloride ingress [25].

4.2.3 Sulfate Attack

Sulfate attack is a chemical deterioration process that impacts the concrete matrix and the embedded steel reinforcement. It occurs when sulfate ions (SO_4^{2-}) from external sources, such as soil, groundwater, or seawater, infiltrate the concrete and react with the hydrated cement compounds. The primary reaction results in the formation of expansive products, such as gypsum ($CaSO_4 \cdot 2H_2O$) and ettringite ($3CaO \cdot Al_2O_3 \cdot 3CaSO_4 \cdot 32H_2O$) [26, 27].

The formation of gypsum and ettringite leads to an expansion within the concrete, generating internal stresses that cause cracking, spalling, and a reduction in structural strength. These cracks created by sulfate attack can serve as pathways for the entry of other harmful agents, like chloride ions, which may trigger or speed up the corrosion of steel reinforcement. When sulfate attack and chloride-induced corrosion occur together, they can be especially damaging [28–30].

Sulfate attacks are more likely to happen in structures exposed to environments with high sulfate levels, such as those in contact with specific soils or industrial waste. The extent of the damage depends on factors like the sulfate ion concentration, concrete permeability, the presence of other harmful substances, and the concrete's overall quality. To reduce the risk of sulfate attack, it is advised to use sulfate-resistant cement, maintain low water-to-cement ratios, and implement proper curing techniques [31, 32].

4.3 Visual Assessment Methods

Visual inspection is the most basic and commonly employed technique for evaluating the state of concrete structures. While it may seem rudimentary compared to advanced diagnostic technologies, visual inspection remains a crucial first step in identifying potential issues, guiding further investigation, and planning maintenance or repairs. It allows inspectors to detect surface-level damage and early signs of deterioration that may indicate deeper, underlying problems [33].

4.3.1 Surface Damage and Deterioration

Surface damage in concrete structures serves as a critical early warning sign of deeper underlying issues such as corrosion, mechanical stress, or environmental degradation, and during visual inspections, various damage types are typically observed, each providing valuable insights into the structure's health. Cracking is one of the most prevalent forms of surface damage, manifesting in several forms, including hairline, shrinkage, or structural cracks, each offering clues about the nature and severity of the underlying problem. For example, vertical cracks may point to settlement or foundational shifts. In contrast, horizontal cracks, particularly those near the level of reinforcement, often signal the onset of corrosion within the embedded steel. The careful assessment of the extent, orientation, and location of these cracks is essential for diagnosing their causes and evaluating their potential impact on the overall structural integrity [34].

Spalling, another common form of surface damage, involves the flaking or chipping away of concrete, which typically results from the expansion of corroded steel reinforcement. As steel corrodes, it produces rust, which occupies more volume than the original metal, exerting pressure on the surrounding concrete and causing pieces of the surface to break away. This not only compromises the esthetic appearance of the structure but also exposes the reinforcement to further corrosive elements, accelerating the degradation process. Therefore, identifying and thoroughly documenting areas of spalling during visual inspections are crucial steps in determining the extent of internal corrosion and planning for necessary repairs to prevent further deterioration [35].

Discoloration of the concrete surface can also be a significant indicator of underlying damage. For instance, rust stains, reddish-brown streaks, or patches are clear signs of corrosion within the embedded steel reinforcement. Additionally, efflorescence, the deposition of white crystalline salts on the surface, often signals moisture movement through the concrete, indicating potential pathways for further damage, such as freeze–thaw cycles or chemical attacks. Inspectors must analyze the patterns and extent of such discoloration to assess the underlying causes and associated risks, ensuring that appropriate remediation measures are taken [36, 37].

Surface erosion, often resulting from chemical attacks, freeze–thaw cycles, or abrasion, is another critical form of damage that gradually leads to the loss of concrete material from the surface. This erosion is particularly concerning in structures exposed to harsh environmental conditions, such as marine or industrial settings. The progressive loss of material not only diminishes the structural thickness of the concrete cover but also increases the vulnerability of the embedded reinforcement to corrosion by reducing the protective barrier between the steel and the external environment. During visual inspections, it is vital to evaluate the depth and extent of surface erosion, as this will determine the urgency and type of protective measures or maintenance or repairs required to maintain the structural stability and prolong the service life of the concrete structure [38, 39].

4.3.2 Signs of Corrosion in Concrete Structures

The visual identification of corrosion in concrete structures is paramount for enabling early intervention and guiding effective maintenance strategies, as corrosion of embedded steel reinforcement typically manifests through specific, sometimes subtle, signs that require careful and systematic observation. Rust staining, one of the most obvious visual indicators of corrosion, results from the oxidation of iron within the steel reinforcement, which produces rust that migrates to the concrete surface. These stains, appearing as brown or reddish-brown streaks, often align with areas where cracks or spalling are present, indicating a direct correlation between surface damage and underlying corrosion. Documenting the location and extent of rust staining during inspections is crucial, as it allows inspectors to assess the progression of corrosion, prioritize areas for more in-depth investigation, and plan targeted remediation efforts to prevent further structural damage [40, 41].

Corrosion-induced cracking is another critical sign, typically following the pattern of the embedded steel reinforcement. Longitudinal cracks, which run parallel to the reinforcement, are common indicators that steel within the concrete is corroding and expanding, exerting pressure along its length and causing the surrounding concrete to crack. In some cases, map cracking or random cracking patterns may also develop, signaling localized areas of corrosion where the reinforcement is undergoing significant deterioration. Careful examination of these cracking patterns is essential during visual inspections, as it helps determine the severity and spread of the corrosion, informing decisions about the necessary repairs or further diagnostic testing [42, 43].

Concrete delamination, a condition where layers of concrete detach from each other, often arises due to the corrosion of embedded reinforcement or poor bonding during construction. Although delamination may not always be immediately visible, it can be detected through sounding techniques, such as tapping the concrete surface with a hammer to identify hollow areas beneath. When delamination is visibly apparent, it often manifests as localized bulging or separation of the concrete surface, which presents a significant risk to the structural stability of the building. If left unaddressed, delamination can lead to further deterioration, making it critical for inspectors to identify and document these areas during visual assessments to ensure timely and effective repair interventions [44].

In advanced stages of corrosion, the protective concrete cover may deteriorate to the point where the steel reinforcement becomes exposed to the environment. Exposed reinforcement is particularly vulnerable to accelerated corrosion due to direct contact with moisture, oxygen, and other corrosive elements. During visual inspections, the condition of the exposed reinforcement and surrounding concrete must be thoroughly assessed to gauge the extent of the corrosion and determine the urgency of repair efforts needed to prevent catastrophic structural failure [45].

Moisture ingress, characterized by signs such as damp patches, water leakage, or the appearance of efflorescence (white crystalline deposits), serves as an indirect but significant indicator of potential corrosion risk. Moisture is a key factor in the corrosion process, as it facilitates the electrochemical reactions that lead to the deterioration of steel reinforcement. The presence of moisture within or on the exterior of concrete structures can significantly elevate the likelihood of reinforcement corrosion, making it imperative for inspectors to examine areas where moisture ingress is evident closely. Identifying and addressing the sources of moisture is crucial to mitigating the risk of further corrosion and ensuring the long-term durability of the structure [46, 47].

While visual inspection alone may not uncover all the complexities of corrosion processes, it remains an indispensable tool in the ongoing maintenance and preservation of concrete structures. The insights gained from visual inspections provide essential first-line information, guiding the deployment of more advanced diagnostic techniques, such as electrochemical testing or non-destructive evaluation methods. Regular and systematic visual inspections are critical to maintaining the longevity, safety, and integrity of concrete infrastructure, as they enable early detection of potential issues and facilitate proactive maintenance, ultimately extending the structure's lifespan and preventing costly repairs or failures [48, 49].

4.4 Electrochemical Assessment Methods

Electrochemical assessment methods are essential for evaluating steel reinforcement's corrosion state within concrete structures. These techniques provide quantitative data on the probability of corrosion, its progression rate, and the overall condition of the embedded steel. Electrochemical methods allow for non-destructive analysis, making them valuable for ongoing monitoring and targeted investigations of potential corrosion hotspots. This section will explore three key electrochemical methods: half-cell potential measurements, linear polarization resistance (LPR), and electrochemical impedance spectroscopy (EIS) [50].

4.4.1 Half-Cell Potential Measurements

Half-cell potential (HCP) measurements are widely used to assess the likelihood of corrosion in steel reinforcement within concrete structures. This method involves placing a reference electrode, typically copper/copper sulfate or silver/silver chloride, on the concrete surface and measuring the potential difference between the reference electrode and the embedded steel reinforcement. The potential difference, known as the half-cell potential, indicates the electrochemical activity of the steel, which is directly correlated with the likelihood of corrosion [51, 52]. The principle behind HCP measurements is that the electrochemical potential of steel in concrete alters when corrosion occurs, allowing for the inference of whether the steel is in a passive (non-corroding) or active (corroding) state, with a more negative potential indicating a higher likelihood of corrosion. The procedure for HCP measurements involves preparing the concrete surface by cleaning it to ensure good electrical contact and then connecting the reference electrode to a high-impedance voltmeter, which is also related to the steel reinforcement [53]. Measurements are taken at various points on the concrete surface, creating a potential map highlighting areas at high corrosion risk. Interpretation of these results is based on standard potential ranges, where potentials more negative than -350 mV (relative to a copper/copper sulfate reference electrode) suggest a high probability of active corrosion. In comparison, potentials more positive than -200 mV indicate a lower likelihood of corrosion, with intermediate values requiring further investigation. HCP measurements are commonly applied in the routine monitoring of large concrete structures, such as bridges and parking garages, where early corrosion detection can prevent costly repairs, making this non-destructive and relatively easy-to-apply method an essential tool in structural health monitoring [54, 55].

4.4.2 Linear Polarization Resistance (LPR)

LPR is an electrochemical method used to directly measure the corrosion rate in steel reinforcement, providing quantifiable data in terms of current density or metal loss over time. Unlike half-cell potential (HCP) measurements, which assess the probability of corrosion, LPR offers a direct assessment by leveraging the direct correlation between the applied electrochemical potential and the corresponding current within a limited potential range close to the corrosion potential [56, 57]. The resulting current response is measured by applying a small, controlled voltage to the steel reinforcement through a working electrode, with a reference and counter electrode employed to close the electrochemical circuit. The slope of the potential-current curve, known as polarization resistance, is then used to calculate the corrosion rate, with higher polarization resistance indicating a lower corrosion rate, signifying better protection of the steel [58]. Conversely, lower polarization resistance suggests a higher corrosion rate, warranting immediate attention and potential intervention. LPR measurements are typically expressed in micrometers per year (μm/year) of metal loss. They are particularly valuable in contexts requiring precise corrosion rate quantification, such as evaluating corrosion inhibitors or protective coatings. Moreover, LPR is instrumental in research settings for assessing the effectiveness of new materials or treatments designed to prevent corrosion. It is a critical tool in applied and experimental corrosion studies [59, 60].

4.4.3 Electrochemical Impedance Spectroscopy (EIS)

EIS is an advanced electrochemical technique that provides in-depth insights into the electrochemical processes at the boundary between the steel reinforcement and the concrete, making it particularly valuable for studying corrosion mechanisms and evaluating the effectiveness of protective coatings or inhibitors [61, 62]. The principle behind EIS is to measure the impedance of the electrochemical system over a spectrum of frequencies, enabling a detailed analysis of the system's behavior in response to an alternating current (AC) signal [63]. The resulting impedance data are typically represented in Nyquist or Bode plots, from which key parameters such as resistance, capacitance, and inductance can be extracted. These parameters reveal critical aspects of the corrosion process, such as charge transfer resistance and protective layers [64]. Table 4.1 compares the electrochemical methods, including EIS, used for corrosion assessment in concrete structures.

The procedure for conducting EIS involves applying a small AC signal to the steel reinforcement and measuring the resulting current across different frequencies. Specialized software is then used to collect and analyze the data, generating impedance spectra that are interpreted to understand the steel's corrosion behavior and the surrounding concrete's protective qualities. Interpreting impedance spectra is typically carried out using equivalent circuit models, representing the system's electrochemical processes [65].

Key parameters like charge transfer resistance, which is inversely related to the corrosion rate, and the capacitance of the double layer, indicating the

Table 4.1 Evaluation of electrochemical methods for corrosion assessment in concrete structures.

Method	Measurement parameter	Principle	Advantages	Limitations	Applications	References
Half-cell potential (HCP)	Corrosion potential (mV)	Measures the potential difference between steel reinforcement and a reference electrode	Simple and non-destructive	Cannot quantify the corrosion rate	Assessment of corrosion risk in large areas of concrete	[66]
Linear polarization resistance (LPR)	Polarization resistance (Ω)	Measures the resistance to polarization to estimate the corrosion rate	Provides a direct corrosion rate	Requires surface access and contact with steel reinforcement	Evaluation of corrosion rate in specific zones	[67]
Electrochemical impedance spectroscopy (EIS)	Impedance (Ω) at varying frequencies	Analyzes the system's impedance across different frequencies to model corrosion behavior	Provides detailed information on corrosion mechanisms	Requires specialized equipment and expertise	Research and detailed evaluation of protective coatings and inhibitors	[68]

(Continued)

Table 4.1 Evaluation of electrochemical methods for corrosion assessment in concrete structures. (*Continued*)

Method	Measurement parameter	Principle	Advantages	Limitations	Applications	References
Galvanostatic pulse measurement	Instantaneous corrosion rate (A/cm^2)	A small current is applied, and the resulting polarization is measured.	Quick measurement and portable equipment	Less sensitive to corrosion initiation	On-site testing for immediate corrosion rate assessment	[69]
Concrete resistivity	The electrical resistance of concrete ($\Omega \cdot m$)	Measures the resistivity of concrete, which influences corrosion likelihood	Useful for assessing the protective capacity of concrete	It does not directly measure the corrosion of steel reinforcement.	Evaluating the quality and durability of concrete over time	[70]

condition of the concrete-steel interface, are critical to assessing the state of the structure. A high charge transfer resistance suggests a low corrosion rate, while a low resistance points to active corrosion. EIS is extensively used in research and development to evaluate the long-term performance of corrosion protection systems, such as coatings, sealants, and inhibitors. Additionally, it is employed in the advanced monitoring of critical infrastructure where a detailed understanding of the corrosion process is essential for guiding maintenance and repair strategies [71–73].

4.5 Non-Destructive Testing (NDT) Techniques

Non-destructive testing (NDT) techniques are essential in assessing the integrity of concrete structures without causing damage [74]. These techniques offer important information about the internal state of structures, allowing for the early detection of potential issues such as corrosion, voids, and other forms of deterioration. The following sections outline key NDT techniques employed in corrosion assessment [75, 76].

4.5.1 Ultrasonic Testing (UT)

UT is a widely used NDT technique that utilizes high-frequency sound waves to detect internal flaws in concrete structures. This technique sends ultrasonic pulses through the concrete, and the reflected waves are examined to detect any anomalies, such as cracks, voids, or delamination. The time it takes for the waves to return to the surface indicates the presence and size of defects, as well as the thickness of the concrete. UT is particularly effective in evaluating the condition of the concrete cover and detecting areas where the steel reinforcement may be compromised due to corrosion [77, 78].

4.5.2 Radiography

Radiography, or X-ray testing, is another powerful NDT technique used in corrosion inspection. This method involves passing X-rays or gamma rays through the concrete to produce an image of its internal structure. Radiography is very effective for identifying voids, cracks, and the placement of steel reinforcement within concrete structures [79, 80]. By analyzing the resulting images, inspectors can identify areas where corrosion might occur, especially in regions where the steel reinforcement is inadequately protected, or where voids could allow for moisture ingress. Despite its effectiveness, radiography is often reserved for critical structures due to the need for specialized equipment and safety precautions [81, 82].

4.5.3 Ground Penetrating Radar (GPR)

GPR is a flexible, non-destructive testing technique that employs radar pulses to capture images of the subsurface in concrete structures [83]. GPR is particularly useful in detecting embedded steel reinforcement, voids, and other anomalies within the concrete. This technique operates by sending electromagnetic waves into the concrete and recording the reflected signals. Variations in the reflected signals indicate differences in the material properties, such as corrosion or voids. GPR is commonly employed to evaluate extensive structures such as bridges and parking garages, where quick, non-intrusive inspection is essential [84].

Table 4.2 Remediation methods for damage due to reinforcement corrosion according to EN 1504-9 [85].

Principle	Methods
P7: Preserving or restoring passivity	M7.1: Increasing cover with additional mortar/concrete
	M7.2: Replacing contaminated or carbonated concrete
	M7.3: Electrochemical realkalization of carbonated concrete
	M7.4: Realkalization of carbonated concrete by diffusion
	M7.5: Electrochemical chloride extraction
P8: Increasing resistivity	M8.1: Hydrophobic impregnation
	M8.2: Impregnation
	M8.3: Coating
	M8.4: Membranes
P9: Cathodic control	M9.1: Saturation
	M9.2: Surface coating
P10: Cathodic protection	M10.1: Impressed current cathodic protection (ICCP)
	M10.2: Galvanic (sacrificial anode) protection (GP)
P11: Control of anodic areas	M11.1: Active coating of the reinforcement
	M11.2: Barrier coating of the reinforcement
	M11.3: Applying corrosion inhibitors in or to the concrete

Table 4.2 provides an overview of repair techniques for addressing reinforcement corrosion damage, serving as a practical guide to complement the non-destructive testing (NDT) methods discussed above. Through the integration of damage identification *via* NDT with suitable remediation methodologies, this synthesis facilitates the connection between inspection outcomes and efficient restoration practices.

4.6 Advanced Technologies in Corrosion Inspection

Integrating advanced technologies in corrosion inspection has significantly enhanced the accuracy, efficiency, and comprehensiveness of corrosion assessments. The following sections explore the cutting-edge techniques transforming the field [86, 87]. Figure 4.2 illustrates the classification of various corrosion assessment and monitoring techniques, providing an overview of the methods used to detect and evaluate corrosion in concrete structures.

4.6.1 Remote Sensing Techniques

Remote sensing techniques have revolutionized corrosion inspection by enabling the monitoring of large and inaccessible structures from a distance. These methods involve using satellites, drones, or ground-based sensors to collect data on the structural health of concrete assets [88, 89]. Remote sensing is especially useful for monitoring extensive infrastructure like pipelines,

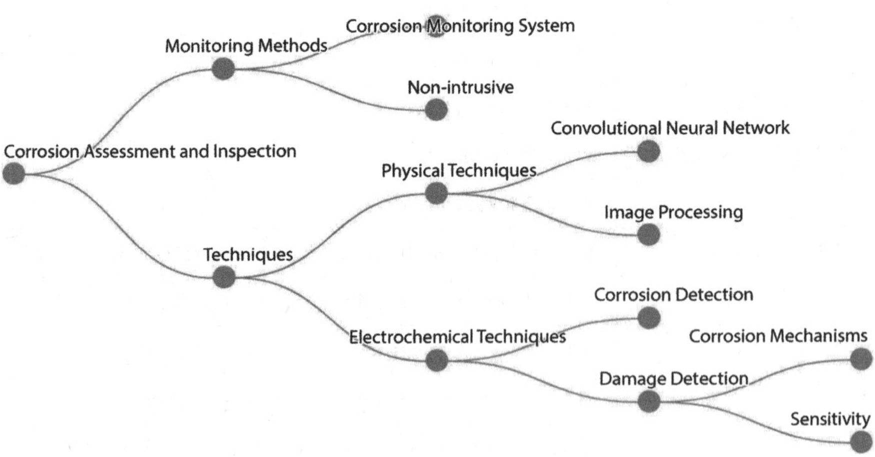

Figure 4.2 Classification of corrosion assessment and monitoring techniques.

bridges, and offshore platforms, where conventional inspection approaches are challenging to implement. The data collected can be analyzed to detect early signs of corrosion, surface degradation, or structural movement, providing critical information for preventive maintenance [90].

4.6.2 Digital Imaging and Analysis

Digital imaging and analysis have become indispensable tools in corrosion inspection, allowing for the detailed visualization and quantification of surface damage [91, 92]. High-resolution cameras and imaging software are used to capture images of concrete structures, which are then analyzed to identify cracks, spalling, and discoloration indicative of corrosion. Advanced image processing algorithms can quantify the extent of damage, track changes over time, and even predict future deterioration. This technology enhances the precision of visual inspections and facilitates the documentation and comparison of structural conditions [93, 94].

4.6.3 Integration of IoT and Data Analytics in Corrosion Monitoring

The integration of the Internet of Things (IoT) and data analytics into corrosion monitoring represents a significant advancement in the proactive management of concrete structures [95]. IoT involves embedding sensors within structures to continuously monitor environmental conditions and corrosion-related parameters, such as humidity, temperature, and electrochemical activity [96]. These sensors transmit real-time data to centralized platforms, where advanced data analytics tools analyze the information to detect trends, identify anomalies, and predict future corrosion risks [97]. This approach enables timely interventions, reducing maintenance costs and extending the lifespan of infrastructure [98].

4.7 Case Studies in Corrosion Assessment

The practical application of corrosion assessment techniques is essential in understanding the complexities of maintaining and extending the lifespan of concrete structures. Real-world case studies provide critical insights into the challenges faced in various environments, the methodologies employed, and the outcomes achieved. The following sections delve into detailed analyses of corrosion assessment in different types of infrastructure, highlighting how specific techniques were applied to address unique corrosion-related challenges [99, 100].

4.7.1 Case Study 1: Assessment in Marine Structures

Marine structures, such as piers, jetties, and offshore platforms, are continuously exposed to harsh saline environments, which significantly accelerate the corrosion of steel reinforcement within concrete. This case study focuses on the comprehensive assessment of a coastal pier, where a combination of ultrasonic testing (UT), ground penetrating radar (GPR), and electrochemical impedance spectroscopy (EIS) was employed to evaluate the extent of corrosion. The pier was particularly vulnerable in the splash zone—the area just above the waterline where alternating wet and dry conditions exacerbate corrosion [101, 102].

Challenges: The primary challenge was accessing and accurately assessing the submerged and splash zones, where corrosion is most aggressive. Traditional visual inspections were insufficient due to the pier's exposure to marine conditions and the difficulty of detecting subsurface damage [103, 104].

Methodology: UT was utilized to detect internal cracks and voids in the concrete, while GPR helped map the location and condition of the steel reinforcement. EIS provided detailed information about the electrochemical properties of the steel–concrete interface, allowing for a deeper understanding of the corrosion mechanisms at play [105, 106].

Outcomes: The assessment revealed significant corrosion in the steel reinforcement within the splash zone, characterized by extensive cracking and spalling of the concrete. The findings necessitated immediate intervention, including removing damaged concrete, applying corrosion-resistant coatings, and installing sacrificial anodes to protect the steel reinforcement. These measures were crucial in extending the service life of the pier and preventing further structural degradation [107, 108].

4.7.2 Case Study 2: Inspection of Bridges and Overpasses

Bridges and overpasses are vital components of transportation infrastructure, and their structural integrity is critical for ensuring public safety. This case study examines the corrosion assessment of a highway overpass that had been in service for several decades. The overpass was showing signs of wear, including visible cracks and surface discoloration, prompting a thorough investigation using NDT techniques such as radiography and LPR [109, 110].

Challenges: The primary challenges included the structure's age, which increased the likelihood of hidden defects, and the need to conduct inspections without disrupting traffic flow. Additionally, the varying environmental conditions to which different parts of the overpass were exposed required a tailored approach to corrosion assessment [111, 112].

Methodology: Radiography was employed to produce detailed images of the steel reinforcement within the concrete, revealing areas where corrosion had compromised the integrity of the structure. LPR measurements were used to quantify the corrosion rate of the steel, providing a clear picture of how quickly the reinforcement was deteriorating [113, 114].

Outcomes: The inspection identified several critical areas of concern, including corrosion-induced cracking near the reinforcement and delamination of the concrete deck. Based on these findings, a targeted maintenance plan was developed, which included the application of corrosion inhibitors to slow down the corrosion process, the installation of cathodic protection systems to safeguard the reinforcement, and the repair of damaged concrete. These interventions were essential in maintaining the overpass's structural integrity and extending its operational life [115, 116].

4.7.3 Case Study 3: Industrial Infrastructure

Industrial facilities, such as chemical plants and refineries, are often subjected to highly corrosive environments due to aggressive chemicals, elevated temperatures, and moisture. This case study explores the corrosion assessment of a concrete containment structure within a chemical plant. The structure was critical for safely containing hazardous materials, making its integrity paramount [117, 118].

Challenges: The containment structure was exposed to various aggressive agents, including acidic and alkaline chemicals, which increased the risk of rapid deterioration. Additionally, the harsh industrial environment made it difficult to conduct traditional inspections, necessitating the use of advanced technologies [119, 120].

Methodology: Remote sensing and digital imaging technologies were employed to thoroughly assess the structure. These techniques allowed inspectors to detect early signs of corrosion in the steel reinforcement, even in areas where the concrete's external appearance remained largely unaffected. The remote sensing data provided a comprehensive overview of the structure's condition, while digital imaging enabled detailed documentation of surface and subsurface damage [121, 122].

Outcomes: The assessment revealed early-stage corrosion in the steel reinforcement, which had not yet manifested in significant visible damage. This early detection was critical, as it allowed for timely interventions that included the application of protective coatings and implementing a monitoring program using embedded sensors to track the ongoing condition of the structure. By addressing the corrosion early, the plant could prevent further degradation, ensuring the continued safe operation of the containment structure and avoiding costly repairs or replacements [123, 124].

4.8 Discussion on Assessment and Inspection Results

The effectiveness of corrosion assessment and inspection lies in the accurate detection of deterioration and the rigorous interpretation and correlation of the gathered data with the overall structural health. This section delves into the critical aspects of interpreting the results from various inspection techniques and understanding how these findings relate to concrete infrastructures' structural integrity and longevity [125, 126].

4.8.1 Interpretation of Data

Interpreting data from corrosion assessments involves a comprehensive analysis of the information obtained through multiple testing methods, such as NDT, electrochemical measurements, and visual inspections. Each technique yields specific types of data, which must be carefully analyzed in the context of the structure's environment, materials, and service history [127].

Data integration: The data collected from different techniques must be integrated to form a holistic understanding of the structure's condition. For instance, the results of half-cell potential (HCP) measurements can provide an overview of areas with a high likelihood of corrosion. Still these findings need to be corroborated with physical evidence from visual inspections or subsurface damage detection using ground penetrating radar (GPR). Similarly, linear polarization resistance (LPR) data, which quantifies the corrosion rate, must be interpreted alongside ultrasonic testing (UT) results that reveal internal cracking or voids, providing a clearer picture of how the corrosion process is affecting the concrete [128].

Thresholds and standards: Interpretation also involves comparing the results against established standards and thresholds to assess the severity of the damage. For example, potential readings from HCP measurements are interpreted based on standard potential ranges, with certain thresholds indicating active corrosion while others suggest a passive state of the steel reinforcement. Similarly, the corrosion rate obtained from LPR is compared against industry benchmarks to determine whether the observed rate is acceptable or requires immediate intervention [129].

Anomalies and outliers: In some cases, the data may show anomalies or outliers that do not fit the expected patterns. These instances require careful consideration, as they could indicate localized issues, such as isolated areas of accelerated corrosion due to specific environmental factors, or they may point to errors in data collection that need to be addressed. Proper interpretation involves recognizing these anomalies, understanding their implications, and determining whether they indicate broader structural issues or isolated problems [130].

4.8.2 Correlation Between Inspection Techniques and Structural Health

The correlation between the results of various inspection techniques and the overall health of a structure is essential for making informed decisions regarding maintenance and repair [131]. By integrating data from complementary methods like ultrasonic testing (UT) and radiography—where UT detects internal flaws such as voids or cracks and radiography provides detailed images of the internal structure, including the condition of steel reinforcement—engineers can help develop a thorough understanding of the structure's physical and electrochemical condition. When these findings are combined with electrochemical assessments such as LPR or EIS, they offer a detailed perspective that improves the precision of the evaluation [132]. This correlation not only identifies existing damage but also aids in predicting future deterioration. For instance, early signs of corrosion detected through half-cell potential (HCP) measurements, when coupled with minor cracking observed during visual inspections, may indicate a future risk of significant structural damage if left unaddressed. Understanding these interrelationships supports a holistic assessment that considers underlying causes, potential risks, and the effectiveness of current protective measures. For example, high corrosion rates from LPR and low charge transfer resistance from EIS might reveal the failure of protective strategies such as coatings or inhibitors, prompting a revaluation of the maintenance approach. Ultimately, this comprehensive correlation directly informs maintenance and repair decisions, enabling engineers to prioritize repairs, allocate resources effectively, and implement targeted preventive measures, thus supporting the development of long-term, proactive maintenance plans that ensure the preservation of structural integrity over time [133].

4.9 Conclusion

This chapter has provided an in-depth examination of corrosion assessment and inspection techniques for concrete structures, highlighting their essential role in preserving structural integrity and extending infrastructure lifespan. The key findings emphasize the diversity of corrosion forms affecting concrete, such as carbonation-induced corrosion, chloride-induced corrosion, and sulfate attack. Each type requires specific assessment techniques for effective identification and mitigation. Visual inspections are a fundamental first step, enabling practitioners to make initial judgments based on surface damage indicators like cracking, spalling, and discoloration. Subtle signs, such as rust staining and delamination, are also critical for identifying deeper issues within the concrete.

Electrochemical assessment methods, including HCP measurements, LPR, and EIS, provide quantifiable data on corrosion likelihood and rate, offering precise evaluations of steel reinforcement conditions within concrete structures. Additionally, NDT techniques, such as UT, radiography, and GPR, are vital for detecting internal defects without damaging the structure, revealing flaws that may not be visible through visual inspections.

Integrating advanced technologies, such as remote sensing, digital imaging, and the Internet of Things (IoT), enhances corrosion assessment accuracy, efficiency, and real-time monitoring capabilities, leading to more informed maintenance decisions. Real-world case studies demonstrate the practical application of these techniques across various infrastructure types, underscoring the importance of tailored approaches based on environmental conditions and structural demands.

4.9.1 Recommendations for Practitioners

In light of these findings, several recommendations are proposed for practitioners engaged in the maintenance and inspection of concrete structures. A multi-method approach is essential, combining visual inspections with electrochemical and NDT techniques to evaluate potential corrosion risks comprehensively. Early corrosion detection is critical to prevent significant structural damage, particularly in high-risk environments like marine or industrial settings. Practitioners should leverage advanced technologies, such as remote sensing and IoT-based monitoring systems, to improve the precision and efficiency of corrosion assessments. Tailoring maintenance and repair strategies to specific corrosion risks is also crucial. For example, structures affected by chloride-induced corrosion may benefit from corrosion inhibitors, while those suffering from carbonation-induced corrosion might require increased concrete cover or protective coatings.

4.9.2 Future Directions in Corrosion Assessment and Inspection

The corrosion assessment and inspection field is poised for significant advancements, driven by ongoing research and technological innovation. Future directions include the development of more accurate predictive models, which would allow for better forecasting of corrosion progression and more effective maintenance planning. Integrating machine learning and artificial intelligence into corrosion assessment processes can revolutionize the field, enabling the analysis of large datasets from various inspection techniques and identifying patterns and risks with greater accuracy and speed.

Advancements in non-invasive techniques are anticipated, potentially leading to more sophisticated methods that provide detailed insights without compromising structural integrity. Improvements in imaging technology, sensors, and data analysis tools could further enhance the effectiveness of corrosion assessments. Additionally, as sustainability becomes increasingly important in construction and maintenance practices, future research may focus on developing eco-friendly corrosion inhibitors and protective coatings. These sustainable solutions would aim to protect concrete structures while minimizing environmental impact.

In conclusion, this chapter has highlighted the importance of a comprehensive and proactive approach to corrosion assessment and inspection in concrete structures. By integrating traditional and advanced techniques, practitioners can ensure the long-term durability and safety of infrastructure, contributing to the sustainability and resilience of the built environment. As the field continues to evolve, embracing new technologies and methodologies will be essential for overcoming the challenges posed by corrosion and maintaining the structural health of critical infrastructure worldwide.

AI disclosure statement: The author confirms that no artificial intelligence (AI)-generated content or AI-assisted writing tools were used in the preparation of this chapter (including summaries and the abstract). All text, analysis, and conclusions are the result of the author's original work.

Bibliography

1. Odili, P.O., Daudu, C.D., Adefemi, A., *et al.*, Integrating Advanced Technologies in Corrosion and Inspection Management for Oil and Gas Operations. *Eng. Sci. Technol. J.*, 5, 597–611, 2024, https://doi.org/10.51594/estj.v5i2.835.

2. Arya, E.K. and Dhanya, B.S., Corrosion control of reinforced concrete structures in construction industry: A review. *IOP Conf. Ser. Mater. Sci. Eng.*, 1114, 012006, 2021, https://doi.org/10.1088/1757-899x/1114/1/012006.
3. Fan, L. and Shi, X., Techniques of corrosion monitoring of steel rebar in reinforced concrete structures: A review. *Struct. Health Monit.*, 21, 1879–1905, 2022, https://doi.org/10.1177/14759217211030911.
4. Foorginezhad, S., Mohseni-Dargah, M., Firoozirad, K., *et al.*, Recent Advances in Sensing and Assessment of Corrosion in Sewage Pipelines. *Process Saf. Environ. Prot.*, 147, 192–213, 2021, https://doi.org/10.1016/j.psep.2020.09.009.
5. Renne, N., Kara De Maeijer, P., Craeye, B., *et al.*, Sustainable Assessment of Concrete Repairs through Life Cycle Assessment (LCA) and Life Cycle Cost Analysis (LCCA). *Infrastructures*, 7, 1–32, 2022, https://doi.org/10.3390/infrastructures7100128.
6. Makul, N., Modern sustainable cement and concrete composites: Review of current status, challenges and guidelines. *Sustain. Mater. Technol.*, 25, e00155, 2020, https://doi.org/10.1016/j.susmat.2020.e00155.
7. Bender, R., Féron, D., Mills, D., *et al.*, Corrosion challenges towards a sustainable society. *Mater. Corros.*, 73, 1730–1751, 2022, https://doi.org/10.1002/maco.202213140.
8. Peng, X., Anyaoha, U., Liu, Z., Tsukada, K., Analysis of Magnetic-Flux Leakage (MFL) Data for Pipeline Corrosion Assessment. *IEEE Trans. Magn.*, 56, 1–15, 2020, https://doi.org/10.1109/TMAG.2020.2981450.
9. Tešić, K., Baričević, A., Serdar, M., Non-destructive corrosion inspection of reinforced concrete using ground-penetrating radar: A review. *Mater. (Basel)*, 14, 1–20, 2021, https://doi.org/10.3390/ma14040975.
10. Coppola, L., Beretta, S., Bignozzi, M.C., *et al.*, The Improvement of Durability of Reinforced Concretes for Sustainable Structures: A Review on Different Approaches. *Mater. (Basel)*, 15, 1–20, 2022, https://doi.org/10.3390/ma15082728.
11. Shafeek, H., Soltan, H.A., Abdel-Aziz, M.H., Corrosion monitoring in pipelines with a computerized system. *Alexandria Eng. J.*, 60, 5771–5778, 2021, https://doi.org/10.1016/j.aej.2021.04.006.
12. Dixit, M. and Gupta, A.K., A review of different assessment methods of corrosion of steel reinforcement in concrete. *Iran. J. Sci. Technol. Trans. Civ. Eng.*, 46, 735–752, 2022, https://doi.org/10.1007/s40996-021-00644-5.
13. Daniyal, M. and Akhtar, S., Corrosion assessment and control techniques for reinforced concrete structures: A review. *J. Build. Pathol. Rehabil.*, 5, 1–20, 2020, https://doi.org/10.1007/s41024-019-0067-3.
14. Rodrigues, R., Gaboreau, S., Gance, J., *et al.*, Reinforced concrete structures: A review of corrosion mechanisms and advances in electrical methods for corrosion monitoring. *Constr. Build. Mater.*, 269, 121240, 2021, https://doi.org/10.1016/j.conbuildmat.2020.121240.

15. Gopal, R. and Sangoju, B., Carbonation-induced corrosion: A brief review on prediction models. *J. Inst. Eng. India Ser. A*, 101, 247–257, 2020, https://doi.org/10.1007/s40030-020-00434-8.
16. Topçu, I.B. and Uzunömeroğlu, A., Properties of corrosion inhibitors on reinforced concrete. *J. Struct. Eng. Appl. Mech.*, 3, 93–109, 2020, https://doi.org/10.31462/jseam.2020.02093109.
17. Li, Z., Jin, Z., Wang, P., Zhao, T., Corrosion mechanism of reinforced bars inside concrete and relevant monitoring or detection apparatus: A review. *Constr. Build. Mater.*, 279, 122432, 2021, https://doi.org/10.1016/j.conbuildmat.2021.122432.
18. Sanjuán, M.A., Menéndez, E., Recino, H., Carbonation resistance of ternary Portland cements made with silica fume and limestone. *Materials*, 17, 1–17, 2024, https://doi.org/10.3390/ma17112705.
19. Vereecken, E., Botte, W., Lombaert, G., Caspeele, R., Assessment of corroded prestressed and posttensioned concrete structures: A review. *Struct. Concr.*, 22, 2556–2580, 2021, https://doi.org/10.1002/suco.202100050.
20. Wang, T., Wu, K., Kan, L., Wu, M., Current understanding on microbiologically induced corrosion of concrete in sewer structures: A review of the evaluation methods and mitigation measures. *Constr. Build. Mater.*, 247, 118539, 2020, https://doi.org/10.1016/j.conbuildmat.2020.118539.
21. Van Steen, C., Nasser, H., Verstrynge, E., Wevers, M., Acoustic emission source characterization of chloride-induced corrosion damage in reinforced concrete. *Struct. Health Monit.*, 21, 1266–1286, 2022, https://doi.org/10.1177/14759217211013324.
22. Leporace-Guimil, B., Conforti, A., Zerbino, R., Plizzari, G.A., Chloride-induced corrosion in reinforced concrete and fiber reinforced concrete elements under tensile service loads. *Cem. Concr. Compos.*, 124, 104245, 2021, https://doi.org/10.1016/j.cemconcomp.2021.104245.
23. Harilal, M., Kamde, D.K., Uthaman, S., *et al.*, Chloride-induced corrosion of fly ash concrete with nanoparticles and corrosion inhibitor. *Constr. Build. Mater.*, 274, 122097, 2021, https://doi.org/10.1016/j.conbuildmat.2020.122097.
24. Feng, W., Tarakbay, A., Ali Memon, S., *et al.*, Methods of accelerating chloride-induced corrosion in steel-reinforced concrete: A comparative review. *Constr. Build. Mater.*, 289, 123165, 2021, https://doi.org/10.1016/j.conbuildmat.2021.123165.
25. Poursaee, A. and Ross, B., The role of cracks in chloride-induced corrosion of carbon steel in concrete—Review. *Corros. Mater. Degrad.*, 3, 258–269, 2022, https://doi.org/10.3390/cmd3020015.
26. Wu, Q., Ma, Q., Huang, X., Mechanical properties and damage evolution of concrete materials considering sulfate attack. *Mater. (Basel)*, 14, 1–19, 2021, https://doi.org/10.3390/ma14092343.
27. Quan, X., Wang, S., Liu, K., *et al.*, Corrosion resistance of engineered cementitious composite (ECC) with high-volume fly ash and low-volume bentonite against sulfate attack and dry-wet cycles. *Constr. Build. Mater.*, 303, 124599, 2021, https://doi.org/10.1016/j.conbuildmat.2021.124599.

28. Sun, D., Huang, C., Cao, Z., *et al.*, Reliability assessment of concrete under external sulfate attack. *Case Stud. Constr. Mater.*, 15, e00690, 2021, https://doi.org/10.1016/j.cscm.2021.e00690.
29. Zhang, G., Wu, C., Hou, D., *et al.*, Effect of environmental pH on phase composition and microstructure of Portland cement paste under sulfate attack. *Composites, Part B*, 216, 108862, 2021, https://doi.org/10.1016/j.compositesb.2021.108862.
30. Zhou, S. and Ju, J.W., A chemo-micromechanical damage model of concrete under sulfate attack. *Int. J. Damage Mech.*, 30, 1213–1237, 2021, https://doi.org/10.1177/1056789521997916.
31. Honglei, C., Zuquan, J., Penggang, W., *et al.*, Comprehensive resistance of fair-faced concrete against sulfate attack in marine environments. *Constr. Build. Mater.*, 277, 122312, 2021, https://doi.org/10.1016/j.conbuildmat.2021.122312.
32. Zhao, G., Shi, M., Guo, M., Fan, H., Degradation mechanism of concrete subjected to external sulfate attack: Comparison of different curing conditions. *Mater. (Basel)*, 13, 1–17, 2020, https://doi.org/10.3390/ma13143179.
33. Ma, L., Ren, C., Wang, J., *et al.*, Self-reporting coatings for autonomous detection of coating damage and metal corrosion: A review. *Chem. Eng. J.*, 421, 127854, 2021, https://doi.org/10.1016/j.cej.2020.127854.
34. Van Steen, C. and Verstrynge, E., Degradation monitoring in reinforced concrete with 3D localization of rebar corrosion and concrete cracking. *Appl. Sci.*, 11, 1–16, 2021, https://doi.org/10.3390/app11156772.
35. Munawar, H.S., Ullah, F., Shahzad, D., *et al.*, Civil infrastructure damage and corrosion detection: Application of machine learning. *Buildings*, 12, 1–27, 2022, https://doi.org/10.3390/buildings12020156.
36. Szczepankowski, A., Przysowa, R., Perczyński, J., Kułaszka, A., Health and durability of protective and thermal barrier coatings monitored in service by visual inspection. *Coatings*, 12, 1–19, 2022, https://doi.org/10.3390/coatings12050624.
37. Reddy, M.S.B., Ponnamma, D., Sadasivuni, K.K., *et al.*, Sensors advancing the capabilities of corrosion detection: A review. *Sens. Actuators, A*, 332, 113086, 2021, https://doi.org/10.1016/j.sna.2021.113086.
38. Khayatazad, M., De Pue, L., De Waele, W., Corrosion detection on steel structures using automated image processing. *Dev. Built Environ.*, 3, 100022, 2020, https://doi.org/10.1016/j.dibe.2020.100022.
39. Chen, H.G., Chow, C.L., Lau, D., Materials, 15, 2022, https://doi.org/10.3390/ma15010354.
40. Nasser, H., Van Steen, C., Vandewalle, L., Verstrynge, E., Experimental assessment of corrosion damage and bending capacity reduction in singly reinforced concrete beams with accelerated corrosion. *Constr. Build. Mater.*, 286, 122773, 2021, https://doi.org/10.1016/j.conbuildmat.2021.122773.
41. Li, C., Chen, Q., Wang, R., *et al.*, Corrosion assessment of reinforced concrete structures exposed to chloride environments in underground tunnels: Theoretical insights and practical data interpretations. *Cem. Concr. Compos.*, 112, 103652, 2020, https://doi.org/10.1016/j.cemconcomp.2020.103652.

42. Lavorato, D., Fiorentino, G., Pelle, A., *et al.*, A corrosion model for the interpretation of cyclic behavior of reinforced concrete sections. *Struct. Concr.*, 21, 1732–1746, 2020, https://doi.org/10.1002/suco.201900232.
43. Melchers, R.E., Long-term durability of marine reinforced concrete structures. *J. Mar. Sci. Eng.*, 8, 1–12, 2020, https://doi.org/10.3390/JMSE8040290.
44. Golewski, G.L., The phenomenon of cracking in cement concretes and reinforced concrete structures: The mechanism of cracks formation, causes of their initiation, types and places of occurrence, and methods of detection—A review. *Buildings*, 13, 1–34, 2023, https://doi.org/10.3390/buildings13030765.
45. Melchers, R.E. and Chaves, I.A., Durability of reinforced concrete bridges in marine environments. *Struct. Infrastruct. Eng.*, 16, 169–180, 2020, https://doi.org/10.1080/15732479.2019.1604769.
46. Senga Kiesse, T., Bonnet, S., Amiri, O., Ventura, A., Analysis of corrosion risk due to chloride diffusion for concrete structures in marine environment. *Mar. Struct.*, 73, 1–12, 2020, https://doi.org/10.1016/j.marstruc.2020.102804.
47. Bojórquez, J., Ponce, S., Ruiz, S.E., *et al.*, Structural reliability of reinforced concrete buildings under earthquakes and corrosion effects. *Eng. Struct.*, 237, 1–11, 2021, https://doi.org/10.1016/j.engstruct.2021.112161.
48. Samson, G., Deby, F., Garciaz, J.L., Lassoued, M., An alternative method to measure corrosion rate of reinforced concrete structures. *Cem. Concr. Compos.*, 112, 103672, 2020, https://doi.org/10.1016/j.cemconcomp.2020.103672.
49. Keßler, S., Probabilistic corrosion condition assessment of a tunnel structure. *Struct. Concr.*, 21, 1345–1355, 2020, https://doi.org/10.1002/suco.201900414.
50. Guo, Z., Guo, R., Yao, G., Bayesian probabilistic model for reinforcement corrosion ratio of reinforcement in concrete prediction based on modified half-cell potential. *J. Civ. Struct. Health Monit.*, 14, 485–500, 2024, https://doi.org/10.1007/s13349-023-00738-w.
51. Almashakbeh, Y., Saleh, E., Al-Akhras, N.M., Evaluation of Half-Cell Potential Measurements for Reinforced Concrete Corrosion. *Coatings*, 12, 1–10, 2022, https://doi.org/10.3390/coatings12070975.
52. Dineshkumar, R., Harikaran, C., Veerapandi, P., Corrosion Assessment in Reinforced Concrete Elements using Half-Cell Potentiometer–A Review. *Measurement*, 9, 12, 2020.
53. Amiri, A.S., Erdogmus, E., Richter-Egger, D., A Comparison between Ultrasonic Guided Wave Leakage and Half-Cell Potential Methods in Detection of Corrosion in Reinforced Concrete Decks. *Signals*, 2, 413–433, 2021, https://doi.org/10.3390/signals2030026.
54. Sarma, C.B., Raja, M., Dwivedi, S.K., Vidyarthi, U.S., Assessment of Corrosion Risk of Reinforcement in Drainage Gallery of Power Station Using Half-Cell Potential Meter. 44–49, 2023, https://doi.org/10.9790/1813-12024449.
55. Machammah, M., Gbinu Samuel, K., Charles, K., Chloride Threshold Ingress Evaluation of Corrosion Probability Using Concrete Electrical Resistivity and Half-Cell Potential Measurements. *Sch. Int. J. Chem. Mater. Sci.*, 204–220, 1–18, 2021, https://doi.org/10.36348/sijcms.2021.v04i07.004.

56. Wu, Z., Yu, H., Ma, H., *et al.*, Rebar Corrosion in Coral Aggregate Concrete: Determination of Chloride Threshold by LPR. *Corros. Sci.*, 163, 1–24, 2020, https://doi.org/10.1016/j.corsci.2019.108238.
57. Wu, Z., Yu, H., Ma, H., *et al.*, Rebar Corrosion Behavior of Coral Aggregate Seawater Concrete by Electrochemical Techniques. *Anti-Corros. Methods Mater*, 67, 59–72, 2020, https://doi.org/10.1108/ACMM-05-2019-2128.
58. Sohail, M.G., Kahraman, R., Alnuaimi, N.A., *et al.*, Electrochemical Behavior of Mild and Corrosion Resistant Concrete Reinforcing Steels. *Constr. Build. Mater.*, 232, 117205, 2020, https://doi.org/10.1016/j.conbuildmat.2019.117205.
59. Durrani, F., Wesley, R., Srikandarajah, V., *et al.*, Predicting Corrosion Rate in Chilled HVAC Pipe Network: Coupon vs Linear Polarisation Resistance Method. *Eng. Fail. Anal.*, 109, 104261, 2020, https://doi.org/10.1016/j.engfailanal.2019.104261.
60. Babu, U.R. and Kondraivendhan, B., Effect of IR Drop on Reinforced Concrete Corrosion Measurements. *IOP Conf. Ser. Mater Sci. Eng.*, 829, 1–5, 2020, https://doi.org/10.1088/1757-899X/829/1/012015.
61. Zhong, X., Li, M., Zhou, B., *et al.*, *In-situ* Electrochemical Impedance Spectroscopy Investigations on the Early Stage Corrosion Behavior of Zircaloy-4 in High-Temperature Lithiated Water. *Corros. Sci.*, 182, 109223, 2021, https://doi.org/10.1016/j.corsci.2020.109223.
62. Wang, S., Zhang, J., Gharbi, O., *et al.*, Electrochemical Impedance Spectroscopy. *Nat. Rev. Methods Prim.*, 1, 1–21, 2021, https://doi.org/10.1038/s43586-021-00039-w.
63. Usman, B.J., Scenini, F., Curioni, M., Corrosion Testing of Anodized Aerospace Alloys: Comparison Between Immersion and Salt Spray Testing using Electrochemical Impedance Spectroscopy. *J. Electrochem. Soc.*, 167, 041505, 2020, https://doi.org/10.1149/1945-7111/ab74e3.
64. Feliu, S., Electrochemical Impedance Spectroscopy for the Measurement of the Corrosion Rate of Magnesium Alloys: Brief Review and Challenges. *Met. (Basel)*, 10, 1–23, 2020, https://doi.org/10.3390/met10060775.
65. Melara, E.K., Mendes, A.Z., Andreczevecz, N.C., *et al.*, Monitoring by Electrochemical Impedance Spectroscopy of Mortars Subjected to Ingress and Extraction of Chloride Ions. *Constr. Build. Mater.*, 242, 118001, 2020, https://doi.org/10.1016/j.conbuildmat.2020.118001.
66. AL-Ameeri, A.S., Rafiq, M.I., Tsioulou, O., Combined Impact of Carbonation and Crack Width on the Chloride Penetration and Corrosion Resistance of Concrete Structures. *Cem. Concr. Compos.*, 115, 103819, 2021, https://doi.org/10.1016/j.cemconcomp.2020.103819.
67. Deeble Sloane, M.J., Betti, R., Marconi, G., *et al.*, Experimental Analysis of a Nondestructive Corrosion Monitoring System for Main Cables of Suspension Bridges. *J. Bridg. Eng.*, 18, 653–662, 2013, https://doi.org/10.1061/(asce)be.1943-5592.0000399.
68. Bouiti, K., Al-Sharabi, H.A., Bensemlali, M., *et al.*, Effect of Temperature on Corrosion Inhibition by Ethanolic Extract of Eriobotrya Japonica Seeds

in Chloride Medium 1M. *EPJ Appl. Phys.*, 97, 1–25, 2022, https://doi.org/10.1051/epjap/2022220070.

69. Zhang, M., Zhang, Y., Ye, L., *et al.*, Hierarchically Constructed Ag Nanowires Shelled with Ultrathin Co-LDH Nanosheets for Advanced Oxygen Evolution Reaction. *Appl. Catal. B Environ.*, 298, 120601, 2021, https://doi.org/10.1016/j.apcatb.2021.120601.
70. Arabzadeh, A., Notani, M.A., Kazemiyan Zadeh, A., *et al.*, Electrically Conductive Asphalt Concrete: An Alternative for Automating the Winter Maintenance Operations of Transportation Infrastructure. *Compos Part B Eng.*, 173, 106985, 2019, https://doi.org/10.1016/j.compositesb.2019.106985.
71. Marzo, F.F., Alberro, M., Manso, A.P., *et al.*, Evaluation of the Corrosion Resistance of Ni(P)Cr Coatings for Bipolar Plates by Electrochemical Impedance Spectroscopy. *Int. J. Hydrogen Energy*, 45, 20632–20646, 2020, https://doi.org/10.1016/j.ijhydene.2020.03.076.
72. De Motte, R., Basilico, E., Mingant, R., *et al.*, A Study by Electrochemical Impedance Spectroscopy and Surface Analysis of Corrosion Product Layers Formed During CO2 Corrosion of Low Alloy Steel. *Corros. Sci.*, 172, 108666, 2020, https://doi.org/10.1016/j.corsci.2020.10866673.
73. Gaona-Tiburcio, C., Montoya-Rangel, M., Cabral-Miramontes, J.A., *et al.*, Corrosion resistance of multilayer coatings deposited by PVD on Inconel 718 using electrochemical impedance spectroscopy technique. *Coatings*, 10, 1–11, 2020, https://doi.org/10.3390/COATINGS10060521.
74. Sharma, K., Analysis of Non-destructive Testing for Improved Inspection and Maintenance Strategies. *E-J. Nondestruct. Test*, 28, 2023, https://doi.org/10.58286/28287.
75. Patel, R., A Review on Non-Destructive Testing (NDT) Techniques: Advances, Researches and Applicability. *Int. J. Curr. Sci. Res. Rev.*, 5, 1342–1355, 2022, https://doi.org/10.47191/ijcsrr/v5-i4-59.
76. Misokefalou, D.E., Papoutsidakis, P.M., Priniotakis, P.G., Non-Destructive Testing for Quality Control in Automotive Industry. *Int. J. Eng. Appl. Sci. Technol.*, 7, 349–355, 2022, https://doi.org/10.33564/ijeast.2022.v07i01.054.
77. Miró, M., Eiras, J.N., Poveda, P., *et al.*, Detecting cracks due to steel corrosion in reinforced cement mortar using intermodulation generation of ultrasonic waves. *Constr. Build. Mater.*, 286, 1–11, 2021, https://doi.org/10.1016/j.conbuildmat.2021.122915.
78. Yilmaz, B., Asokkumar, A., Jasiūnienė, E., Kažys, R.J., Air-coupled, contact, and immersion ultrasonic non-destructive testing: Comparison for bonding quality evaluation. *Appl. Sci.*, 10, 1–22, 2020, https://doi.org/10.3390/app10196757.
79. Sastry, C.C., Abeens, M., Pradeep, N., Manickam, M.A.M., Microstructural analysis, radiography, tool wear characterization, induced residual stress and corrosion behavior of conventional and cryogenic trepanning of DSS 2507. *J. Mech. Sci. Technol.*, 34, 2535–2547, 2020, https://doi.org/10.1007/s12206-020-0529-1.
80. Licata, M., Parker, H.M.O., Aspinall, M.D., *et al.*, Fast neutron and γ-ray backscatter radiography for the characterization of corrosion-born defects

in oil pipelines. *EPJ Web Conf.*, 225, 06009, 2020, https://doi.org/10.1051/epjconf/202022506009.

81. Jamshidi, V. and Davarnejad, R., Photon backscatter radiography application for the simulation of corrosion detection inside a pipeline: A novel proposal for 360° corrosion consideration in the pipelines. *Appl. Radiat. Isot.*, 176, 109844, 2021, https://doi.org/10.1016/j.apradiso.2021.109844.
82. May, Z., Alam, M.K., Nayan, N.A., Recent Advances in Nondestructive Method and Assessment of Corrosion Undercoating in Carbon–Steel Pipelines. *Sensors*, 22, 1–30, 2022, https://doi.org/10.3390/s22176654.
83. Guo, Y., Liu, G., Jing, G., *et al.*, Ballast fouling inspection and quantification with ground penetrating radar (GPR). *Int. J. Rail Transp.*, 11, 151–168, 2023, https://doi.org/10.1080/23248378.2022.2064346.
84. Shokri, N.H., Amin, Z.M., Seli, V.A., Non-Destruction Method for Detecting Corroded Underground Pipe Using Ground Penetrating Radar. *IOP Conf. Ser. Earth Environ. Sci.*, 540, 0–9, 2020, https://doi.org/10.1088/1755-1315/540/1/012027.
85. BS EN 1504, European Standard EN 1504: A simplified, illustrated guide for all involved in concrete repair. *Concr. Repair Prod. Syst.*, 1, 48, 2009.
86. Thomas, D. and Gündel, M., Hyperspectral imaging systems for corrosion detection from remotely operated vehicles. *Ce/Papers*, 6, 934–938, 2023, https://doi.org/10.1002/cepa.213287.
87. Meribout, M., Mekid, S., Kharoua, N., Khezzar, L., Online monitoring of structural materials integrity in process industry for I4.0: A focus on material loss through erosion and corrosion sensing. *Meas. J. Int. Meas. Confed.*, 176, 109110, 2021, https://doi.org/10.1016/j.measurement.2021.109110.
88. Feroz, S. and Dabous, S.A., UAV-based remote sensing applications for bridge condition assessment. *Remote Sens.*, 13, 1–38, 2021, https://doi.org/10.3390/rs13091809.
89. Gagliardi, V., Tosti, F., Bianchini Ciampoli, L., *et al.*, Satellite Remote Sensing and Non-Destructive Testing Methods for Transport Infrastructure Monitoring: Advances, Challenges and Perspectives. *Remote Sens.*, 15, 1–29, 2023, https://doi.org/10.3390/rs15020418.
90. Li, L., Chakik, M., Prakash, R., A review of corrosion in aircraft structures and graphene-based sensors for advanced corrosion monitoring. *Sensors*, 21, 1–31, 2021, https://doi.org/10.3390/s21092908.
91. Robuschi, S., Tengattini, A., Dijkstra, J., *et al.*, A closer look at corrosion of steel reinforcement bars in concrete using 3D neutron and X-ray computed tomography. *Cem. Concr. Res.*, 144, 106439, 2021, https://doi.org/10.1016/j.cemconres.2021.106439.
92. Ribeiro, D., Santos, R., Shibasaki, A., *et al.*, Remote inspection of RC structures using unmanned aerial vehicles and heuristic image processing. *Eng. Fail. Anal.*, 117, 104813, 2020, https://doi.org/10.1016/j.engfailanal.2020.104813.

93. Majhi, S., Mukherjee, A., George, N.V., *et al.*, Corrosion monitoring in steel bars using Laser ultrasonic guided waves and advanced signal processing. *Mech. Syst. Signal Process.*, 149, 107176, 2021, https://doi.org/10.1016/j.ymssp.2020.107176.
94. Xia, D.H., Song, S., Tao, L., *et al.*, Review-material degradation assessed by digital image processing: Fundamentals, progresses, and challenges. *J. Mater. Sci. Technol.*, 53, 146–162, 2020, https://doi.org/10.1016/j.jmst.2020.04.033.
95. Sunny, A.I., Zhao, A., Li, L., Kanteh Sakiliba, S., Low-cost IoT-based sensor system: A case study on harsh environmental monitoring. *Sens. (Switzerland)*, 21, 1–12, 2021, https://doi.org/10.3390/s21010214.
96. Peng, Z., Jian, J., Wen, H., *et al.*, Distributed fiber sensor and machine learning data analytics for pipeline protection against extrinsic intrusions and intrinsic corrosions. *Opt. Express*, 28, 27277–27289, 2020, https://doi.org/10.1364/oe.397509.
97. Mabrouki, J., Azrour, M., Dhiba, D., *et al.*, IoT-based data logger for weather monitoring using Arduino-based wireless sensor networks with remote graphical application and alerts. *Big Data Min. Anal.*, 4, 25–32, 2021, https://doi.org/10.26599/BDMA.2020.9020018.
98. Cheng, J.C.P., Chen, W., Chen, K., Wang, Q., Data-driven predictive maintenance planning framework for MEP components based on BIM and IoT using machine learning algorithms. *Autom. Constr.*, 112, 103087, 2020, https://doi.org/10.1016/j.autcon.2020.103087.
99. Restrepo, C.E., Simonoff, J.S., Zimmerman, R., Causes, cost consequences, and risk implications of accidents in US hazardous liquid pipeline infrastructure. *Int. J. Crit. Infrastruct. Prot.*, 2, 38–50, 2009, https://doi.org/10.1016/j.ijcip.2008.09.001.
100. Blustein, G., Romagnoli, R., Jaén, J.A., *et al.*, Zinc basic benzoate as eco-friendly steel corrosion inhibitor pigment for anticorrosive epoxy-coatings. *Colloids Surf. A Physicochem. Eng. Asp.*, 290, 7–18, 2006, https://doi.org/10.1016/j.colsurfa.2006.04.043.
101. Qiao, G., Guo, B., Ou, J., Corrosion monitoring of reinforced concrete structures, in: *Corrosion of Reinforced Concrete Structures: Mechanics, Monitoring, Control, and Beyond*, pp. 91–139, 2023, https://doi.org/10.1016/B978-0-12-819548-2.00004-4.
102. Borah, M.M., Dey, A., Sil, A., Reliability assessment of a marine bridge structure considering Indian climatic conditions under time-variant loads. *Innov. Infrastruct. Solut.*, 7, 2–3, 2022, https://doi.org/10.1007/s41062-022-00758-3.
103. Valdez, B., Ramirez, J., Eliezer, A., *et al.*, Corrosion assessment of infrastructure assets in coastal seas. *J. Mar. Eng. Technol.*, 15, 124–134, 2016, https://doi.org/10.1080/20464177.2016.1247635.
104. Choi, J.Y., Lee, K.Y., Chung, J.S., Kim, S.H., Rail pad corrosion effects on the dynamic behavior of direct fixation track systems in marine environments. *Appl. Sci.*, 10, 1–16, 2020, https://doi.org/10.3390/app10072245.
105. Ben Seghier, M.E.A., Corriea, J.A.F.O., Jafari-Asl, J., *et al.*, On the modeling of the annual corrosion rate in main cables of suspension bridges using combined

soft computing model and a novel nature-inspired algorithm. *Neural Comput. Appl.*, 33, 15969–15985, 2021, https://doi.org/10.1007/s00521-021-06199-w.

106. Melchers, R.E. and Jeffrey, R.J., Probabilistic models for steel corrosion loss and pitting of marine infrastructure. *Reliab. Eng. Syst. Saf.*, 93, 423–432, 2008, https://doi.org/10.1016/j.ress.2006.12.006.
107. Karbhari, V.M. and Seible, F., Fiber reinforced composites - advanced materials for the renewal of civil infrastructure. *Appl. Compos. Mater.*, 7, 95–124, 2000, https://doi.org/10.1023/a:1008915706226.
108. Leng, Y., Lu, Z.H., Li, C.Q., Zhao, Y.G., Time-variant probabilistic assessment of corrosion initiation of marine concrete structures considering maximum phenomenon. *Constr. Build. Mater.*, 272, 121967, 2021, https://doi.org/10.1016/j.conbuildmat.2020.121967.
109. Chen, S., Duffield, C., Miramini, S., *et al.*, Life-cycle modelling of concrete cracking and reinforcement corrosion in concrete bridges: A case study. *Eng. Struct.*, 237, 112143, 2021, https://doi.org/10.1016/j.engstruct.2021.112143.
110. Liu, X., Han, W., Yuan, Y., *et al.*, Corrosion fatigue assessment and reliability analysis of short suspender of suspension bridge depending on refined traffic and wind load condition. *Eng. Struct.*, 234, 111950, 2021, https://doi.org/10.1016/j.engstruct.2021.111950.
111. Alamri, A.H., Localized corrosion and mitigation approach of steel materials used in oil and gas pipelines – An overview. *Eng. Fail. Anal.*, 116, 104735, 2020, https://doi.org/10.1016/j.engfailanal.2020.104735.
112. Furuya, K., Kitagawa, M., Nakamura, S.I., Suzumura, K., Corrosion mechanism and protection methods for suspension bridge cables. *Struct. Eng. Int. J. Int. Assoc. Bridg Struct. Eng.*, 10, 189–193, 2000, https://doi.org/10.2749/101686600780481518.
113. Lazorenko, G., Kasprzhitskii, A., Nazdracheva, T., Anti-corrosion coatings for protection of steel railway structures exposed to atmospheric environments: A review. *Constr. Build. Mater.*, 288, 123115, 2021, https://doi.org/10.1016/j.conbuildmat.2021.123115.
114. Fujino, Y. and Siringoringo, D.M., Recent research and development programs for infrastructures maintenance, renovation and management in Japan. *Struct. Infrastruct. Eng.*, 16, 3–25, 2020, https://doi.org/10.1080/15732479.2019.1650077.
115. Tang, K. and Wilkinson, S., Corrosion resistance of electrified railway tunnels made of steel fibre reinforced concrete. *Constr. Build. Mater.*, 230, 117006, 2020, https://doi.org/10.1016/j.conbuildmat.2019.117006116.
116. Shi, X., Fay, L., Yang, Z., *et al.*, Corrosion of Deicers to Metals in Transportation Infrastructure: Introduction and Recent Developments. *Corros. Rev.*, 27, 23–52, 2011, https://doi.org/10.1515/corrrev.2009.27.1-2.23.
117. Prasad, A.R., Kunyankandy, A., Joseph, A., Corrosion inhibition in oil and gas industry: Economic considerations, in: *Corrosion Inhibition in Oil and Gas Industry*, pp. 135–150, 2020, https://doi.org/10.1002/9783527822140.ch5.

118. Frau, A.F., Pernites, R.B., Advincula, R.C., A conjugated polymer network approach to anticorrosion coatings: Poly(vinylcarbazole) electrodeposition. *Ind. Eng. Chem. Res.*, 49, 9789–9797, 2010, https://doi.org/10.1021/ie100813t.
119. Chemestry, I., Mechanical performance of geopolymeric mortars based on Tunisian calcined clay, fly ash, and metakaolin, 2015.
120. Komary, M., Komarizadehasl, S., Tošić, N., *et al.*, Low-Cost Technologies Used in Corrosion Monitoring. *Sensors*, 23, 1–28, 2023, https://doi.org/10.3390/s23031309.
121. Chu, J.H., Lee, J., Chang, C.C., *et al.*, Antimicrobial characteristics in Cu-containing Zr-based thin film metallic glass. *Surf Coat. Technol.*, 259, 87–93, 2014, https://doi.org/10.1016/j.surfcoat.2014.05.019.
122. He, X., Li, J., Meng, Q., *et al.*, Enhanced adsorption capacity of sulfadiazine on tea waste biochar from aqueous solutions by the two-step sintering method without corrosive activator. *J. Environ. Chem. Eng.*, 9, 104898, 2021, https://doi.org/10.1016/j.jece.2020.104898.
123. Quraishi, M.A., Chauhan, D.S., Saji, V.S., Heterocyclic Organic Corrosion Inhibitors: Principles and Applications, 2020.
124. Idumah, C.I., Obele, C.M., Emmanuel, E.O., Hassan, A., Recently Emerging Nanotechnological Advancements in Polymer Nanocomposite Coatings for Anti-corrosion, Anti-fouling and Self-healing. *Surf. Interfaces*, 21, 100734, 2020, https://doi.org/10.1016/j.surfin.2020.100734.
125. Taleb-Berrouane, M., Khan, F., Hawboldt, K., Corrosion risk assessment using adaptive bow-tie (ABT) analysis. *Reliab. Eng. Syst. Saf.*, 214, 107731, 2021, https://doi.org/10.1016/j.ress.2021.107731126.
126. Qin, G. and Cheng, Y.F., A review on defect assessment of pipelines: Principles, numerical solutions, and applications. *Int. J. Press Vessel Pip*, 191, 104329, 2021, https://doi.org/10.1016/j.ijpvp.2021.104329.
127. Kelly, E.R., Cronk, R., Kumpel, E., *et al.*, How we assess water safety: A critical review of sanitary inspection and water quality analysis. *Sci. Total Environ.*, 718, 137237, 2020, https://doi.org/10.1016/j.scitotenv.2020.137237.
128. Arzaghi, E., Chia, B.H., Abaei, M.M., *et al.*, Pitting corrosion modelling of X80 steel utilized in offshore petroleum pipelines. *Process Saf. Environ. Prot.*, 141, 135–139, 2020, https://doi.org/10.1016/j.psep.2020.05.024.
129. Guillal, A., Ben Seghier, M.E.A., Nourddine, A., *et al.*, Probabilistic investigation on the reliability assessment of mid- and high-strength pipelines under corrosion and fracture conditions. *Eng. Fail. Anal.*, 118, 104891, 2020, https://doi.org/10.1016/j.engfailanal.2020.104891.
130. Hawari, A., Alkadour, F., Elmasry, M., Zayed, T., A state of the art review on condition assessment models developed for sewer pipelines. *Eng. Appl. Artif. Intell.*, 93, 103721, 2020, https://doi.org/10.1016/j.engappai.2020.103721.
131. Azimi, M., Eslamlou, A.D., Pekcan, G., Data-driven structural health monitoring and damage detection through deep learning: State-of-the-art review.

132. Dong, W., Huang, Y., Lehane, B., Ma, G., XGBoost algorithm-based prediction of concrete electrical resistivity for structural health monitoring. *Autom. Constr.*, 114, 103155, 2020, https://doi.org/10.1016/j.autcon.2020.103155.
133. Dong, C.Z. and Catbas, F.N., A review of computer vision–based structural health monitoring at local and global levels. *Struct. Heal Monit.*, 20, 692–743, 2021, https://doi.org/10.1177/1475921720935585.
134. Abdelkhalek, S. and Zayed, T., Comprehensive Inspection System for Concrete Bridge Deck Application: Current Situation and Future Needs. *J. Perform. Constr. Facil*, 34, 1–16, 2020, https://doi.org/10.1061/(asce)cf.1943-5509.0001484.

5

Impact of Concrete Corrosion on Environmental Sustainability

Abhinay Thakur[1*], Valentine C. Anadebe[2], Elyor Berdimurodov[3,4,5], Abdelkader Zarrouk[6,7], Omar Dagdag[8] and Ashish Kumar[9]

[1]*Division of Research and Development, Lovely Professional University, Phagwara, Punjab, India*

[2]*Department of Chemical Engineering, Alex Ekwueme Federal University Ndufu Alike, Abakaliki, Ebonyi State, Nigeria*

[3]*Chemical & Materials Engineering, New Uzbekistan University, Tashkent, Uzbekistan*

[4]*University of Tashkent for Applied Sciences, Tashkent, Uzbekistan*

[5]*Faculty of Chemistry, National University of Uzbekistan, Tashkent, Uzbekistan*

[6]*Laboratory of Materials, Nanotechnology and Environment, Faculty of Sciences, Mohammed V University in Rabat, Agdal-Rabat, Morocco*

[7]*Research Centre, Manchester Salt & Catalysis, Manchester, United Kingdom*

[8]*Department of Mechanical Engineering, Gachon University, Seongnam, Republic of Korea*

[9]*Nalanda College of Engineering, Bihar Engineering University, Department of Science, Technology and Technical Education, Government of Bihar, Bihar, India*

Abstract

Modern infrastructure's structural integrity and environmental sustainability are seriously threatened by concrete corrosion. The endurance and functionality of concrete structures, which are crucial for infrastructure and urban development, are jeopardized by various corrosion processes, including carbonation and corrosion generated by chloride. Concrete deteriorating structures frequently require more maintenance and repair work, which adds to the resource consumption and waste production associated with construction. This increases the depletion of raw materials and energy supplies and the environmental impact of infrastructure projects. Moreover, deteriorating concrete can leak toxic materials into the

**Corresponding author*: thakurabhinay96@gmail.com

Ruby Aslam, Qihui Wang, Zhitao Yan and Afroz Aslam (eds.) Corrosion in Concrete Structures: Integrating Advanced Technologies and Sustainable Practices, (119–162) © 2026 Scrivener Publishing LLC

environment, affecting the soil and water quality. The numerous effects of concrete corrosion on sustainability and the environment are examined in this chapter, with particular attention to how these effects affect ecological systems and resource management. It emphasizes how crucial it is to use sustainable methods while managing and building concrete to preserve the environment and prolong the life of infrastructure systems.

Keywords: Corrosion, sustainability, infrastructure, durability, environmental, management

5.1 Introduction

In today's infrastructure, concrete is a key component used to build anything from roads and dams to buildings and bridges. Because of its innate strength, adaptability, and longevity, it is a vital option for many different structural applications. Concrete is not immune to degradation despite its extensive use and essential function. The endurance and efficacy of concrete are seriously threatened by corrosion, which raises serious concerns [1–3]. The leading cause of concrete corrosion is the steel reinforcement deterioration ingrained in the concrete matrix. Concrete buildings lack tensile strength and are only robust in compression until steel reinforcement or rebar is added. When corrosion begins to occur, concrete's structural integrity is compromised, resulting in problems including cracking, spalling, and a reduced ability to support weight. Concrete corrodes due to various processes, each of which is impacted by a unique set of environmental conditions. For example, carbonation-induced corrosion happens when atmospheric carbon dioxide combines with concrete's calcium hydroxide to generate calcium carbonate. The protective alkaline environment surrounding the steel reinforcement is diminished by this reaction, which lowers the pH of the concrete and increases its susceptibility to rust. Although carbonation happens gradually, its effects might become noticeable over time, especially in places where CO_2 concentrations are high [4, 5]. Concrete corrosion that is generated by chloride is another common and harmful type. Chlorides seep into the concrete and get to the steel reinforcing; they are frequently added by seawater or de-icing salts. These chlorides cause a corrosive reaction by rupturing the passive oxide layer that typically shields the steel. This kind of corrosion is widespread, particularly in coastal areas or areas where de-icing solutions are used in the winter. Sulfurates from soil or groundwater react with concrete components to generate expansive compounds that cause the concrete to swell and crack.

This process is known as sulphate-induced corrosion. High amounts of sulfate present a risk for this kind of corrosion.

Furthermore, although less often, microbiological corrosion is caused by bacteria or fungi that release acidic byproducts, exacerbating concrete deterioration. Microbial corrosion is still a relatively new field of study, but it is becoming more and more of a problem for the longevity of concrete. Comprehending these corrosion pathways is essential to creating successful mitigation and prevention plans. Researchers and engineers can create more robust concrete mixtures, put protective measures in place, and set up maintenance schedules to increase the lifespan of concrete structures by pinpointing the precise causes of each form of corrosion. In addition to preserving concrete's structural integrity, addressing these corrosion mechanisms is essential to guaranteeing the infrastructure's general security and durability [6–8]. The failure mechanism of a three-coat zinc-rich coating system in an environmental setting, as illustrated in Figure 5.1, involves degradation progressing from the outermost to the innermost layer. This system comprises three layers: a polyurethane-based topcoat, a midcoat made of a generic polymer, and a bottom layer containing over 80% zinc by weight in a dry film blended with epoxy.

The topcoat acts as a hydrophobic barrier, repelling water by reducing the water's contact angle with the surface, thereby slowing the penetration of ions or gases. It also absorbs ultraviolet light (hv), shielding the underlying layers from UV-induced deterioration. Despite its flexibility and resistance to abrasions, this layer's malleability allows ions and gases to gradually diffuse through it. Beneath the topcoat lies the midcoat, typically made of a rigid, highly crystalline polymer that operates below its glass transition temperature (Tg). These polymers, often polyesters or polyamides, provide

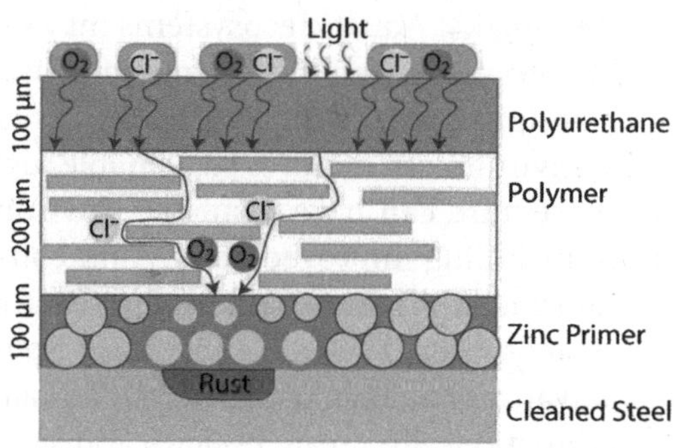

Figure 5.1 Failure mechanism of a three-layer zinc-rich coating system: sacrificial zinc primer, polymer midcoat, and polyurethane topcoat. Adapted from [11] under CCBY 4.0.

an additional barrier to the passage of ions and gases, further slowing the corrosion process. However, the crystalline structure of this layer makes it prone to failure through delamination or adhesion loss. The bottom layer, rich in zinc, contains either organic (epoxy) or inorganic (silicate) fillers. Zinc, being more anodic than iron, corrodes preferentially, protecting the steel substrate through galvanic action [9, 10]. As zinc corrodes, its volume expands significantly, potentially compromising the integrity of the crystalline midcoat and facilitating the diffusion of corrosive agents to the steel. Once the zinc layer is depleted, the exposed steel begins to corrode, forming iron oxide (FeO), which can flake off, taking the coating with it and resulting in complete coating failure.

It is impossible to exaggerate the significance of environmental sustainability in concrete construction. Concrete manufacture and maintenance have a substantial impact on the environment that goes beyond obvious structural issues. Concrete requires a lot of resources to produce; significant amounts of raw materials like cement, aggregates, and water are needed. Damage caused by corrosion frequently necessitates frequent maintenance and repairs, which adds to the cost of resources [12, 13]. The depletion of natural resources is accelerated by this increased material use, which also worsens the environment.

Additionally, building waste is produced during the demolition and restoration of corroded concrete structures, adding to landfill trash and environmental damage. The environmental effect of concrete is further worsened by the energy required for its manufacture, transportation, and preservation. A significant carbon footprint is associated with the lifecycle of concrete structures, including the repair and maintenance phases. Resolving corrosion-related problems frequently requires more energy inputs, which might increase overall environmental impact. Furthermore, deteriorating concrete can leak toxic materials into the environment, lowering the soil and water quality. Aquatic ecosystems may suffer if corroded materials are released into marine environments in coastal places where chloride-induced corrosion is common. Reducing concrete corrosion's adverse effects on the environment requires sustainable methods and solutions. Concrete infrastructure can have a smaller environmental impact owing to innovations, including improved design and construction methods, corrosion-resistant materials and coatings, and routine maintenance schedules. Significant steps toward sustainability have been made possible by material science advances, such as using recycled materials and more durable concrete combinations. To create policies and standards that support sustainable practices and stimulate the uptake of cutting-edge technologies, policymakers and business executives must work together.

5.2 Types of Concrete Corrosion

5.2.1 Carbonation-Induced Corrosion

One common type of concrete deterioration is carbonation-induced corrosion, which is brought on by the reaction of carbon dioxide (CO_2) with the alkaline elements in the concrete matrix. Calcium hydroxide, a byproduct of cement hydration, gives concrete its inherent alkaline properties. The implanted steel reinforcement is naturally shielded from damage by this high pH environment, which keeps a passive oxide coating on its surface [14]. However, the concrete's pH is lowered when CO_2 from the atmosphere seeps through and combines with calcium hydroxide to make calcium carbonate. The passive protective layer surrounding the steel reinforcement becomes less effective as the pH decreases. When exposed to moisture, this lack of protection renders the steel more prone to rusting. The long-term nature of the carbonation process makes it a slow-moving but sneaky danger to concrete buildings. The rate at which carbonation advances is influenced by variables like concrete porosity, ambient CO_2 concentration, and mix quality. Carbonation-induced corrosion is especially dangerous since it frequently goes undiagnosed until serious harm has been done. Surface discoloration or small fissures are some of the early indicators of carbonation that are often undetectable and easily missed [15, 16]. Eventually, rust expands due to the corrosion of the steel reinforcement, which causes cracking and spalling in the concrete. This jeopardizes the concrete's structural integrity and calls for expensive maintenance and repairs.

Several tactics are used to mitigate corrosion caused by carbonation. The longevity of concrete can be increased by making sure the mix is properly designed, with a sufficient amount of cement and a lower ratio of water to cement. Furthermore, adding additional cementitious ingredients to concrete, like slag or fly ash, might strengthen its resistance to carbonation. Regular inspection and maintenance procedures are crucial to managing carbonation-induced corrosion's long-term effects. These procedures include monitoring for carbonation depth and evaluating structural conditions.

5.2.2 Chloride-Induced Corrosion

Another important type of concrete corrosion is chloride-induced corrosion, which is more common near coasts or in places where de-icing salts are utilized. Chlorides can pass through concrete and get to the embedded steel reinforcement. Usually, they come from sources like saltwater or road salts [17–19]. Chlorides cause a corrosive reaction in steel by rupturing the passive oxide layer that shields the reinforcing. The process starts when the chlorides move through the concrete matrix, which is frequently made easier by moisture. Localized corrosion results from the chloride ions interfering with the passive film when they come into touch with the steel reinforcement. Chloride-induced corrosion can be more aggressive and quick than carbonation-induced corrosion, especially in settings with a significant concentration of chloride. Concrete spalling, rust discoloration, and surface cracking are common signs of chloride-induced corrosion. Rust expands as corrosion worsens, applying internal pressure to the surrounding concrete, resulting in more harm. Bridges, highways, and maritime constructions are among the structures that are most vulnerable to this kind of corrosion because of their exposure to extreme weather conditions [20, 21]. There are multiple steps involved in preventing corrosion caused by chloride. One helpful tactic is using low-permeability concrete, which is attained by adding more cementitious ingredients and lowering the water-to-cement ratio. Furthermore, adding sealants or protective coatings can prevent chloride intrusion into the concrete. Cathodic protection systems and chloride extraction techniques can be used to control and lessen continuous corrosion in existing structures.

5.2.3 Other Corrosion Mechanisms

Concrete structures are also at risk from several different corrosion mechanisms, while carbonation and chloride-induced corrosion are the ones most frequently mentioned. These consist of the alkaline-silica reaction, microbiological corrosion, and sulphate-induced corrosion.

- Sulfurates in soil or groundwater react with concrete's constituents to produce expanding compounds, which causes sulphate-induced corrosion [22]. Ettringite or gypsum are commonly produced by this reaction, resulting from sulfates interacting with calcium aluminate phases in the cement. Concrete swells, cracks, and eventually deteriorates due to the production of these expansive compounds. In regions where soil or groundwater concentrations of sulfate are high, sulfate-induced corrosion is frequently a concern.
- Though less frequent, microbial corrosion is becoming a growing source of worry. Concrete can deteriorate because some fungi and bacteria can colonize its surface and produce

acidic byproducts. These bacteria have the ability to change the chemical environment in concrete, which speeds up corrosion [23, 24]. Although microbial corrosion research is still in its early stages, it poses a significant threat to the longevity of concrete, particularly in areas where microbial activity is common.

- Another important cause of concrete deterioration is the alkaline-silica reaction (ASR). ASR is the result of a reaction between the alkaline hydroxides in the cement paste and reactive silica in the aggregate materials, which forms a gel-like substance. As a result of this gel's absorption of moisture and gradual expansion, internal pressure and concrete cracks result. ASR has the potential to seriously harm concrete structures, impairing both their structural soundness and aesthetic appeal. Using non-reactive materials and regulating the cement's alkali level are two preventive techniques for ASR.

5.3 Impact on Structural Integrity and Concrete Durability

Strong and long-lasting, concrete is highly regarded as crucial to the structural integrity of buildings, bridges, and other infrastructure, but different types of corrosion can seriously impair these qualities, resulting in a host of structural problems that impact concrete structures' usability and safety. Corrosion has a significant negative influence on structural integrity in several ways, including reducing load-bearing capacity, making structures more prone to failure and requiring costly repairs. Concrete's structural integrity is primarily affected by corrosion of the embedded steel reinforcement. Concrete buildings naturally lack tensile strength; they are strong in compression but weak in tension [25, 26]. Reinforcement, usually composed of steel rebar, gives concrete structures tensile strength. Rust forms on the steel reinforcement as a result of a chemical reaction that occurs when corrosion begins. Because this rust takes up more volume than the original steel, it creates expanding forces that press against the nearby concrete. Rust builds up and causes internal strains in the concrete, which lead to cracking and spalling. These fissures worsen the situation by allowing more moisture and corrosive substances to enter the concrete and reducing its structural strength. A critical effect of corrosion is a reduction in load-bearing capability. The cross-sectional area of the corroding steel reinforcement decreases, which causes a loss of strength. The concrete structure's overall load-bearing capacity may decline due to this strength loss, making it less able to withstand the imposed loads [27, 28]. Severe circumstances may jeopardize the system's structural integrity, possibly resulting in a partial or total breakdown. This is especially troubling for vital infrastructure, such as high-rise buildings and bridges, where structural failure can have disastrous effects.

The increased vulnerability of structures to structural failure is another effect of corrosion on structural integrity [29]. Damage caused by corrosion frequently worsens over time, making it difficult to identify and treat before it reaches a dangerous point. Because corrosion occurs gradually, structural problems might not be noticed until considerable damage has been done. The corrosion deterioration weakens the concrete and lowers its ability to support weight, making the structure more susceptible to outside pressures like strong winds, seismic activity, or large loads. This greater susceptibility can raise the possibility of structural failure, especially in areas that frequently experience harsh weather. Another significant effect of corrosion on structural integrity is the need for substantial repairs and maintenance. Repairing the damage caused by corrosion to concrete can be difficult and expensive. The damaged areas are usually removed to repair corroded concrete, the underlying steel reinforcement is treated, and fresh concrete is applied. This procedure takes a lot of time and energy and interferes with the structure's ability to operate while it is being repaired [30, 31]. The degree of the damage may, in some circumstances, need partial or complete reconstruction, which adds to the logistical and financial strain. Besides these immediate effects, corrosion can impact concrete buildings' structural integrity and visual appeal. Spalling, rust discoloration, and surface cracking can reduce a structure's esthetic appeal and possibly lower its value and usefulness. Visible deterioration of public infrastructure, such as highways and bridges, can also prompt public attention and perhaps regulatory actions due to safety and maintenance issues.

One important factor that impacts concrete structures' lifetime and long-term performance is their durability. Different types of corrosion significantly impact concrete's resistance to deterioration, the ability to endure harsh weather, and the ability to retain structural integrity. Corrosion has a variety of consequences on the durability of concrete, including both direct and indirect effects that can drastically shorten the life of concrete structures. A principal result of corrosion on the longevity of concrete is the diminished ability to withstand environmental influences [32–34]. Environmental factors, such as moisture, temperature swings, and chemical exposure, can affect concrete. Because corrosion weakens the internal structure of concrete and lessens its protective qualities, it can make it less resilient to environmental stresses. For instance, moisture may be able to enter the concrete more deeply due to corrosion-induced cracking, which would result in further degradation and decreased durability. The faster moisture infiltration in areas with high humidity or frequent rains can exacerbate a cycle of continuous degradation. The effect of corrosion on concrete's resistance to chemical attacks is another way that corrosion affects durability. Concrete is frequently exposed to various substances, including salts, acids, and alkalis, which can hasten its degradation. Because corrosion modifies concrete's chemical composition and structural integrity, it can make it less resistant to various chemical agents. This is one example of how corrosion-induced rust formation can further weaken the durability of concrete by generating more byproducts that react with it. The overall performance and longevity of the concrete might be adversely affected by this degradation, which can be especially severe in industrial settings or locations with heavy chemical exposure. Saura-Gómez *et al.* [35] demonstrated that the valuable life of reinforced concrete structures is significantly impacted by reinforcement corrosion, especially in environments where carbonation occurs. Carbonation is a chemical process in which carbon dioxide from the atmosphere penetrates the concrete and reacts with calcium hydroxide, lowering the pH of the concrete. As the pH drops, the passive protective layer around the steel reinforcement is weakened, making the steel vulnerable to corrosion. This degradation can severely affect the stability, durability, and safety of buildings, necessitating repairs much earlier than anticipated. In the Spanish Mediterranean region, environmental exposure conditions are classified such as non-aggressive and normal, which are considered relatively mild and less likely to cause severe corrosion. According to Spanish legislation, structures in these environments should have a longer service life. However, in practice, the research has revealed that specific structural components, notably columns at ground level, are highly susceptible to carbonation. These areas are exposed to moisture, which enhances the carbonation process, leading to premature corrosion. The study evaluated 17 columns from 10 buildings located in non-aggressive environments, which were expected to conform to the standard Spanish model for carbonation rates. Surprisingly, 14 of the columns displayed carbonation levels higher than those predicted by the model, indicating that corrosion was occurring at a faster rate than expected. The only exceptions were three columns, which had carbonation levels in line with regulatory expectations. This significant deviation suggests that the current Spanish EHE 08 regulations may be inadequate for certain exposure conditions and need revision. Several factors may contribute to the accelerated carbonation observed in these columns. One key issue is the influence of moisture, particularly in columns that are in contact with the ground, where capillary action can transport water into the concrete, accelerating the carbonation process. Furthermore, differences in electrochemical potential between the steel reinforcement and the surrounding materials can create conditions that promote corrosion. The current model fails to account for these influences, leading to inaccurate predictions of service life.

The need to update the EHE 08 regulation becomes evident in light of these findings. Future revisions should incorporate considerations for moisture transport through capillarity and the electrochemical interactions between materials to provide a more comprehensive and realistic assessment of the durability of reinforced concrete structures. By addressing these issues, engineers can better predict the useful life of such structures, ensuring that they remain safe and functional for their intended design life. The corrosion pathologies observed in this study highlight the visible damage caused by reinforcement corrosion. For example (Figure 5.2), in building 8, pillar 21, cracks parallel to the main reinforcements were observed, marking the propagation phase of corrosion and in building 9, pillar 1, generalized corrosion indicated the end of the corrosion propagation phase and the onset of structural failure. These examples underscore the urgency of addressing reinforcement corrosion through improved standards and preventive measures to ensure the long-term durability of reinforced concrete structures.

(a) (b)

Figure 5.2 Corrosion pathologies: (a) Cracks along reinforcements during corrosion propagation in Building 8, Pillar 21; (b) Generalized corrosion and end of service life in Nuilding 9, Pillar 1. Adapted from [35] under CCBY 4.0.

The impact of corrosion on load-bearing capacity is another important factor affecting the longevity of concrete. As was already established, corrosion weakens the steel reinforcement by reducing its cross-sectional area. This decrease in load-bearing capacity may impact the concrete structure's overall durability and reduce its ability to sustain applied loads over time. The reduced ability to support loads may further impact the concrete's durability and performance, which may result in early failure or the need for repeated repairs. Alterations in the thermal characteristics of concrete can also result from damage caused by corrosion. Although corrosion can change the material's thermal behavior, concrete buildings are made to endure temperature changes and thermal loads. Rust production and subsequent expansion can impact the concrete's insulating and thermal conductivity qualities. This may result in problems including thermal cracking, decreased energy effectiveness, and heightened vulnerability to temperature-related harm. In severe circumstances, temperature variations may impact the concrete's structural integrity, thus jeopardizing its longevity. One potential consequence of corrosion on the long-term durability of concrete is a diminished ability to withstand mechanical forces. Concrete constructions are exposed to various mechanical forces, such as impact, vibration, and loading [36]. Because corrosion damages the internal structure of concrete and lowers its load-bearing capability, it can make it less resilient to these forces. Decreased mechanical resistance may result in problems like elevated deflection, diminished stability, and heightened vulnerability to structural harm. These factors can significantly affect the concrete's durability and general performance over time. Apart from the immediate consequences of corrosion on the longevity of concrete, secondary effects associated with upkeep and restoration exist. Damage caused by corrosion frequently needs extensive repairs, which can impair the structure's operation and raise expenditures. Because each intervention may require removing and replacing pieces of the concrete, the necessity for periodic maintenance and repair might further affect the overall durability of the concrete. This cycle of continuous maintenance may shorten the concrete structure's useful life and raise operating expenses.

5.4 Environmental Consequences

5.4.1 Resource Consumption and Waste Generation

Concrete has an environmental impact at every stage, from production to removal. Making concrete uses a lot of resources because it needs large amounts of raw materials like water, aggregates, and cement. The environmental impact of corrosion on concrete structures is increased since it requires more resources for upkeep, repair, and replacement over time. Accelerating corrosion-induced concrete deterioration necessitates these actions, raising material and energy consumption. Numerous resource-intensive procedures are needed to upkeep and repair deteriorated concrete structures [37, 38]. First, specific tests and inspections are necessary to detect and evaluate corrosion damage, and these procedures take time and money. Concrete parts that have corroded are frequently removed and replaced as part of repair procedures once the extent of the damage has been assessed. New materials, such as cement, aggregates, and reinforcement steel, must usually be extracted in order to complete this process. These materials require more energy and resources to produce, which increases their overall environmental effect.

Furthermore, creating construction debris is a common outcome of fixing damaged concrete buildings. The amount of waste material that needs to be managed grows when decaying concrete is removed. This waste is disposed of frequently in landfills, which degrades the environment. Major environmental issues are still associated with the total impact of increasing trash generation and its management, even though some concrete waste can be recycled or used for other purposes.

Concrete constructions have a lifespan encompassing both their final demolition and their earliest phases of construction and restoration [39–41]. The necessity for demolition and replacement when a structure reaches the end of its helpful life increases waste production and resource consumption. Large volumes of debris are created during the demolition of corroded concrete structures; this waste needs to be appropriately managed and disposed of. The requirement for new building materials to replace the decaying structures exacerbates the demolition process' environmental impact. In summary, during a concrete structure's lifetime, corrosion causes an increase in resource consumption and waste production. Corroded concrete eventually necessitates maintenance, repair, and replacement, which uses more energy and resources and degrades the environment. Concrete infrastructure's environmental impact is made worse by the creation and handling of construction trash.

5.4.2 Release of Harmful Substances

Numerous dangerous compounds may be released into the environment by deteriorating concrete structures, which could majorly impact ecological systems and public health. Rust, chemical byproducts, and other contaminants can be released from concrete when it corrodes, potentially hurting the environment. Rust is one of the main toxic chemicals emitted during concrete deterioration. One of the byproducts of the corrosion of the steel reinforcement in concrete is rust, also known as iron oxide [42–45]. Surface cracking and spalling result from the steel reinforcement's expansion and pressure on the surrounding concrete as it corrodes. Eventually, this rust may be released into the environment by direct exposure or as part of the construction waste produced during repairs. The quality of soil and water can be impacted by the discharge of rust particles, which can lead to pollution. Apart from rust, corrosion can result in additional chemical byproducts that could pose a threat to the environment. For instance, chlorinated compounds may arise as a result of the corrosion of concrete buildings exposed to chloride ions [46–54]. These substances have the ability to contaminate water and soil through leaching. The biodiversity and ecosystem health of plants and animals may be negatively impacted by the emission of these compounds.

When concrete structures corrode due to sulfate, toxic byproducts may also be released. Expanding compounds like ettringite or gypsum are formed when sulfates react with cement constituents, a process known as sulfate-induced corrosion [55–57]. These substances have the ability to change the chemical makeup of the environment, which may have an impact on the quality of the soil and water. The possible leakage of dangerous substances during the destruction of concrete structures deteriorated by age is another cause for concern. Any pollutants found in the concrete, like heavy metals or industrial chemicals, may be released into the environment during the dismantling of structures. To protect the environment and maintain public safety, these hazardous compounds must be appropriately managed. Concrete corrosion creates dangerous chemical releases that seriously risk human health and the environment. In order to reduce the dangers associated with corrosion and its negative effects on ecological systems and public health, it is imperative to implement effective management measures, such as the appropriate disposal of construction waste and environmental quality monitoring.

5.4.3 Impact on Soil and Water Quality

One of the most significant environmental concerns is how concrete corrosion affects the soil and water quality. Damage from corrosion can cause pollutants to leak out, negatively affecting soil and water resources, agriculture, ecosystems, and human health. Rust and other corrosion-related byproducts can deteriorate soil quality and impact fertility. The physical and chemical characteristics of soil can be changed by rust particles, which can build up in the soil due to concrete structures that corrode [58, 59]. The texture and structure of the soil may alter due to iron oxide content, which could impact the soil's capacity to retain water and drain it. Given that plants depend on healthy soil for nutrients and appropriate root development, these alterations may affect plant growth and agricultural productivity. Other chemical byproducts of corrosion and rust can also impact soil quality. For instance, sulfate byproducts and chlorinated compounds can change the pH and chemical makeup of the soil, which may affect plant health and soil fertility. The availability of nutrients and microbial activity, both critical for preserving soil health and promoting plant growth, can be affected by pH changes in the soil. The buildup of pollutants in the soil as a result of concrete deterioration may potentially indirectly affect nearby ecosystems. Contamination of soil can result in toxic materials being transferred to plants and animals, upsetting food chains and perhaps having an effect on wildlife populations. For example, hazardous materials may be absorbed and accumulated by plants growing in polluted soil [60, 61]. These compounds can then be transferred to herbivores and, eventually, carnivores within the ecosystem.

Another important environmental problem is the effect of concrete corrosion on water quality. Pollution from deteriorating concrete can seep into nearby rivers, lakes, and groundwater. Water resources may get contaminated due to rust particles and chemical byproducts released from corroded concrete, which may impact aquatic life and water quality. Concrete deterioration can release rust particles into waterways through leaching or runoff. These particles can cause water to become more turbid, impairing the water's clarity and visual appeal. Reduced light penetration and altered photosynthesis in aquatic plants are further ways that elevated turbidity can harm aquatic life. The chemical byproducts of corrosion, like sulfate and chlorinated compounds, can harm water quality more. For instance, fish and other aquatic invertebrates may be poisoned by chlorinated chemicals. These substances impact aquatic organisms' health and reproduction ability, which can upset aquatic ecosystems. Similarly, sulfate by-products may help create acid-sulfate soils, which raise the acidity of water bodies and may endanger aquatic life. An additional concern is the potential for groundwater contamination from concrete corrosion. For many communities, groundwater is an essential supply of drinking water, and contaminants can harm human health [62, 63]. The quality and safety of drinking water sources may be impacted by pollutants that seep into groundwater, such as heavy metals or chemical byproducts from corroded concrete.

5.5 Sustainability Challenges

5.5.1 Lifecycle of Concrete Structures

Concrete structures have a lifecycle of multiple phases, each adding to the overall environmental impact. These stages comprise raw material extraction and processing, concrete manufacture and construction, the building's operational phase, and the final concrete demolition and disposal. The distinct sustainability issues associated with each phase impact the concrete infrastructure's overall environmental footprint. The extraction and preparation of raw materials is the first step in the lifecycle of concrete. Various raw materials are needed to produce concrete, including water, aggregates (sand, gravel, and crushed stone), and cement. The exploitation of these materials may have a substantial negative influence on the environment. For example, aggregate quarrying can result from soil erosion, habitat degradation, and biodiversity loss [64]. Heavy equipment and transportation are frequently used in the extraction process, which increases energy consumption, dust emissions, and noise pollution. One of the main ingredients of concrete is cement, which requires a lot of resources to produce. In order to generate cement, limestone and other ingredients must be heated in a kiln to create clinker, which is subsequently processed into cement powder. Significant amounts of energy, mostly from fossil fuels, are needed for this process, which also produces a lot of CO_2 emissions [65–67]. One of the leading industrial sources of CO_2 emissions worldwide, cement production has a significant impact on climate change.

Concrete is created by mixing the raw elements after they have been removed and processed. In order to attain the necessary qualities, cement, aggregates, water, and occasionally additives are combined during the concrete mixing process. Producing concrete requires a lot of energy because it must be mixed, transported, and placed. The energy needed for these procedures adds to the manufacturing of concrete's overall carbon footprint. There are sustainability issues throughout the building phase as well [68, 69]. Concrete use in buildings frequently necessitates using significant amounts of energy and materials. Waste can be produced by construction operations, such as extra concrete, packing materials, and other trash. There is constant work to be done in managing this garbage and reducing its environmental impact. In addition, the construction process may cause disturbances to nearby populations and ecosystems, such as increased traffic, dust, and noise. A concrete structure's operational phase impacts sustainability, which is determined by various aspects, including durability, energy consumption, and maintenance. Continuous maintenance is necessary for concrete structures to handle problems like corrosion, surface wear, and cracking. Using extra materials and resources is a common component of maintenance tasks, increasing their overall environmental impact. The energy consumption related to the structure is also included in the operational period. For instance, concrete buildings could need lighting, air conditioning, and heating, all of which use energy. The energy efficiency of concrete structures can be influenced by their design and construction. For example, high thermal mass can help control indoor temperature and lessen the need for mechanical heating and cooling. However, a structure's overall energy performance depends on several variables, including design, occupancy patterns, and insulation.

Demolition and disposal are part of a concrete structure's end-of-life phase. Waste from the demolition process, such as rebar and other components and pieces of concrete, can be rather large. Proper management is crucial to reduce this waste's negative effects on the environment. Concrete waste can frequently be recycled or used for other purposes. Still the success of these initiatives depends on the accessibility of recycling facilities and the calibre of the recovered materials [70, 71]. Concrete waste disposal at landfills can exacerbate environmental issues, such as contaminated water and soil. Concrete deterioration in landfills can discharge toxic materials into the environment, including chemical byproducts and heavy metals. The requirement to strike a balance between resource recovery and the environmental impact of disposal methods makes handling concrete waste even more difficult.

5.5.2 Resource Depletion and Energy Use

The usage and manufacture of concrete significantly impact energy use and resource depletion. One of the world's most extensively utilized building materials, concrete places a significant demand on energy and natural resources [72, 73]. These issues must be resolved to advance sustainability and lessen the adverse environmental effects of concrete infrastructure. Substantial resources are used in the raw material extraction process that produces concrete. Limestone is a limited resource that is used extensively in the making of cement, for example. Large-scale limestone extraction may result in the depletion of natural resources and environmental problems like habitat damage and soil erosion. Depletion of natural resources and disturbance of ecosystems can also result from the exploitation of aggregates for concrete, such as sand and gravel. The sustainability of resources is another issue brought up by using natural aggregates in manufacturing concrete [74]. Although aggregates are necessary to give concrete strength and stability, their extraction and usage can result in the loss of important resources. Demand for aggregates has surpassed supply in some areas, increasing competition for these resources and raising the possibility of environmental disputes. Energy is used extensively in the creation of concrete, especially in the cement-making process. In order to produce cement, resources like limestone must be heated in a kiln, which uses a lot of energy. In the past, fossil fuels have been the main source of this energy, which has led to greenhouse gas emissions and climate change. Given that a substantial amount of the total carbon footprint of concrete is attributed to the energy used in cement manufacture, this energy use is a major sustainability concern. Energy is used in the mixing, transporting, and laying of concrete in addition to the manufacture of cement [75, 76]. Vehicle fuel is needed for the transportation of both raw materials and completed concrete, which increases energy consumption and emissions. The energy needed for these procedures increases concrete infrastructure's total environmental impact. In order to address sustainability concerns, concrete production efficiency and energy consumption reduction initiatives are essential. For example, using alternative fuels and energy sources is one innovation in cement production that can assist in lowering the industry's carbon impact. Concrete mixing and transportation procedures can be made more energy-efficient to reduce energy use.

Adopting innovative technology and sustainable practices is necessary to address the energy consumption and resource depletion connected with manufacturing concrete. Recycled aggregates, supplementary cementitious materials (SCMs), and industrial byproducts are examples of alternative materials that can be used in concrete mixtures to assist in lessening the need for natural resources. SCMs can partially substitute cement in concrete, reducing the overall demand for cement and the carbon footprint [77, 78]. Examples of these materials are fly ash, slag, and silica fume. Natural resource conservation and waste reduction can be achieved by recycling concrete waste and using recycled aggregates into new concrete mixes. In order to produce concrete more sustainably, recycled concrete aggregates can be utilized in place of natural aggregates. Energy consumption and emissions can be decreased by implementing energy-efficient technology in cement production and concrete mixing. Utilizing waste heat recovery systems in kiln technology, for instance, can enhance the energy efficiency of cement production. The need for regular repairs and replacements can be reduced by designing concrete structures with lifetime and durability in mind. Sustainability can be increased by using materials resistant to corrosion, improving mix designs, and implementing maintenance procedures that increase the longevity of concrete structures.

5.6 Mitigation Strategies and Recent Advancements

The longevity and safety of concrete structures are significantly impacted by concrete corrosion. Various mitigating techniques are used to improve the resilience and durability of concrete in order to overcome this issue. Using corrosion-resistant coatings and additives, adhering to design and construction standards, and putting early detection and maintenance procedures into practice are some of these measures. The use of coatings and additives resistant to corrosion is essential in the fight against corrosion in concrete. By strengthening the concrete's natural resistance to degradation or creating a barrier against corrosive substances, they protect concrete structures. These safeguards are intended to increase concrete's longevity and decrease maintenance requirements. Concrete is protected on its surface by corrosion-resistant coatings, including polyurethane, epoxy, and silane/siloxane sealers [79, 80]. Epoxy coatings are well known for their outstanding durability, chemical resistance, and adherence. They are perfect for places subjected to harsh conditions since they form a strong barrier against moisture and chemicals. Conversely, polyurethane coatings provide robustness and flexibility, which are especially helpful for surfaces that are subjected to movement or mechanical stress. These coatings provide long-lasting protection by resisting UV deterioration and abrasion [81]. Concrete is penetrated by silane and siloxane sealers, which offer a water-repellent barrier that lowers moisture and chloride infiltration.

Cementitious coatings provide a protective layer that strengthens the concrete's resilience to exposure to chemicals and water when placed as a slurry or overlay. Corrosion-inhibiting chemicals are put into the concrete mix in addition to surface coatings to offer internal protection. Chemicals known as corrosion inhibitors, such as amine-based and volatile corrosion inhibitors, either neutralize corrosive elements or provide a protective coating on the steel reinforcement. While VCIs volatilize and move through the concrete to preserve the steel, surface-active agents coat the metal with a protective layer. By adding extra binding phases and decreasing permeability, supplementary cementitious materials (SCMs) such as fly ash, slag, and silica fume improve the characteristics of concrete and lessen the danger of corrosion [82, 83]. The appropriate application, compatibility with the concrete, and environmental factors that the concrete is subjected to all affect how effective these coatings and additives are. Coatings require routine inspection and maintenance to guarantee their continuous effectiveness and to repair any wear or damage.

It is essential to follow design and construction guidelines to reduce the dangers of concrete corrosion. These standards offer recommendations and best practices for building long-lasting, corrosion-resistant concrete buildings. Longevity and safety are enhanced by using efficient design and construction techniques in concrete structures. Reducing the likelihood of corrosion is dependent primarily on design considerations. The cover depth or concrete thickness between the steel reinforcement and the surface is an important design consideration. Steel is shielded from moisture and corrosive substances by an adequate cover depth [84, 85]. Increasing the cover depth can offer more protection. However, design rules set minimum cover depths based on exposure conditions. Another important consideration is the design of the concrete mix; a well-thought-out mix with low permeability and the right amounts of cement, aggregates, and water helps prevent the entry of moisture and chloride. Concrete's resistance to corrosion can be further increased by adding SCMs and corrosion inhibitors. In the design stage, proper drainage and ventilation avoid moisture buildup and the accumulation of corrosive agents. Concrete movement may be controlled and major fissures that could let moisture in are prevented with control joints and crack control techniques. Ensuring the endurance of concrete structures requires careful consideration of construction procedures. Ensuring adequate mixing, placing, and curing of concrete is ensured *via* quality control during construction [86–88]. Testing building materials and techniques regularly aids in identifying and resolving potential durability-affecting problems. Achieving the right strength and longevity for concrete requires proper curing, which keeps the temperature and moisture content just right for hydration. Maintaining the integrity of concrete requires keeping things like chemicals, oils, and chlorides out of the mix. Joints and connections made during construction must be carefully planned and carried out to preserve structural integrity and avoid weak spots that could lead to corrosion.

In order to control concrete corrosion and maintain the structural integrity of concrete structures, early identification and maintenance are essential. The long-term durability of the structure can be ensured by timely intervention and routine monitoring to resolve corrosion-related problems before they cause substantial harm. Inspection and monitoring techniques are crucial to spot corrosion indicators and evaluate the state of the concrete. The most basic technique is visual inspection, which identifies obvious symptoms such as rust staining, cracking, and spalling. Inspectors examine exposed reinforcement, joints, and concrete surfaces to spot early warning indicators of damage. Non-destructive testing (NDT) methods, such as radiography, acoustic emissions, and ultrasonic testing, reveal interior features like cracks and delaminations without causing harm [89, 90]. The corrosion activity of steel reinforcement is assessed using electrochemical techniques, such as half-cell potential measurement and corrosion rate monitoring, which aid in determining the efficacy of corrosion control strategies. Proactive management is made possible by environmental monitoring of parameters, including temperature, humidity, and chloride levels, which reveal elements that contribute to corrosion. Concrete structure preservation is aided by maintenance and rehabilitation techniques, which address corrosion-related problems. In order to restore corroded concrete, surface damage must be addressed, the steel reinforcement underneath must be treated, and fresh concrete must be applied. Patching, resurfacing, and using repair mortars are examples of common repair methods. It is imperative to properly prepare the surface, which includes cleaning the steel reinforcement and removing any loose or degraded concrete. In addition, reinforcing buildings, applying fresh coatings or sealants, and using specialty repair materials are examples of rehabilitation measures. Extending the service life of concrete structures and reducing the impacts of corrosion need effective maintenance techniques, such as routine inspections and prompt repairs. For instance, in an experiment, Yang *et al.* [91] introduced a novel approach by creating molecularly engineered epoxy nanocomposite coatings incorporating corrosion inhibitor-functionalized nanoparticles. Specifically, a siloxane-based corrosion inhibitor precursor was designed and chemically grafted onto silica nanoparticles, forming the functionalized nanofiller, referred to as NPInh. The findings revealed that modifying the silica nanoparticles with the corrosion inhibitor precursor significantly improved the uniform dispersion of the nanofiller within the epoxy matrix. This enhanced distribution of the NPInh nanoparticles played a crucial role in strengthening both the mechanical properties and the corrosion resistance of the coatings. The nanoparticles contributed to mechanical reinforcement, while the corrosion inhibitor component provided a shielding effect, thereby preventing corrosion. The coatings also exhibited optimized strain durability, impact resistance, and adhesion strength. Electrochemical measurements further demonstrated the exceptional performance of these nanocomposite coatings. The coatings blended with NPInh reduced the corrosion current density to 10^{-10} A/cm^2 and achieved a remarkable protection efficiency of 99.9% after exposure to a 3.5 wt% NaCl solution. In addition, salt spray testing showed that these coatings provided robust durability, withstanding over 1,000 h of salt spray exposure without significant degradation. This study suggests that molecularly engineered nanocomposite coatings, through improved nanoparticle dispersion and enhanced barrier properties, represent a promising new direction for corrosion protection. These coatings effectively combine mechanical reinforcement with corrosion inhibition, offering superior protection for mild steel in corrosive environments. Similarly, Falzone *et al.* [92] proposed an innovative solution using calcium aluminate cement (CAC) formulations that include NO_3^--AFm (aluminate ferrite monosulfate) phases as a functional coating. These coatings function as anion-exchange systems, capturing chloride ions and simultaneously releasing nitrate ions (NO_3^-), which act as corrosion inhibitors. This dual action binds harmful Cl^- ions and passivates the steel reinforcement, providing a protective effect. To test the viability of this approach, the study evaluated Cl^- binding isotherms and ion-diffusion coefficients for a range of hydrated CAC formulations mixed with calcium nitrate ($Ca(NO_3)_2$). This data was then incorporated into a multi-species Nernst–Planck model, which was solved using finite element analysis (FEM) for a typical bridge-deck structure exposed to seawater. The results demonstrated that CAC coatings containing calcium nitrate can significantly delay the onset of steel corrosion when compared to conventional ordinary Portland cement (OPC)-based systems, even with the same concrete cover thickness. The delay factor (df), which quantifies the corrosion initiation time, showed improvements ranging from five to 10 times that of OPC systems. The research also examined how hindered ionic diffusion through these coatings further contributes to corrosion mitigation. In particular, the presence of NO_3^- ions aids in the (re)passivation of the steel reinforcement, restoring its protective layer and reducing the likelihood of chloride-induced corrosion. The study highlights the importance of controlling Cl^-/OH^- and Cl^-/NO_3^- molar ratios in the concrete solution as critical thresholds for delaying corrosion. In summary, the development of CAC-based coatings with anion-capture capabilities offers a promising new route for mitigating steel corrosion in concrete infrastructure. By incorporating both chloride-binding and nitrate-releasing functionalities, these coatings not only protect the steel from chloride attack but also actively enhance the corrosion resistance of reinforced concrete structures, potentially extending their service life in aggressive environments.

Kim *et al.* [93] explored the effectiveness of electrochemical deposition treatment (EDT) as a comprehensive rehabilitation method for addressing corrosion-induced deterioration in reinforced concrete across various severity levels. Corrosion in reinforced concrete structures occurs in three main stages: initiation (de-passivation of steel), propagation (active corrosion), and acceleration (formation of surface-breaking cracks). Each of these stages presents different challenges for repair and rehabilitation. To test the feasibility of EDT, the researchers conducted a series of accelerated corrosion experiments that simulated these stages of corrosion. Concrete samples with varying levels of damage were treated using EDT, which involved using a magnesium chloride ($MgCl_2$) solution as the electrolyte. The experiments were divided into three phases, each focusing on a specific aspect of the EDT process. In the first phase, the study assessed the effect of different concentrations of $MgCl_2$ (ranging from 0 to 3.0 M) on corrosion prevention and mitigation during the early stages of corrosion. In the second phase, they investigated the influence of charging time in the EDT process (from 0 to 7 days) on the rehabilitation performance. The third phase examined the configuration of pre- and post-treatment processes, including the application of a sodium hydroxide (NaOH) solution following EDT, to enhance the crack-repairing capabilities of the method (Figure 5.3). The results indicated that EDT is highly effective in both preventing and mitigating corrosion during the initiation and propagation phases of corrosion. It also showed promising results in repairing surface-breaking cracks that form during the acceleration phase of corrosion. Specifically, higher concentrations of $MgCl_2$ and longer charging times in EDT were associated with improved corrosion resistance in the treated concrete. Furthermore, applying NaOH in the post-treatment phase significantly enhanced the ability of EDT to repair cracks, resulting in better overall rehabilitation performance.

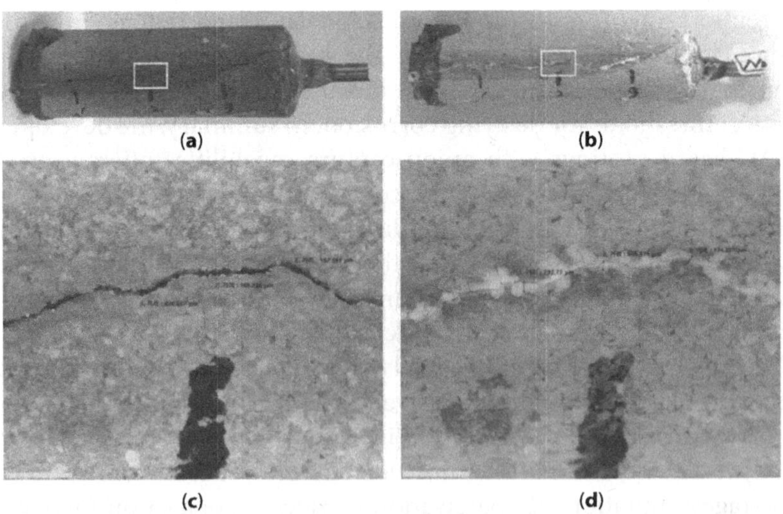

Figure 5.3 Optical microscope images of a surface-breaking crack in concrete sample TP3-1: (a) before and (b) after rehabilitation *via* EDT, with (c, d) close-up views of the highlighted areas from (a) and (b). Adapted from [93] under CCBY 4.0.

Liu *et al.* [94] focused on creating an innovative nanocomposite anti-corrosion filler designed to enhance the protection of steel in challenging environments. This filler was synthesized through a hydrothermal process, where cerium oxide was grown on the surface of graphene oxide, achieving a specific mass ratio of 4:1. The choice of graphene oxide is significant due to its excellent mechanical properties and high surface area, which, when combined with cerium oxide, can significantly improve corrosion resistance. After synthesizing the filler, it was integrated into pure epoxy resin at a concentration of 0.5%. This careful formulation led to the development of a nanocomposite epoxy coating tailored to provide superior protective qualities against corrosive agents. To evaluate the effectiveness of this newly developed coating, a series of comprehensive tests were conducted, focusing on key performance metrics such as surface hardness, adhesion strength, and anti-corrosion efficacy. The testing specifically targeted Q235 low-carbon steel, a material commonly used in construction, and subjected it to conditions simulating seawater and concrete pore solutions. These environments are notorious for their corrosive potential, making them ideal for assessing the coating's performance. After a 90-day exposure period, the findings revealed that the nanocomposite coating, particularly when combined with a corrosion inhibitor, exhibited the lowest corrosion current density recorded at Icorr = 1.001×10^{-9} A/cm^2. This remarkable result translates to an outstanding protection efficiency of 99.92%, underscoring the coating's ability to shield the steel substrate from corrosion effectively. Figure 5.4 presents a series of scanning electron microscopy (SEM) images that illustrate the effects of corrosion on the CeO_2-GO/epoxy (EP) coating when exposed to a 60% corrosive solution over a period of 60 days. Figure 5.4(a) reveals the overall condition of the corroded surface, showcasing the extent of deterioration that has occurred. The surface morphology indicates significant degradation, highlighting the vulnerabilities of the coating in harsh environments. Figure 5.4(b) focuses on the corrosion products formed during this exposure, illustrating the byproducts of the corrosion process that have accumulated on the surface. These products can provide insight into the chemical reactions taking place between the coating and the corrosive medium. Figure 5.4(c), a close-up view of a locally corroded area, is shown, emphasizing specific regions where corrosion has intensified. This detail underscores the non-uniform nature of corrosion, which can vary significantly across the surface. Figure 5.4(d) depicts an area where corrosion amplification has occurred, indicating regions where the protective capabilities of the coating have been compromised, leading to accelerated material loss. Finally, Figure 5.4(e) presents the optical microscopic morphology of the coating

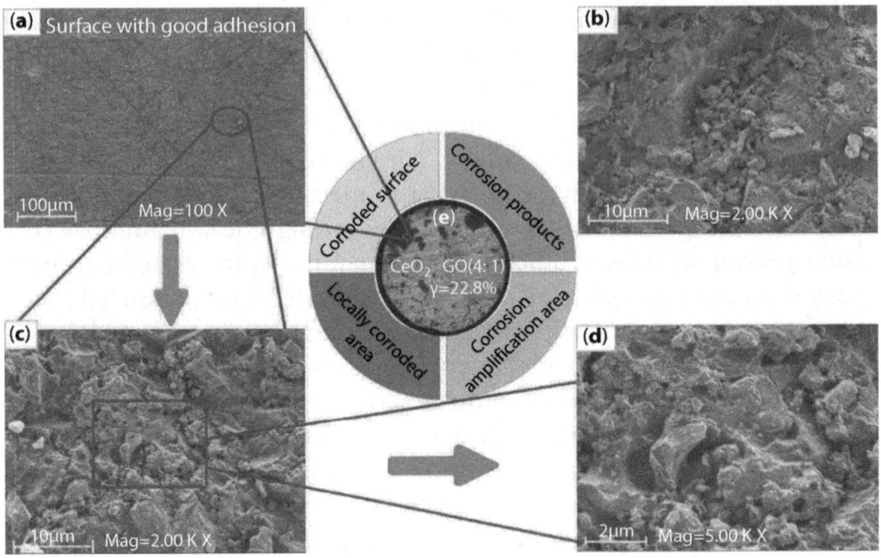

Figure 5.4 SEM images of the CeO_2-GO/epoxy coating after 60 days in a 60% corrosive solution: (a) corroded surface, (b) corrosion products, (c) locally corroded areas, (d) corrosion amplification zone, and (e) optical microscopic morphology post-exposure. Adapted from [94] under CCBY 4.0.

after 60 days of exposure to the corrosive solution. This view complements the SEM images by providing additional context on the structural changes and surface characteristics developed due to prolonged exposure. Together these images give a comprehensive overview of the corrosion behavior of the CeO_2-GO/EP coating in a challenging environment, revealing important insights into its durability and performance.

Huang *et al.* [95] introduced limestone calcined clay cement (LC^3) which stands out as a viable alternative, offering significant reductions in CO_2 emissions and resource consumption within the cement manufacturing sector. This innovation is especially pertinent as the construction industry seeks to minimize its ecological footprint. In addition to LC^3, incorporating recycled concrete aggregate (RCA), seawater, and sea sand provides further sustainability benefits, particularly in coastal regions where these materials are readily available and can reduce the need for transport, thereby lowering emissions. Moreover, fiber-reinforced polymer (FRP) bars present an ideal solution in environments prone to corrosion. These bars possess superior corrosion resistance compared to traditional steel reinforcement, making them particularly suitable for structures exposed to harsh conditions. This study examines a novel green concrete structure that combines

LC^3 with seawater, sea sand, RCA, and FRP bars, aiming to assess its performance in terms of sustainability while also evaluating its life-cycle costs and environmental impacts. The results of this investigation reveal that the LC^3 concrete exhibits markedly lower environmental impacts than traditional concrete. Specifically, CO_2 emissions and energy consumption are reduced by 32%, 41.7%, 23.3%, and 28%, respectively. Although the upfront costs associated with LC^3 concrete and CFRP bars are higher, these initial investments are offset by their longer service life and reduced maintenance needs throughout their lifespan. When comparing the life-cycle costs and environmental impacts of the LC^3 concrete structure to those of conventional reinforced concrete (RC) structures, significant reductions are observed: total life-cycle costs decrease by 39%.

In comparison, CO_2 emissions and energy consumption are reduced by 30.5% to 65.4% and 18.5% to 58.9%, respectively. However, it is important to note that the replacement ratio of LC^3 can negatively influence environmental indices related to the concrete and the overall structure. This underscores the complexity of balancing cost-effectiveness with environmental performance. Despite these challenges, the economic and ecological advantages associated with LC^3 concrete position it as a compelling solution for enhancing sustainability within the construction industry. The findings from this research advocate for the broader adoption of LC^3 concrete, emphasizing its potential to significantly reduce the environmental footprint of construction practices while contributing to the overarching goal of sustainable development. Similarly, Sun *et al.* [96] explored the electrical resistivity (ER) of hybrid fiber-reinforced concrete (HFRC) containing both basalt microfibers (BF) and basalt macrofibers (BMF) through a combination of experimental investigations and data-driven methods. The experimental phase assessed the influence of various factors, such as BF and BMF content, fiber hybridization, curing duration, and the impact of seawater corrosion, on the electrical resistivity of HFRC. Following the experimental work, a machine learning model was developed using a large dataset (1,171 entries from the literature and 540 experimental datasets) to predict the ER of HFRC. The results from both the experimental and data-driven approaches highlighted that the optimal fiber content for maximizing ER was 0.1% for BF, 2.5% for BMF, and 0.1% BF combined with 2.5% BMF in the hybrid fiber system. Notably, the hybrid fiber system outperformed the single-fiber systems in terms of electrical resistivity. Extending the curing time significantly enhanced the ER of HFRC, whereas prolonged exposure to seawater corrosion reduced the corrosion resistance, as evidenced by a lower ER. The study also developed a highly accurate predictive model using the sooty tern optimization XGBoost

(STOA-XGBoost) algorithm, outperforming other machine learning models. With a mean residual of 0.00479 and a standard deviation of 3.0964, the model demonstrated impressive predictive accuracy, as indicated by performance metrics such as RMSE (0.538), MSE (0.047), MAPE (0.505), MAE (0.710), and R^2 (0.980). Further analysis using Shapley additive explanations (SHAP) identified environmental humidity and curing time as the most influential factors affecting HFRC-ER. The findings suggest that minimizing environmental humidity and extending curing times are key strategies for enhancing the ER of hybrid fiber-reinforced concrete. This research presents a novel approach to optimizing HFRC performance and offers effective methodologies for nondestructive performance evaluation, particularly in marine environments where corrosion is a critical concern.

Han *et al.* [97] investigated latex-modified concrete (LMC), which offers superior flexibility and durability, making it an ideal candidate for high-temperature pavement construction. The research aims to optimize LMC for improved workability and performance, especially in bridge construction, where demanding conditions require durable and workable concrete mixes. Latex is integrated into the concrete mix because it enhances the material's physical properties and flexibility. The study focuses on improving the concrete's performance under harsh summer conditions by evaluating the role of nonionic emulsifiers, such as alkyl ethylene oxide (AEO) and alkyl phenyl ethylene oxide (APEO). These emulsifiers are crucial for stabilizing latex in the concrete mix, and the research found that AEO emulsifiers with alkyl chains of 9 to 13 carbon atoms and 20 to 40 ethylene oxide units provided the best improvement in workability. This means the latex-modified concrete could be mixed and applied more efficiently, even under high temperatures, ensuring a more uniform and consistent final product. In addition to using nonionic emulsifiers, the study explored the effect of adding superplasticizers to enhance the fluidity and workability of the latex-modified concrete. A key finding was that incorporating 0.5 wt% of superplasticizer per cement unit significantly increased the ease of mixing, pouring, and compacting the concrete, which is vital for on-site construction. The improved fluidity helped achieve a smoother and more uniform surface and contributed to better compaction, reducing the likelihood of air voids and defects.

Furthermore, the research highlighted that the combination of latex, nonionic emulsifiers, and superplasticizers improved workability and enhanced the concrete's compressive strength. This is particularly important for bridge construction, where concrete is subjected to substantial mechanical loads over its lifespan. The higher compressive strength ensures that the concrete can withstand long-term stresses, making it more durable due to environmental factors such as heat, moisture, and traffic loads. In conclusion, this study demonstrates that when optimized with the right blend of nonionic emulsifiers and superplasticizers, latex-modified concrete provides significant advantages in summer pavement construction. The findings suggest that by enhancing workability and durability, LMC can overcome many of the challenges associated with high-temperature environments, leading to higher-quality concrete pavements with fewer defects and longer service lives. The research offers a practical solution for improving concrete performance, particularly in bridge and other infrastructure projects where durability and workability are critical concerns.

Furthermore, concrete is indispensable in modern construction, yet its environmental footprint, particularly its carbon dioxide (CO_2) emissions and susceptibility to corrosion, poses significant sustainability challenges. Developing carbon-capturing concrete, which actively sequesters CO_2 during production and enhances durability, provides an innovative approach to mitigate these issues [98–100]. Carbon-capturing concrete addresses environmental and infrastructure-related concerns by integrating sustainability with structural resilience. The carbon-capturing process relies on the reaction between CO_2 and calcium hydroxide, a byproduct of cement hydration. This reaction forms calcium carbonate, which strengthens the concrete matrix. Carbonation curing, a controlled process exposing fresh concrete to CO_2, accelerates this reaction, enhancing both material performance and carbon sequestration.

Additionally, incorporating supplementary cementitious materials (SCMs) such as fly ash, slag, silica fume, and recycled aggregates increases the reactive surface area for CO_2 absorption, further boosting the material's environmental benefits. From an environmental perspective, carbon-capturing concrete offers a dual advantage: reducing the carbon footprint of cement production while offsetting emissions through CO_2 sequestration. Cement manufacturing contributes approximately 8% of global CO_2 emissions, mainly due to calcination [101–104]. These emissions can be significantly curtailed by integrating SCMs, partially replacing clinker and employing carbonation curing. Furthermore, using recycled materials minimizes the demand for virgin resources, conserving natural aggregates and reducing the environmental degradation associated with extraction processes. In addition to its environmental advantages, carbon-capturing concrete enhances structural performance and durability. The formation of calcium carbonate densifies the concrete matrix, improving its compressive strength and reducing permeability. This makes the material more resistant to ordinary corrosion pathways, such as chloride ingress and carbonation-induced corrosion. Moreover, the dense microstructure protects steel reinforcements, mitigating corrosion and reducing the need for frequent repairs. As a result, the overall lifecycle costs of infrastructure are lowered, aligning with the principles of sustainable construction. Corrosion is a leading cause of structural degradation and environmental harm, as it often necessitates resource-intensive repairs and generates construction waste. Carbon-capturing concrete addresses this by enhancing durability and reducing susceptibility to environmental stressors. Controlled carbonation during curing does not compromise the alkalinity required to protect steel reinforcements, and the reduced porosity minimizes chloride and moisture ingress. These properties are especially valuable in marine and coastal environments, where chloride-induced corrosion is prevalent. Case studies [105–111] have demonstrated the practical benefits of carbon-capturing concrete. For example, CarbonCure, a commercially available technology, has been adopted in various projects to inject CO_2 into ready-mix concrete, reducing emissions by up to 10% without compromising strength. Similarly, marine structures constructed with carbon-capturing formulations have shown enhanced resistance to chloride-induced corrosion, extending their operational life. Urban infrastructure projects in cities like Toronto and Singapore have also successfully integrated this technology, showcasing its scalability and effectiveness in reducing urban carbon footprints.

Despite its potential, carbon-capturing concrete faces challenges that must be addressed to achieve widespread adoption. Initial production costs are higher due to specialized materials and processes, and retrofitting existing facilities for carbonation curing requires significant investment. Furthermore, the absence of standardized guidelines for testing and implementation can hinder the adoption of this material. Overcoming these barriers will require collaborative efforts from policymakers, researchers, and industry stakeholders. The future of carbon-capturing concrete lies in continued innovation and support. Policy incentives, such as subsidies or tax benefits, can encourage its adoption, while research should focus on optimizing carbonation curing and integrating advanced materials like nanotechnology to enhance CO_2 absorption. Comprehensive lifecycle assessments can highlight long-term benefits, and training programs for engineers and builders will ensure proper application and maximize impact.

5.7 Challenges and Future Directions

The problems we confront and the paths we must take going forward in the field of concrete corrosion mitigation become more intricate and varied.

The need for comprehensive policy and regulatory frameworks and technical advancements drives the continuous search for more efficient solutions. Innovative technologies that promise to improve concrete structures' performance and longevity are reshaping the concrete corrosion mitigation field. These cutting-edge technologies present fresh approaches to overcome the drawbacks of conventional techniques and present chances to increase the efficacy and efficiency of corrosion control strategies [112–114]. The application of cutting-edge nanomaterials is among the most promising developments. At the molecular level, nanotechnology has created new opportunities for improving the characteristics of concrete. To increase concrete's resistance to corrosion, nano-sized additives, including nano-silica, nano-titanium dioxide, and nano-clay, are added to the mix. These nanoparticles can strengthen the link between cement and aggregates, decrease permeability, and improve the densification of the concrete matrix. It has been demonstrated that adding these components can greatly increase the long-term durability of concrete. Using self-healing concrete is another cutting-edge technique. This novel method includes adding bacteria or microcapsules to the concrete mixture as healing agents. These substances are triggered when cracks appear, starting a healing process that closes the fractures and preserves the concrete's structural integrity [115, 116]. By repairing damage before it worsens and becomes more severe, self-healing concrete can lower maintenance requirements and increase the lifespan of structures. Smart sensors and monitoring tools will significantly influence concrete corrosion control. Embedded sensors in these technologies can continually and instantly monitor the state of concrete constructions. Temperature, moisture content, and corrosion rates are just a few variables that sensors may measure. This data is crucial for the early diagnosis of corrosion and deterioration. The management of concrete infrastructure can be improved by combining these sensors with data analytics and predictive maintenance systems to enable more targeted and proactive interventions.

Furthermore, computational modeling and simulation developments are enhancing our comprehension of the corrosive processes affecting concrete. The evaluation of alternative mitigation measures under varying conditions and the simulation of corrosion scenarios are made possible by modern computer tools [1, 117, 118]. This feature enables the construction of more precise and trustworthy performance forecasts for concrete structures and the optimization of design and maintenance procedures.

The frameworks of laws and regulations governing the construction and upkeep of concrete must change along with the technological environment. Effective policies and regulations are necessary to guarantee that infrastructure projects are managed sustainably and that the most recent developments in corrosion mitigation are applied consistently. Several important suggestions can help resolve present issues and direct the next advancements in the control of concrete corrosion. First, new norms and rules that consider newly developed materials and technologies are required [119, 120]. New corrosion-resistant coatings, additives, and repair methods must be incorporated into the requirements as they become available. Regulatory agencies should collaborate extensively with researchers and industry specialists to guarantee that standards are grounded on the most recent scientific discoveries and technological developments. This will ensure the efficient application of modern technologies while preserving the durability and safety of concrete constructions. Secondly, it is imperative to provide more funding for research and development to mitigate concrete corrosion. Academic institutions, industrial players, and government agencies should collaborate to support and encourage research that fills knowledge gaps and looks for new approaches. Research funding can result in the creation of corrosion-fighting materials, technologies, and techniques that are more successful overall, extending the life of infrastructure and promoting environmental sustainability. Thirdly, experts engaged in designing, building, and maintaining concrete should get specialized education and training [121, 122]. For new technologies to be implemented successfully, engineers, architects, and construction workers must have an up-to-date understanding of corrosion mitigation techniques and best practices. Opportunities for professional development and continuing education can keep workers abreast of developments in their fields. Regulations should also promote sustainable building techniques when building with concrete. The overall environmental effect of concrete infrastructure can be decreased with the support of policies that encourage the use of recycled materials, energy-efficient technologies, and environmentally friendly additives. Grants or tax credits are examples of incentives that can be used to push industrial stakeholders to incorporate environmental concerns into their initiatives. Lastly, a focus on long-term performance tracking and maintenance scheduling is necessary [123, 124]. To guarantee that concrete structures are maintained in excellent condition for their valuable lives, regulations should mandate the installation of reliable monitoring systems and maintenance plans. Using real-time data from smart sensors and monitoring technologies, proactive maintenance tactics can help find problems early on and fix them before they cause severe degradation.

5.8 Conclusion

Concrete deterioration, caused by processes including carbonation and chloride-induced reactions, presents serious threats to environmental health and infrastructure longevity. Beyond just structural issues, corrosion impacts resource consumption, waste production, and the discharge of toxic compounds into soil and water. Utilizing corrosion-resistant coatings and additives, adhering to design and construction standards, and putting early detection and maintenance procedures into practice are all part of the multifaceted strategy that goes into mitigating concrete corrosion. Coatings resistant to corrosion, such as polyurethane, epoxy, and silane/siloxane sealers, are essential for shielding concrete surfaces from acidic substances. Corrosion-inhibiting additives, such as volatile corrosion inhibitors and surface-active chemicals, improve the concrete's ability to withstand internal deterioration. The key to reducing corrosion hazards is to use efficient design and construction techniques, such as ensuring enough cover depth, maximizing the design of the concrete mix, and putting suitable drainage in place. Furthermore, maintaining the integrity and prolonging the service life of concrete structures depend heavily on early identification using sophisticated inspection techniques and prompt maintenance interventions.

Future concrete corrosion and sustainability issues underscore the necessity of ongoing innovation and adaptability. Novel technologies present encouraging alternatives, like self-healing concrete that fixes damage on its own, nanomaterials that improve concrete's characteristics at the molecular level, and smart sensors that offer real-time monitoring. These developments have the potential to completely change the way we maintain infrastructure and mitigate corrosion, providing more practical and efficient ways to increase the lifespan of concrete structures. However, technological progress on its own is insufficient. Newly revised legislative and regulatory frameworks that support sustainable activities and consider these recent advancements are desperately needed. To ensure that new technologies are applied consistently and successfully, regulatory standards must be updated to reflect the most recent advancements in materials and procedures. Adopting sustainable practices, continuing professional education, and funding research and development are all crucial parts of a comprehensive strategy to combat concrete deterioration. To manage infrastructure throughout its service life, thorough maintenance planning and performance monitoring are also essential.

AI disclosure statement: The author confirms that no artificial intelligence (AI)-generated content or AI-assisted writing tools were used in the preparation of this chapter (including summaries and the abstract). All text, analysis, and conclusions are the result of the author's original work.

References

1. Owais, M. and Idriss, L.K., Modeling green recycled aggregate concrete using machine learning and variance-based sensitivity analysis. *Constr. Build Mater [Internet]*, 440, March, 137393, 2024, Available from: https://doi.org/10.1016/j.conbuildmat.2024.137393.
2. Raj, A., Yamkasikorn, P., Wangtawesap, R., Win, T.T., Ngamkhanong, C., Jongvivatsakul, P., *et al.*, Effect of Graphene Quantum Dots (GQDs) on the mechanical, dynamic, and durability properties of concrete. *Constr. Build Mater [Internet]*, 441, 137597, 2024, Available from: https://www.sciencedirect.com/science/article/pii/S0950061824027399.
3. Surehali, S., Singh, A., Biligiri, K.P., A state-of-the-art review on recycling rubber in concrete: Sustainability aspects, specialty mixtures, and treatment methods. *Dev. Built Environ. [Internet]*, 14, 100171, 2023, Available from: https://www.sciencedirect.com/science/article/pii/S2666165923000534.
4. Villalba, P., Sánchez-Garrido, A.J., Yepes, V., Life cycle evaluation of seismic retrofit alternatives for reinforced concrete columns. *J. Clean. Prod. [Internet]*, 455, 142290, 2024, Available from: https://www.sciencedirect.com/science/article/pii/S0959652624017384.
5. Hafez, H., Teirelbar, A., Tošić, N., Ikumi, T., de la Fuente, A., Data-driven optimization tool for the functional, economic, and environmental properties of blended cement concrete using supplementary cementitious materials. *J. Build Eng. [Internet]*, 67, 106022, 2023, Available from: https://www.sciencedirect.com/science/article/pii/S2352710223002012.
6. Qian, R., Li, Q., Fu, C., Zhang, Y., Wang, Y., Jin, X., Atmospheric chloride-induced corrosion of steel-reinforced concrete beam exposed to real marine-environment for 7 years. *Ocean Eng. [Internet]*, 286, 115675, 2023, Available from: https://www.sciencedirect.com/science/article/pii/S0029801823020590.
7. Ran, T., Pang, J., Liu, Y., Zou, J., Cai, F., Improving concrete fatigue resistance with COVID-19 rubber gloves: An innovative sustainable approach. *Case Stud. Constr. Mater [Internet]*, 18, e01914, 2023, Available from: https://www.sciencedirect.com/science/article/pii/S2214509523000931.
8. Zuo, W., Zhao, X., Luo, J., He, Z., Zhao, J., Investigating the impact of carbonation curing concentrations on the mechanical properties and microstructure of recycled concrete. *J. Build Eng. [Internet]*, 96, 110465, 2024, Available from: https://www.sciencedirect.com/science/article/pii/S2352710224020333.

9. Liu, J., Shi, Y., Guo, A., Mechanical properties of concrete-filled double-skin steel tubular columns subjected to axial loading with non-uniform corrosion. *Struct. [Internet]*, 64, 106597, 2024, Available from: https://www.sciencedirect.com/science/article/pii/S2352012424007495.
10. Feng, G., Jin, Z., Jiang, Y., Wang, X., Zhu, D., Localized corrosion propagation of steel in cracked mortar and long-term corrosion of steel reinforcement in cracked concrete in seawater environment. *Corros Sci. [Internet]*, 228, 111793, 2024, Available from: https://www.sciencedirect.com/science/article/pii/S0010938X23008363.
11. Adsetts, J.R., Ebrahimi, N., Zhang, J., Jalaei, F., Noël, J.J., Steel Bridge-Coating Systems and Their Environmental Impacts: Current Practices and Future Trends. *Coatings.*, 13, 5, 850, 2023.
12. Nguyen, V.M., Phung, T.B., Pham, D.T., Ho, L.S., Mechanical properties and durability of concrete containing coal mine waste rock, F-class fly ash, and nano-silica for sustainable development. *J. Eng. Res. [Internet]*, 11, 3, 75–86, 2023, Available from: https://www.sciencedirect.com/science/article/pii/S2307187723000986.
13. Müller, C., How standards support sustainability of cement and concrete in Europe. *Cem. Concr. Res. [Internet]*, 173, 107288, 2023, Available from: https://www.sciencedirect.com/science/article/pii/S0008884623002028.
14. Lu, J. and Choon, S.K.L., Assessment of Surface Treatment Systems for Protecting Concrete Structure. *Biogeotechnics [Internet]*, 100134, 2024, Available from: https://www.sciencedirect.com/science/article/pii/S2949929124000664.
15. Elshinawy, A., Yousry Elshikh, M.M., Kaloop, M.R., El-Demerdash, W.E., Elemam, W.E., An experimental investigation on mechanical characteristics of steel-fiber-reinforced volcanic concrete. *Next Mater [Internet]*, 6, 100308, 2025, Available from: https://www.sciencedirect.com/science/article/pii/S2949822824002053.
16. Qian, J., Zheng, Y., Dong, Y., Wu, H., Guo, H., Zhang, J., Sustainability and resilience of steel shape memory alloy reinforced concrete bridge under compound earthquakes and functional deterioration within entire life-cycle. *Eng. Struct. [Internet]*, 271, 114937, 2022, Available from: https://www.sciencedirect.com/science/article/pii/S0141029622010136.
17. Wang, C., Yuan, J., Lu, B., Zhang, Y., Ma, Z., Mesoscopic 3D simulation and *in-situ* 4D CT investigation on the mechanical behaviors of high-toughness recycled aggregate concrete. *Constr. Build Mater [Internet]*, 442, 137560, 2024, Available from: https://www.sciencedirect.com/science/article/pii/S0950061824027028.
18. Etcheverry, J.M., Laveglia, A., Villagran-Zaccardi, Y.A., De Belie, N., A technical-environmental comparison of hybrid and blended slag cement-based recycled aggregate concrete tailored for optimal field performance. *Dev. Built Environ. [Internet]*, 17, 100370, 2024, Available from: https://www.sciencedirect.com/science/article/pii/S2666165924000516.

19. Zhu, P., Chen, X., Liu, H., Wang, Z., Chen, C., Li, H., Recycling of waste recycled aggregate concrete in freeze-thaw environment and emergy analysis of concrete recycling system. *J. Build Eng. [Internet]*, 96, 110377, 2024, Available from: https://www.sciencedirect.com/science/article/pii/S2352710224019454.
20. Vandecruys, E., Hendriks, M.A.N., van de Velde, M., Lombaert, G., Verstrynge, E., Dynamic stiffness parameter assessment of cracked reinforced concrete beams: A numerical and experimental study. *Eng. Struct. [Internet]*, 318, 118758, 2024, Available from: https://www.sciencedirect.com/science/article/pii/S0141029624013208.
21. Jacobsen, OEK., Kristoffersen, M., Dey, S., Børvik, T., Sustainable shielding: Ballistic performance of low-carbon concrete. *Constr. Build Mater [Internet]*, 415, 135103, 2024, Available from: https://www.sciencedirect.com/science/article/pii/S0950061824002447.
22. Bayat, H., Banar, R., Nikravan, M., Farnood, P., Ramezanianpour, A.M., Kashani, A., Durability, mechanical, workability, and environmental assessment of self-consolidating concrete containing blast furnace slag and natural zeolite. *J. Build Eng. [Internet]*, 86, 108737, 2024, Available from: https://www.sciencedirect.com/science/article/pii/S235271022400305X.
23. Xiong, L., Li, S., Jiang, K., He, J., Kong, F., Development and assessment of sustainable steel-concrete composite beams with novel demountable shear connections. *Soil Dyn. Earthq. Eng. [Internet]*, 180, 108606, 2024, Available from: https://www.sciencedirect.com/science/article/pii/S0267726124001581.
24. Li, J., Jia, C., Guo, L., Behavior of eccentrically loaded slender concrete-filled steel tubes with defect representing random local corrosion. *Eng. Struct. [Internet]*, 318, 118766, 2024, Available from: https://www.sciencedirect.com/science/article/pii/S0141029624013282.
25. Li, S., Fu, C., Shi, W., He, Q., Wang, J., Chloride penetration in reinforced slag-based concrete beams under combined effects of loads and drying–wetting cycles. *Mag. Concr. Res. [Internet]*, 75, 19, 1012–26, 2023, Available from: https://www.sciencedirect.com/science/article/pii/S1751763X23000417.
26. Ouyang, X., Chen, J., Li, J., Wu, K., Ma, Y., Fu, J., Production of sustainable lime-based brick using carbonated recycled concrete fines: Mechanical, mineralogical and microstructure properties. *Constr. Build Mater [Internet]*, 411, 134385, 2024, Available from: https://www.sciencedirect.com/science/article/pii/S095006182304103X.
27. Barzegar, M., Goracci, G., Martauz, P., Dolado, J.S., Sustainable geopolymer concrete for thermoelectric energy harvesting. *Constr. Build Mater [Internet]*, 411, 134398, 2024, Available from: https://www.sciencedirect.com/science/article/pii/S0950061823041168.
28. Golafshani, E.M., Behnood, A., Kim, T., Ngo, T., Kashani, A., Metaheuristic optimization based- ensemble learners for the carbonation assessment of recycled aggregate concrete. *Appl. Soft Comput. [Internet]*, 159, 111661, 2024, Available from: https://www.sciencedirect.com/science/article/pii/S1568494624004356.

29. Bozorgmehr Nia, S. and Nemati Chari, M., Applied development of sustainable-durable high-performance lightweight concrete: Toward low carbon footprint, durability, and energy saving. *Results Mater [Internet]*, 20, 100482, 2023, Available from: https://www.sciencedirect.com/science/article/pii/S2590048X23001206.
30. Bahmani, H. and Mostofinejad, D., Comparative analysis of environmental, social, and mechanical aspects of high-performance concrete with calcium oxide-activated slag reinforced with basalt, and recycled PET fibers. *Case Stud. Constr. Mater [Internet]*, 20, e02895, 2024, Available from: https://www.sciencedirect.com/science/article/pii/S2214509524000469.
31. Hossain, M.A., Datta, S.D., Akid, A.S.M., Sobuz, M.H.R., Islam, M.S., Exploring the synergistic effect of fly ash and jute fiber on the fresh, mechanical and non-destructive characteristics of sustainable concrete. *Heliyon [Internet]*, 9, 11, e21708, 2023, Available from: https://www.sciencedirect.com/science/article/pii/S2405844023089168.
32. Xu, C., Li, Q., Wang, P., Fan, Q., Kong, Z., Wang, L., *et al.*, Theoretical porosity design, mechanical properties, and durability of large-pore sandy recycled concrete. *Case Stud. Constr. Mater [Internet]*, 21, e03655, 2024, Available from: https://www.sciencedirect.com/science/article/pii/S2214509524008064.
33. Gerengi, H., Maraşlı, M., Rizvi, M., Kohen, B., Taşkıran, I., Protection of steel hooks embedded in glass-fiber-reinforced concrete against macrocell corrosion. *Pet Res. [Internet]*, 9, 2, 317–26, 2024, Available from: https://www.sciencedirect.com/science/article/pii/S2096249524000012.
34. Soltanianfard, M.A., Abuhishmeh, K., Hojat Jalali, H., Shah, S.P., Sustainable concrete made with wastewater from different stages of filtration. *Constr. Build Mater [Internet]*, 409, 133894, 2023, Available from: https://www.sciencedirect.com/science/article/pii/S0950061823036127.
35. Saura-Gómez, P., Rizo-Maestre, C., Echarri-Iribarren, V., The Useful Life of Reinforced Concrete Structures with Reinforcement Corrosion Due to Carbonation in Non-Aggressive and Normal Exposures in the Spanish Mediterranean. *Mater. (Basel)*, 15, 3, 745, 2022.
36. Li, M., The Effect of Environmental Conditions on Electrochemical Corrosion Behavior of AISI 1018 Low-carbon Steel in Concrete Pore Solution. *Int. J. Electrochem. Sci. [Internet]*, 16, 8, 21088, 2021, Available from: https://www.sciencedirect.com/science/article/pii/S1452398123032297.
37. Jia, Z., Cunha, S., Aguiar, J., Study on phase change materials integration in concrete: Form-stable PCM and direct addition. *Process Saf. Environ. Prot. [Internet]*, 189, 1293–302, 2024, Available from: https://www.sciencedirect.com/science/article/pii/S0957582024008401.
38. Alwash, S.S., AL-Ameeri, A.S., Mattar, S.G., Assessing the effect of corrosion on optimised low carbon concrete. *Mater Today Proc. [Internet]*, 2023, Available from: https://www.sciencedirect.com/science/article/pii/S2214785323025877.
39. Bischof, P., Mata-Falcón, J., Kaufmann, W., Fostering innovative and sustainable mass-market construction using digital fabrication with concrete. *Cem.*

Concr. Res. [Internet], 161, 106948, 2022, Available from: https://www.sciencedirect.com/science/article/pii/S000888462200240X.

40. Liu, Z., Chin, C.S., Xia, J., Lu, J., Wang, X., Exploring the sustainability of concrete with fly ash, recycled coarse aggregate and biomineralisation method by life cycle assessment. *J. Clean. Prod. [Internet]*, 406, 137077, 2023, Available from: https://www.sciencedirect.com/science/article/pii/S0959652623012350.
41. Yao, J.J. and Chu, S.H., Durability of sustainable marine sediment concrete. *Dev. Built Environ. [Internet]*, 13, 100118, 2023, Available from: https://www.sciencedirect.com/science/article/pii/S2666165922000527.
42. Rezaeicherati, F., Memarzadeh, A., Esmailpour, A., Fallahnejad, H., Ghorbanzadeh, A., Nematzadeh, M., Durability evaluation and environmental implications of blended cement with colloidal nano-silica for use in recycled fine aggregate concrete: Experimental and theoretical study. *Constr. Build Mater [Internet]*, 402, 132926, 2023, Available from: https://www.sciencedirect.com/science/article/pii/S0950061823026429.
43. Yildizel, S.A., Uzun, M., Armagan, K., Ozbakkaloglu, T., Evaluation and optimization of red pine needle-reinforced roller-compacted concrete by bioinspired algorithms. *Compos. Commun. [Internet]*, 51, 102073, 2024, Available from: https://www.sciencedirect.com/science/article/pii/S245221392400264X.
44. Varma, D.A., Joseph, L., Madhavan, M.K., Jayanarayanan, K., Pegoretti, A., Strength, durability and finite element analysis of hybrid jute/basalt fiber reinforced polymer confined concrete column under axial compression. *Results Eng. [Internet]*, 22, 102281, 2024, Available from: https://www.sciencedirect.com/science/article/pii/S259012302400536X.
45. Karthik, S., Srinivasnaik, M., Kasirajan, G., Kumar, R.S., Govindarajan, S., Sethu Ramalingam, P., Evaluation of the mechanical performance and durability of concrete using reinforced cement concrete in the partial replacement of coarse and fine aggregates. *Mater Today Proc. [Internet]*, 2024, Available from: https://www.sciencedirect.com/science/article/pii/S2214785324004188.
46. Sharma, S., Ganjoo, R., Thakur, A., Kumar, A., Investigation of corrosion performance of expired Irnocam on the mild steel in acidic medium. *Mater Today Proc. [Internet]*, 66, 540–3, 2022, Available from: https://www.sciencedirect.com/science/article/pii/S2214785322040731.
47. Sharma, S., Ganjoo, R., Kr. Saha, S., Kang, N., Thakur, A., Assad, H., *et al.*, Investigation of inhibitive performance of Betahistine dihydrochloride on mild steel in 1 M HCl solution. *J. Mol. Liq. [Internet]*, 347, 118383, 2022, Available from: https://www.sciencedirect.com/science/article/pii/S0167732221031081.
48. Thakur, A., Kaya, S., Kumar, A., Recent Innovations in Nano Container-Based Self-Healing Coatings in the Construction Industry. *Curr. Nanosci. [Internet]*, 18, 2, 203–16, 2022, Available from: https://www.sciencedirect.com/science/article/pii/S1573413722000037.
49. Sharma, D., Thakur, A., Sharma, M.K., Kumar, A., Jakhar, K., Kumar, S., *et al.*, A convenient synthesis, electrochemical profiling, and morphological studies of a pyridine-based 1,3,4-oxadiazole hybrid: A promising study

for corrosion mitigation of mild steel in strongly acidic environment. *Inorg. Chem. Commun. [Internet]*, 158, 111554, 2023, Available from: https://www.sciencedirect.com/science/article/pii/S1387700323011668.

50. Verma, C., Thakur, A., Ganjoo, R., Sharma, S., Assad, H., Kumar, A., *et al.*, Coordination bonding and corrosion inhibition potential of nitrogen-rich heterocycles: Azoles and triazines as specific examples. *Coord. Chem. Rev. [Internet]*, 488, 215177, 2023, Available from: https://www.sciencedirect.com/science/article/pii/S0010854523001662.
51. Assad, H., Saha, S.K., Kang, N., Kumar, S., Sharma, P.K., Dahiya, H., *et al.*, Electrochemical and computational insights into the utilization of 2, 2- dithio bisbenzothiazole as a sustainable corrosion inhibitor for mild steel in low pH medium. *Environ. Res. [Internet]*, 242, 117640, 2024, Available from: https://www.sciencedirect.com/science/article/pii/S0013935123024441.
52. Thakur, A., Kumar, A., Kaya, S., Benhiba, F., Sharma, S., Ganjoo, R., *et al.*, Electrochemical and computational investigations of the Thysanolaena latifolia leaves extract: An eco-benign solution for the corrosion mitigation of mild steel. *Results Chem. [Internet]*, 6, 101147, 2023, Available from: https://www.sciencedirect.com/science/article/pii/S2211715623003867.
53. Thakur, A., Dagdag, O., Berisha, A., Ebenso, E.E., Kumar, A., Sharma, S., *et al.*, Mechanistic insights into the corrosion inhibition of mild steel by eco-benign Asphodelus Tenuifolius aerial extract in acidic environment: Electrochemical and computational analysis. *Surf Coat. Technol. [Internet]*, 480, 130568, 2024, Available from: https://www.sciencedirect.com/science/article/pii/S0257897224001981.
54. Sharma, D., Thakur, A., Sharma, M.K., Sharma, R., Kumar, S., Sihmar, A., *et al.*, Effective corrosion inhibition of mild steel using novel 1,3,4-oxadiazole-pyridine hybrids: Synthesis, electrochemical, morphological, and computational insights. *Environ. Res. [Internet]*, 234, 116555, 2023, Available from: https://www.sciencedirect.com/science/article/pii/S0013935123013592.
55. Xu, C., Li, Q., Wang, P., Guo, Y., Preparation and performance research of ecological concrete using waste wood. *Case Stud. Constr. Mater [Internet]*, 20, e03221, 2024, Available from: https://www.sciencedirect.com/science/article/pii/S2214509524003723.
56. Subedi, J., Wu, Y., Bazli, M., Rajabipour, A., Hassanli, R., Shahbazi, R., Bond performance between sand coated hybrid glass-carbon FRP tubes and seawater sea sand concrete after exposure to elevated temperatures. *Constr. Build Mater [Internet]*, 446, 137984, 2024, Available from: https://www.sciencedirect.com/science/article/pii/S095006182403126X.
57. Jena, B., Zade, N.P., Sarkar, P., Karak, S.K., Sustainable integration of coal mine overburden as a substitute for natural sand in concrete to enhance its mechanical and durability properties. *Constr. Build Mater [Internet]*, 411, 134488, 2024, Available from: https://www.sciencedirect.com/science/article/pii/S0950061823042071.

58. Jia, Z., Aguiar, J., Cunha, S., Jesus, C., Castro, F., Durability properties of lightweight concrete with ceramic mould casting waste. *Mag. Concr. Res. [Internet]*, 76, 11, 548–56, 2024, Available from: https://www.sciencedirect.com/science/article/pii/S1751763X24000175.
59. Valizadeh Kiamahalleh, M., Gholampour, A., Ngo, T.D., Ozbakkaloglu, T., Mechanical, durability and microstructural properties of waste-based concrete reinforced with sugarcane fiber. *Struct. [Internet]*, 67, 107019, 2024, Available from: https://www.sciencedirect.com/science/article/pii/S2352012424011718.
60. Kanwal, M., Adnan, F., Khushnood, R.A., Jalil, A., Khan, H.A., Wattoo, A.G., *et al.*, Biomineralization and corrosion inhibition of steel in simulated bio-inspired self-healing concrete. *J. Build Eng. [Internet]*, 82, 108224, 2024, Available from: https://www.sciencedirect.com/science/article/pii/S235271022302404X.
61. Xia, B., Xiao, J., Li, S., Sustainability-based reliability design for reuse of concrete components. *Struct. Saf. [Internet]*, 98, 102241, 2022, Available from: https://www.sciencedirect.com/science/article/pii/S0167473022000510.
62. Ding, J., Xiao, L., Mei, G., Easin, A., Reverse-seepage and saturation action anti-corrosion tech (RS-AAT) for corrosion resistance and self-healing of cracked concrete in seawater immersion. *Appl. Ocean Res. [Internet]*, 143, 103874, 2024, Available from: https://www.sciencedirect.com/science/article/pii/S0141118723004157.
63. Jin, Z., Mao, S., Zheng, Y., Liang, K., Pre-treated corn straw fiber for fiber-reinforced concrete preparation with high resistance to chloride ions corrosion. *Case Stud. Constr. Mater [Internet]*, 19, e02368, 2023, Available from: https://www.sciencedirect.com/science/article/pii/S221450952300548X.
64. Lu, Z., Zhao, C., Zhao, J., Shi, C., Xie, J., Bond durability of FRP bars and seawater–sea sand–geopolymer concrete: Coupled effects of seawater immersion and sustained load. *Constr. Build Mater [Internet]*, 400, 132667, 2023, Available from: https://www.sciencedirect.com/science/article/pii/S0950061823023838.
65. Khan, M. and Javed, M.F., Towards sustainable construction: Machine learning based predictive models for strength and durability characteristics of blended cement concrete. *Mater Today Commun. [Internet]*, 37, 107428, 2023, Available from: https://www.sciencedirect.com/science/article/pii/S2352492823021190.
66. Yang, X., Lu, J., Wan, X., Wang, C., Wang, D., Liu, S., Mechanical and microstructural properties of municipal solid waste incinerator bottom ash (MSWIBA) concrete exposed to salt erosion and drying-wetting cycles. *J. Build Eng. [Internet]*, 96, 110482, 2024, Available from: https://www.sciencedirect.com/science/article/pii/S2352710224020503.
67. Gómez-Luna, G.F., Lopez-Calvo, H.Z., Bremner, T.W., Castro-Borges, P., Montes-García, P., Performance of fly ash and CNI as corrosion prevention methods for steel reinforcement embedded in cracked HPC concrete exposed to a natural marine environment. *Constr. Build Mater [Internet]*,

428, 136260, 2024, Available from: https://www.sciencedirect.com/science/article/pii/S0950061824014016.

68. Yuan, F., Liu, P., Li, H., Wu, Y., Experimental investigations on the flexural behavior of compression-cast seawater sea-sand concrete beams reinforced with CFRP bars. *Constr. Build Mater [Internet]*, 445, 137754, 2024, Available from: https://www.sciencedirect.com/science/article/pii/S0950061824028964.
69. Habib, N., Saqib, M., Najeh, T., Gamil, Y., Eco-Transformation of construction: Harnessing machine learning and SHAP for crumb rubber concrete sustainability. *Heliyon [Internet]*, 10, 5, e26927, 2024, Available from: https://www.sciencedirect.com/science/article/pii/S240584402402958X.
70. Li, C., Bi, K., Hao, H., Li, C., Comparative studies on the seismic performances of precast segmental columns with different concrete, reinforcement and tendon materials. *Struct. [Internet]*, 67, 106963, 2024, Available from: https://www.sciencedirect.com/science/article/pii/S2352012424011159.
71. Merachtsaki, D., Tsiaras, S., Peleka, E., Zouboulis, A., Selection of magnesium hydroxide coatings for corrosion mitigation in concrete sewer pipes by using multiple criteria decision analysis. *Environ. Sustain Indic. [Internet]*, 13, 100168, 2022, Available from: https://www.sciencedirect.com/science/article/pii/S2665972721000696.
72. Xi, B., Huang, Z., Al-Obaidi, S., Ferrara, L., Healing capacity of Ultra High Performance Concrete under sustained through crack tensile stresses and aggressive environments. *Cem. Concr. Compos. [Internet]*, 145, 105355, 2024, Available from: https://www.sciencedirect.com/science/article/pii/S0958946523004298.
73. Xiao, S.H., Cai, Y.J., Xie, Z.H., Guo, Y.C., Zheng, Y., Lin, J.X., Bond durability between steel-FRP composite bars and concrete under seawater corrosion environments. *Constr. Build Mater [Internet]*, 419, 135456, 2024, Available from: https://www.sciencedirect.com/science/article/pii/S095006182400597X.
74. Zhu, M., Li, X., Deng, J., Peng, S., Electrochemical impedance based interfacial monitoring for concrete beams strengthened with CFRP subjected to wetting–drying cycling and sustained loading. *Constr. Build Mater [Internet]*, 366, 130238, 2023, Available from: https://www.sciencedirect.com/science/article/pii/S0950061822038946.
75. Pradhan, S., Chang Boon Poh, A., Qian, S., Impact of service life and system boundaries on life cycle assessment of sustainable concrete mixes. *J. Clean. Prod. [Internet]*, 342, 130847, 2022, Available from: https://www.sciencedirect.com/science/article/pii/S0959652622004851.
76. Mirshahmohammad, M., Rahmani, H., Maleki-Kakelar, M., Bahari, A., Effect of sustained service loads on the self-healing and corrosion of bacterial concretes. *Constr. Build Mater [Internet]*, 322, 126423, 2022, Available from: https://www.sciencedirect.com/science/article/pii/S0950061822001167.
77. Liu, J., Li, Z., Zhang, W., Jin, H., Xing, F., Tang, L., The impact of cold-bonded artificial lightweight aggregates produced by municipal solid waste

incineration bottom ash (MSWIBA) replace natural aggregates on the mechanical, microscopic and environmental properties, durability of sustainable concrete. *J. Clean. Prod. [Internet]*, 337, 130479, 2022, Available from: https://www.sciencedirect.com/science/article/pii/S0959652622001226.

78. Wang, Y., Liao, J., Liu, Z., Service life prediction and environmental impact evaluation of recycled aggregate concrete with supplementary cementitious materials. *Constr. Build Mater [Internet]*, 395, 132270, 2023, Available from: https://www.sciencedirect.com/science/article/pii/S0950061823019864.
79. Elmenshawy, Y., Elmahdy, M.A.R., Moawad, M., Elshami, A.A., Ahmad, S.S.E., Nagai, K., Investigating the bacterial sustainable self-healing capabilities of cracks in structural concrete at different temperatures. *Case Stud. Constr. Mater [Internet]*, 20, e03188, 2024, Available from: https://www.sciencedirect.com/science/article/pii/S2214509524003395.
80. Abushanab, A. and Alnahhal, W., Life cycle cost analysis of sustainable reinforced concrete buildings with treated wastewater, recycled concrete aggregates, and fly ash. *Results Eng. [Internet]*, 20, 101565, 2023, Available from: https://www.sciencedirect.com/science/article/pii/S2590123023006928.
81. Bheel, N., Nadeem, G., Almaliki, A.H., Al-Sakkaf, Y.K., Dodo, Y.A., Benjeddou, O., Effect of low carbon marble dust powder, silica fume, and rice husk ash as tertiary cementitious material on the mechanical properties and embodied carbon of concrete. *Sustain. Chem. Pharm. [Internet]*, 41, 101734, 2024, Available from: https://www.sciencedirect.com/science/article/pii/S2352554124003097.
82. Guo, J., Chan, T.M., Wang, Y., Precast concrete pavement applications, design and joint load transfer characteristics. *Struct. [Internet]*, 69, 107176, 2024, Available from: https://www.sciencedirect.com/science/article/pii/S2352012424013286.
83. Rawat, A., Karade, S.R., Thapliyal, P.C., Mechanism of inhibitors in control of corrosion of steel in concrete. *Mater Today Proc. [Internet]*, 2023, Available from: https://www.sciencedirect.com/science/article/pii/S2214785323035897.
84. Zhao, P., Xu, G., Wang, Q., Wen, C., Impact of sustained load on damage characteristics of reinforced concrete beams under the combined action of salt freeze-thaw cycles and corrosion. *Constr. Build Mater [Internet]*, 273, 121744, 2021, Available from: https://www.sciencedirect.com/science/article/pii/S095006182033748X.
85. Bakde, S., Suryawanshi, P., Murkute, S., Bharti, R., Kumar Shaw, S., Ali Khan, H., Impacts of fibre and wastage material on the sustainable concrete: A comprehensive review. *Mater Today Proc. [Internet]*, 2023, Available from: https://www.sciencedirect.com/science/article/pii/S2214785323010118.
86. Yang, W., Huang, Y., Tang, Z., Xiong, X., Li, C., Zhong, X., *et al.*, Impermeability performance and corrosion resistance mechanism of basalt fiber recycled concrete under the coastal tidal environment. *Constr. Build Mater [Internet]*, 411, 134510, 2024, Available from: https://www.sciencedirect.com/science/article/pii/S0950061823042290.

87. Guo, Q., Zhang, T., Liu, J., Study of endogenous chloride ion content on the behavior of steel bar depassivation in sea sand concrete. *J. Build Eng. [Internet]*, 97, 110710, 2024, Available from: https://www.sciencedirect.com/science/article/pii/S2352710224022782.
88. Taheri-Shakib, J. and Al-Mayah, A., Effect of corrosion pit distribution of rebar on pore, and crack characteristics in concrete. *Cem. Concr. Compos. [Internet]*, 148, 105476, 2024, Available from: https://www.sciencedirect.com/science/article/pii/S0958946524000489.
89. Chen, D., Guo, W., Wu, B., Shi, J., Service life prediction and time-variant reliability of reinforced concrete structures in harsh marine environment considering multiple factors: A case study for Qingdao Bay Bridge. *Eng. Fail Anal. [Internet]*, 154, 107671, 2023, Available from: https://www.sciencedirect.com/science/article/pii/S1350630723006258.
90. Wattanavichien, P. and Iwanami, M., Investigation of the mechanical, microstructure, and durability properties of concrete with fine uniform and non-uniform polyethylene terephthalate (PET) aggregates. *Clean Mater [Internet]*, 13, 100267, 2024, Available from: https://www.sciencedirect.com/science/article/pii/S2772397624000510.
91. Yang, W., Feng, W., Liao, Z., Yang, Y., Miao, G., Yu, B., *et al.*, Protection of mild steel with molecular engineered epoxy nanocomposite coatings containing corrosion inhibitor functionalized nanoparticles. *Surf. Coat. Technol.*, 406, 126639, September 2020, 2021.
92. Falzone, G., Balonis, M., Bentz, D., Jones, S., Sant, G., Anion capture and exchange by functional coatings: New routes to mitigate steel corrosion in concrete infrastructure. *Cem. Concr. Res. [Internet]*, 101, September 2016, 82–92, 2017, Available from: http://dx.doi.org/10.1016/j.cemconres.2017.08.021.
93. Kim, J.K., Yee, J.J., Kee, S.H., Electrochemical deposition treatment (Edt) as a comprehensive rehabilitation method for corrosion-induced deterioration in concrete with various severity levels. *Sensors*, 21, 18, 6287, 2021.
94. Liu, X., Wu, Z., Lyu, Y., Li, T., Yang, H., Liu, Y., *et al.*, Corrosion Resistance of CeO2-GO/Epoxy Nanocomposite Coating in Simulated Seawater and Concrete Pore Solutions. *Polym. (Basel)*, 15, 12, 2602, 2023.
95. Huang, X., Jiao, Z., Xing, F., Sui, L., Hu, B., Zhou, Y., Performance assessment of LC3 concrete structures considering life-cycle cost and environmental impacts. *J. Clean. Prod. [Internet]*, 436, October 2023, 140380, 2024, Available from: https://doi.org/10.1016/j.jclepro.2023.140380.
96. Sun, Z., Li, Y., Han, T., Su, L., Zhu, X., He, J., *et al.*, Performance evaluation of hybrid fiber-reinforced concrete based on electrical resistivity: Experimental and data-driven method. *Constr. Build. Mater.*, 446, 137992, August, 2024.
97. Han, H., Han, K.R., Cheong, I.W., Optimizing latex-modified concrete for enhanced workability and durability in pavement construction. *Constr. Build Mater [Internet]*, 444, August, 137722, 2024, Available from: https://doi.org/10.1016/j.conbuildmat.2024.137722.

98. Rampini, M.C., Zani, G., Colombo, M., di Prisco, M., Advancing construction techniques: Textile reinforced concrete shells and high-performance fiber-reinforced concrete beams for partially prefabricated elevated slabs. *Struct. [Internet]*, 70, 107710, 2024, Available from: https://www.sciencedirect.com/science/article/pii/S2352012424018630.
99. Hason, M.M. and Al-Janabi, M.A.Q., Behavior of zero cement reinforced concrete slabs under monotonic and impact loads: Experimental and numerical investigations. *Case Stud. Constr. Mater [Internet]*, 21, e03789, 2024, Available from: https://www.sciencedirect.com/science/article/pii/S2214509524009409.
100. ALMisned, G., Susoy, G., Sen, B.D., AH, A., Tekin, H.O., Evaluating the efficacy of multiple concrete compositions in gamma ray and fast neutron shielding. *J. Build Eng. [Internet]*, 98, 111205, 2024, Available from: https://www.sciencedirect.com/science/article/pii/S2352710224027736.
101. Alfouneh, M., Ben Seghier, M.E.A., Keshtegar, B., Tabatabaie Shojah, S.A., A new formulation for predicting the ultimate capacities of FRP-confined concrete using advanced machine learning framework: Developed structural reliability analysis. *Struct. [Internet]*, 68, 107158, 2024, Available from: https://www.sciencedirect.com/science/article/pii/S2352012424013109.
102. Xiao, X., Santoso, H.S., Unluer, C., Yang, E.H., The use of waste concrete sludge as alkali source to recover reactive MgO from reject brine. *Desalin. [Internet]*, 600, 118470, 2025, Available from: https://www.sciencedirect.com/science/article/pii/S0011916424011810.
103. Amin, M.N., Ahmad, A., Khan, K., Qadir, M.T., Precision assessment of the machine learning tools for the strength optimization of environmental-friendly lightweight foam concrete. *J. Environ. Manage [Internet]*, 373, 123462, 2025, Available from: https://www.sciencedirect.com/science/article/pii/S0301479724034480.
104. Farahzadi, L., Tellnes, L.G.F., Shafei, B., Kioumarsi, M., Life-cycle environmental assessment of ultra-high-performance concrete with sustainable materials and fiber substitutions. *Clean Eng. Technol. [Internet]*, 23, 100846, 2024, Available from: https://www.sciencedirect.com/science/article/pii/S2666790824001265.
105. Chen, Y., Zeng, J., Jia, J., Jabli, M., Abdullah, N., Elattar, S., *et al.*, A fusion of neural, genetic and ensemble machine learning approaches for enhancing the engineering predictive capabilities of lightweight foamed reinforced concrete beam. *Powder Technol. [Internet]*, 440, 119680, 2024, Available from: https://www.sciencedirect.com/science/article/pii/S003259102400322X.
106. Lin, Y., Yuan, Y., Wan, S., Ding, B., Yang, C., Zong, Z., *et al.*, Flexural behaviour of reinforced one-part geopolymer concrete beams. *J. Build Eng. [Internet]*, 96, 110435, 2024, Available from: https://www.sciencedirect.com/science/article/pii/S2352710224020035.
107. Ucak, K., Fromme, P., Duffour, P., Chen, L., Bai, Y., Influence of replacing Portland cement with anionic asphalt emulsion on vibration damping, mechanical and microstructural properties of concrete. *Case Stud. Constr.*

Mater [Internet], 21, e03901, 2024, Available from: https://www.sciencedirect.com/science/article/pii/S2214509524010520.

108. Pfeiffer, O.P., Gong, K., Severson, K.A., Chen, J., Gregory, J.R., Ghosh, S., *et al.*, Bayesian design of concrete with amortized Gaussian processes and multi-objective optimization. *Cem. Concr. Res. [Internet]*, 177, 107406, 2024, Available from: https://www.sciencedirect.com/science/article/pii/S0008884623003216.
109. Le Nguyen, K., Uddin, M., Pham, T.M., Generative artificial intelligence and optimisation framework for concrete mixture design with low cost and embodied carbon dioxide. *Constr. Build Mater [Internet]*, 451, 138836, 2024, Available from: https://www.sciencedirect.com/science/article/pii/S0950061824039783.
110. Spartali, H., Hegger, J., Beckmann, B., Chudoba, R., Closed-form design concept for fully utilized FRP-reinforced rectangular concrete cross-sections accompanied by experimental validation on carbon-reinforced beams. *Struct. [Internet]*, 61, 106042, 2024, Available from: https://www.sciencedirect.com/science/article/pii/S2352012424001942.
111. Mohtasham Moein, M., Rahmati, K., Mohtasham Moein, A., Rigby, S.E., Saradar, A., Karakouzian, M., Utilizing construction and demolition waste in concrete as a sustainable cement substitute: A comprehensive study on behavior under short-term dynamic and static loads *via* laboratory and numerical analysis. *J. Build Eng. [Internet]*, 97, 110778, 2024, Available from: https://www.sciencedirect.com/science/article/pii/S2352710224023465.
112. Lu, D., Jiang, X., Qu, F., Huo, Y., Mitigating sulfate ions migration in concrete: A targeted approach to address recycled concrete aggregate's impact. *J. Clean. Prod. [Internet]*, 442, 141135, 2024, Available from: https://www.sciencedirect.com/science/article/pii/S0959652624005821.
113. Ortolan, T.L.P., Borges, P.M., Silvestro, L., da Silva, S.R., Possan, E., de Oliveira Andrade, J.J., Durability of concrete incorporating recycled coarse aggregates: carbonation and service life prediction under chloride-induced corrosion. *Constr. Build Mater [Internet]*, 404, 133267, 2023, Available from: https://www.sciencedirect.com/science/article/pii/S0950061823029847.
114. Zeng, W., Ayough, P., Wang, Y.H., Elchalakani, M., Numerical simulation of axially loaded circular concrete-filled steel tubular short columns with localized corrosion. *Eng. Struct. [Internet]*, 312, 118243, 2024, Available from: https://www.sciencedirect.com/science/article/pii/S0141029624008058.
115. Merino-Maldonado, D., Antolín-Rodríguez, A., Blanco, S., Ma Morán-del Pozo, J., García-González, J., Juan-Valdés, A., Improving the performance of recycled concrete by biodeposition of biogenic silica as a surface coating. *Constr. Build Mater [Internet]*, 408, 133781, 2023, Available from: https://www.sciencedirect.com/science/article/pii/S0950061823034992.
116. Bandeira Barros, L., Knockaert, M., Tenório Filho, J.R., Towards a more sustainable construction industry: Bridging the gap between technical progress and commercialization of self-healing concrete. *Constr. Build Mater [Internet]*, 403, 133094, 2023, Available from: https://www.sciencedirect.com/science/article/pii/S0950061823028118.

117. Song, L., Wei, M., Zhang, X., Li, C., Dong, Z., Wang, Y., Flexural performance of reinforced concrete beams repaired with BFRP grid-TSUHDC. *Constr. Build Mater [Internet]*, 443, 137748, 2024, Available from: https://www.sciencedirect.com/science/article/pii/S0950061824028903.
118. Shankar Biswal, U. and Dinakar, P., Evaluating corrosion resistance of recycled aggregate concrete integrating ground granulated blast furnace slag. *Constr. Build Mater [Internet]*, 370, 130676, 2023, Available from: https://www.sciencedirect.com/science/article/pii/S0950061823003872.
119. Lehner, P. and Hrabová, K., Evaluation of degradation and mechanical parameters and sustainability indicators of zeolite concretes. *Constr. Build Mater [Internet]*, 371, 130791, 2023, Available from: https://www.sciencedirect.com/science/article/pii/S0950061823005020.
120. Singh, A., Ansari, K.R., AbdelRahim, S.K., Ali, I.H., EL Ibrahimi, B., Alanazi, A.K., *et al.*, Potential application of Ginkgo biloba extract as a green corrosion inhibitor for carbon steel reinforcement in chloride-polluted simulated concrete pore solution. *Process Saf. Environ. Prot. [Internet]*, 186, 819–32, 2024, Available from: https://www.sciencedirect.com/science/article/pii/S0957582024003550.
121. Xing, W., Tam, V.W.Y., Le, K.N., Hao, J.L., Wang, J., Life cycle assessment of sustainable concrete with recycled aggregate and supplementary cementitious materials. *Resour. Conserv. Recycl. [Internet]*, 193, 106947, 2023, Available from: https://www.sciencedirect.com/science/article/pii/S0921344923000848.
122. Mohamed, A.M., Tayeh, B.A., Majeed, S.S., Aisheh, Y.I.A., Salih, M.N.A., Fresh, hardened, durability and microstructure properties of seawater concrete: A systematic review. *J. CO2 Util. [Internet]*, 83, 102815, 2024, Available from: https://www.sciencedirect.com/science/article/pii/S2212982024001501.
123. Robles, K.P.V., Gucunski, N., Kee, S.H., Evaluation of steel corrosion-induced concrete damage using electrical resistivity measurements. *Constr. Build Mater [Internet]*, 411, 134512, 2024, Available from: https://www.sciencedirect.com/science/article/pii/S0950061823042319.
124. Mojapelo, K.S., Kupolati, W.K., Ndambuki, J.M., Sadiku, E.R., Ibrahim, I.D., Maepa, C., Sustainable usage and the positive environmental impact of wastewater dry sludge-based concrete. *Results Mater [Internet]*, 16, 100336, 2022, Available from: https://www.sciencedirect.com/science/article/pii/S2590048X2200084X.

6

Role of Anti-Corrosion Coatings and Sealants in Preventing Concrete Corrosion

Humira Assad[1,2], Richika Ganjoo[2], Shveta Sharma[2], Praveen Kumar Sharma[2], Femiana Gapsari[3] and Ashish Kumar[4*]

[1]*Department of Metallurgical Engineering and Materials Science, Indian Institute of Technology Bombay, Mumbai, India*
[2]*Department of Chemistry, School of Chemical Engineering and Physical Sciences, Lovely Professional University, Punjab, India*
[3]*Department of Mechanical Engineering, Faculty of Engineering, Brawijaya University, Malang, Indonesia*
[4]*Nalanda College of Engineering, Bihar Engineering University, Department of Science Technology and Technical Education, Government of Bihar, India*

Abstract

Concrete is a broadly utilized building matter that is highly appraised for its durability, strength, and adaptability. However, the sturdiness and integrity of concrete assemblies are seriously threatened by their sensitivity to corrosion, especially in harsh settings. The key worry is the corrosion of the embedded steel reinforcement in concrete, of which the upshot can be reduced load-bearing capacity, structural deterioration, and eventual failure. Concrete corrosion can be initiated and spread by several variables, comprising climatic conditions, carbonation, and chloride infiltration. New materials science developments provide fresh approaches to increased durability, and recent breakthroughs in anti-corrosion technology are revolutionizing the industry. Case studies and real-world implementations highlight the valuable advantages of these precautionary measures. Sealants and anti-corrosion coatings have become essential as a defensive tactic to address these problems. By acting as barriers, these materials keep corrosive substances out of the concrete and protect its structural integrity. The current chapter explores the complex function of sealants and anti-corrosion coatings in preventing concrete

**Corresponding author*: drashishchemlpu@gmail.com

Ruby Aslam, Qihui Wang, Zhitao Yan and Afroz Aslam (eds.) Corrosion in Concrete Structures: Integrating Advanced Technologies and Sustainable Practices, (163–186) © 2026 Scrivener Publishing LLC

corrosion in this regard. The basic mechanisms of concrete corrosion and the variables that aggravate it are first examined. Subsequently, the chapter offers an in-depth analysis of several kinds of anti-corrosion sealants and coatings, together with their characteristics and modes of operation. There is also a discussion of practical topics like performance evaluation and application strategies. The final section of the chapter highlights the continued development of anti-corrosion techniques in maintaining the sustainability of concrete infrastructure by examining potential future trends and research areas.

***Keywords*:** Concrete, aggressive environments, corrosion, protective measures, coatings, sealants

6.1 Introduction

Modern building relies heavily on concrete, which is prized for its durability and adaptability. On the other hand, corrosion is a notable issue, especially when steel reinforcement is present [1, 2]. In addition to compromising concrete's structural integrity, corrosion can result in serious safety risks and expensive repairs [3–6]. In this regard, sealing and anti-corrosion coatings are essential to prevent the deterioration of concrete buildings [7]. These safeguards act as barriers against substances such as chemicals, moisture, and other environmental stresses that might cause or hasten corrosion [8, 9] as shown below in Figure 6.1.

Coatings like epoxy and polyurethane stand out among many anti-corrosion solutions because of their remarkable protective qualities [11, 12]. Epoxy coatings are well known for having firm, impermeable surfaces that keep out chemicals and water, which make them perfect for severe settings. Their superior mechanical strength and adhesion guarantee long-lasting performance [10]. Conversely, polyurethane coatings are more flexible and resistant to UV rays, enabling them to adjust to concrete movement without cracking. Their longevity in outdoor settings where temperature changes are common is significantly increased by this adaptability [13]. Sealants, namely, those based on silane and siloxane, offer an additional layer of protection that is distinct but complementary. By completely saturating the concrete's pores, these penetrating sealants lower the material's permeability to water and chloride ions. Sealants work by forming a hydrophobic surface that stops moisture from penetrating the concrete and let it breathe, reducing the possibility of moisture buildup that could cause more harm. Because of these combined qualities, sealants work especially well in areas where seawater or deicing salts are present [14, 15]. These sealants and coatings offer concrete protection through a variety of techniques.

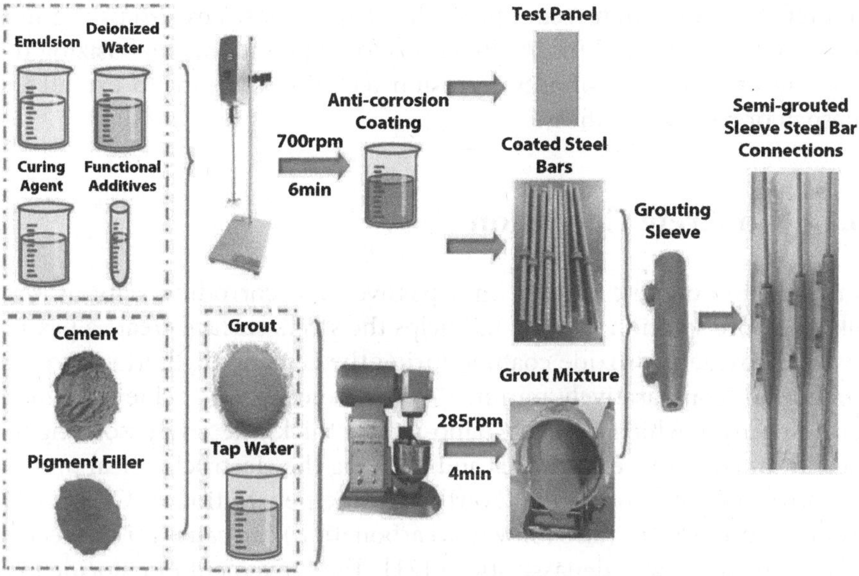

Figure 6.1 Preparation process for samples and coated test panels [10].

They primarily lower the danger of corrosion by erecting a physical barrier that separates the steel reinforcement from its surroundings.

Furthermore, many coatings are designed to resist particular substances that can degrade them, offering specialized protection depending on the use [16]. This defense is further strengthened by the sealants' hydrophobic qualities, which lower water absorption and lessen the likelihood of freeze–thaw cycles. Furthermore, flexible coatings such as polyurethane allow for the normal movement of concrete, preserving a continuous layer of protection that deters cracking. In general, the integrity and longevity of concrete structures depend on the function of anti-corrosion coatings and sealants in reducing concrete corrosion [17]. They are better than other corrosion prevention techniques because they can serve as barriers to keep out moisture, chemicals, and other corrosive agents. These preventive measures will become increasingly important as technological breakthroughs continue to occur since they will guarantee the robustness and sustainability of our built environment, which will eventually secure infrastructure for coming generations [18, 19]. Investing in anti-corrosion solutions is essential to strengthen concrete's resistance to the problems presented by a changing environment, not just a preventive one. This chapter extensively examines different anti-corrosion sealants and coatings, examining their properties and workings. It also covers practical issues, such as application and efficacy evaluation techniques. The chapter closes by discussing new directions in study and future directions for exploration, emphasizing the ongoing development of anti-corrosion techniques meant to extend the lifespan of concrete edifices.

6.2 Concrete Corrosion

The steel in concrete is often in a passive, non-corroding state. A very alkaline pore solution (pH > 13.5) helps the steel's surface create a passive hydrated oxide/hydroxide coating during the cement's hydration process. Decades of comparatively minimal maintenance can be achieved by this shielding layer, which is merely nano-meters thick, effectively isolating the steel from corrosive electrolytes and delaying the electrochemical process of corrosion initiation [20]. Nevertheless, the penetration of Cl and CO_2 from the atmosphere (also known as carbonation) can cause this protective layer to break down (depassivation) [21]. The movement of chloride ions through the concrete pore network, either by diffusion or capillary action, occurs rapidly and has harmful effects. On the other hand, the slow penetration of carbon dioxide into the concrete leads to chemical reactions with $Ca(OH)_2$ and other cement hydration byproducts, gradually lowering the pH of the pore solution to about 9.0, which breaks down the protective passive layer around the reinforcing steel. All above-ground outdoor constructions have abundant water and oxygen available simultaneously, which can lead to considerable corrosion, as shown in Figure 6.2.

On the other hand, corrosion happens more quickly when exposed to chlorides. Steel corrodes in concrete due to an electrochemical process that produces a current flow that dissolves metal when oxygen and moisture are present. In typical practice, carbonation is mild and relatively easy to avoid in contrast to corrosion caused by chloride [23]. Concrete may get contaminated by soluble chlorides found in seawater, groundwater, or de-icing salts through ion diffusion in water or capillary absorption. Additionally, utilizing contaminated materials or adding water when making concrete might introduce chlorides into the mixture. Chemical admixtures can occasionally contain excessive amounts of chlorides. For example, adding calcium chloride as a set accelerator to freshly mixed concrete has produced premature structural issues with steel corrosion [24]. Steel passivity has been predicted to be compromised by chloride ion (Cl^-) concentrations as low as 0.6 kg per cubic meter of concrete by mass, even though Cl^- has minimal impact on the pore solution pH [25]. According to the European standard EN 206, the allowable chloride ion concentrations are 0.2%–0.4% by the

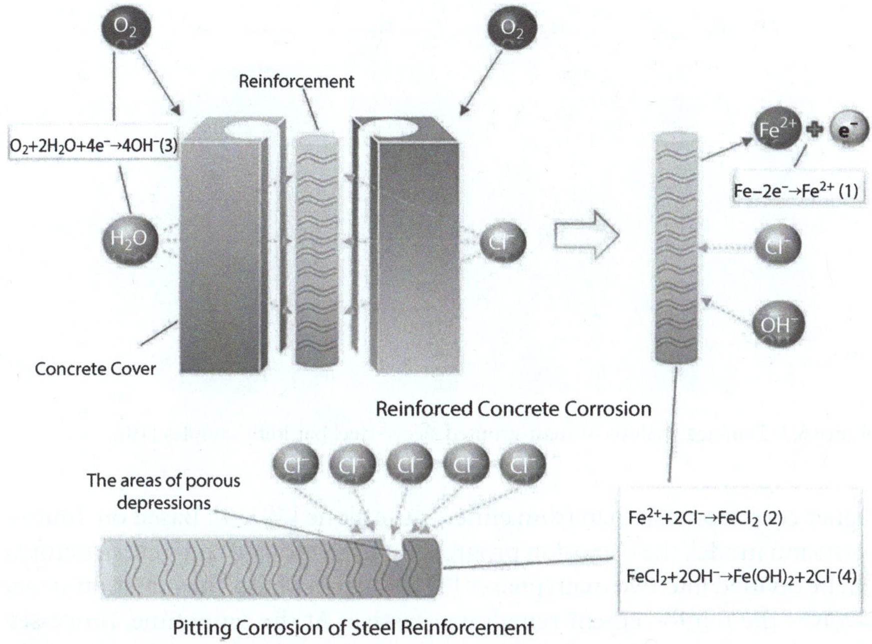

Figure 6.2 Mechanism of corrosion reaction in concrete [22].

mass of binder for reinforced concrete and 0.1%–0.2% for pre-stressed concrete [26]. Chlorides can adhere to the hydrated cement mix by physical or chemical binding. Several researchers have linked corrosion risk to chloride levels. The amount of total chloride is typically used to express the chloride content of concrete. It should be understood that the overall chloride content does not primarily cause corrosion. The free chlorides, instead of bound chlorides, are the only ones that can conceivably erode the passive coating on the surface of the reinforcing steel and consequently commence corrosion. It is believed that corrosion begins when the crucial chloride content, often known as the "chloride threshold," is reached at the implanted steel surface. Corrosion cells arise between the anodic zones, where active corrosion occurs, and the cathodic regions, which maintain passivity, as a result of the local rupture of the passive film that arises once the chloride threshold is exceeded [27]. As is evident, chlorides help accelerate the start of corrosion in this process by acting as a catalyst.

It should be emphasized that corrosion resulting from carbonation develops at a considerably slower pace than that caused by chloride penetration [28], the corrosion process will be more complex when carbonation and chloride ingress are combined, and the combined effect will result in a much

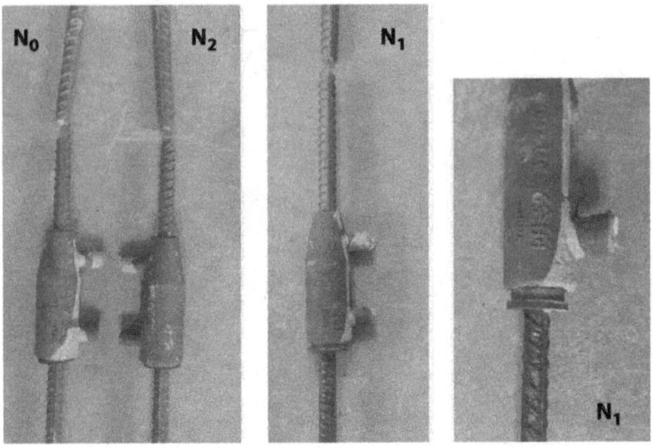

Figure 6.3 Damage analysis of semi-grouted sleeve steel bar joint samples [10].

higher corrosion risk than from either cause alone [29, 30]. Based on Tuutti's corrosion model, the corrosion progression in reinforced concrete structures can be divided into two main phases [31]. The first phase, the initiation stage, involves the reinforcement remaining passive. At the same time, processes like carbonation or chloride infiltration into the concrete occur, potentially leading to a loss of passivity. Oxygen and water near the steel influence the second phase of corrosion progression. It begins with the steel's depassivation and can continue until the concrete layer begins to break internally or externally and spalls as shown in Figure 6.3. The period of time before this kind of failure is commonly known as the reinforced concrete element's service life, and it is established by adding up the duration of these two stages.

6.3 Overview of Anti-Corrosion Coatings and Sealant Systems

Concrete structures are protected from various types of damage by using coatings or sealer systems as surface treatments. By preventing water and several harmful substances from penetrating, such as chloride salts used for de-icing or anti-icing, sulphate and chloride salts from seawater, sulphate ions from soil, and harmful gases (such carbon dioxide) from the atmosphere, they shield concrete [32]. Concrete can currently be protected using a variety of sealers and coatings [33, 34]. Protective sealers and coatings are typically employed to safeguard concrete bridge and building structures from environmental deterioration due to abrasion, freeze–thaw cycles, and moisture fluctuations. They aim to lessen several types of concrete damage, including cracking, delamination, popout, surface scaling, and spalling [35, 36]. Moreover, corrosion damage and chemical attacks brought on by anti-icing/de-icing corrosive chemicals are reduced or eliminated by applying sealers and coatings [36].

6.3.1 Types

The following coating and sealer products are typically applied to concrete surfaces to shield them from field-related damage. The sealer and coating systems' protective mechanisms are the basis for this classification [37, 38].

- Penetrating sealers
  Low-viscosity materials are "flood coated" to penetrate concrete by about 1–3 mm on average. Most of these are based on silicate or silane/siloxane. Water penetration into concrete is decreased using silane/siloxane-based sealers, which provide a hydrophobic covering on the pore-wall of capillary pores. On the other hand, calcium hydroxide, or portlandite, is a byproduct of cement hydration in concrete. Penetrating sealers based on silicate can react chemically with this substance to produce more C-S-H (calcium silicate hydrate) gel, which fills capillary pores and cracks in the concrete, making it denser and possibly stronger [38].
- Surface sealers
  These are low-solid, medium-viscosity materials that create a thin layer on the surface of the concrete but do not penetrate its capillary pores. These sealers must stick to the concrete surface when applied to create a thin layer. They mainly consist of materials based on urethane, epoxy, or methyl methacrylate.
- Barrier or surface coatings
  High-viscosity/high-solid materials coat concrete surfaces in a comparatively thick layer that acts as a barrier to protect concrete surfaces rather than seeping into the capillary pores of the material. These coatings must have high adhesion with the concrete substrates to work at their best in field settings. They are generally acrylic, epoxy, or urethane-based polymeric compounds. Bituminous or cement-based materials can also be used for these coatings.
- Impregnating polymers
  Ultra-low-viscosity compounds restrict the capillary pores in concrete and are applied using specific drying and suction

procedures to ensure deep penetration. These mostly consist of materials based on high molecular weight methacrylate, acrylonitrile, or styrene. Based on the primary constituent elements included in the coating composition, the previously described barrier or surface coating systems are further divided into the following categories [39].

The nature, characteristics, and/or quality of the sealer or coating, the environmental conditions during application and service, and the condition of the underlying concrete (e.g., moisture content, smoothness, roughness), the effectiveness of the substrate preparation methods, and the application method all influence the product's selection and performance. The chosen sealants or coatings should possess the following qualities:

- superior ability to apply sealant to newly exposed, old, or previously coated surfaces
- superior ability to minimize or completely stop moisture infiltration; superior ability to transfer moisture vapor from concrete (breathability)
- superior resistance to salt scaling
- superior chemical resistance
- superior resistance to ultraviolet (UV) rays
- enhanced infiltration into the concrete outer layer
- superior skid resistance
- minimal toxicity
- suitable volatility for the intended application [40]

When choosing which sealer or coating products to use on concrete structures, remember the following particular criteria:

- concrete substrate type and condition
- exposure type
- concrete protection type
- anticipated longevity
- service condition
- sealer or coating installation
- surface treatment expenses

These elements will affect how well sealers or coatings work in the standard tools used to put coating and sealer systems, which are airless spray guns, rollers, brushes, brooms, or squeegees [40]. According to Attanayaka *et al.* [35], the best way to apply a penetrating sealer correctly is to flood the area. According to Ho and Harrison [41], surface coatings work better when they are thicker, providing more fortification against carbonation. Additionally, they claimed that a coating applied at the same coverage rate on an abrasive-blasted concrete substrate is less effective than a coating applied on an unprepared concrete surface. This is because, in order to obtain a uniform coating thickness, abrasive blasting usually results in a rough surface profile that calls for a higher coverage rate. It denotes that for a specific coverage rate, the thickness of the coating is reduced when applied to a concrete surface treated with abrasive blasting. Nevertheless, an enhanced coverage rate is typically advised when using a coating over the abrasively blasted concrete substrate because abrasive blasting is still helpful in achieving the necessary level of coating adhesion.

6.3.2 Criteria for Selecting Coatings and Sealants

Coatings, sealants, and adhesives are essential in manufacturing processes, offering protection, functionality, and esthetic appeal to various products. Choosing the right materials is critical to ensuring product quality, durability, and compliance with industry standards. This document outlines the key guidelines that manufacturers should consider when selecting paints, coatings, sealants, and adhesives for specific applications. The foremost consideration when choosing these materials is substrate compatibility. The selected paint, coating, sealant, or adhesive must be compatible with the surface it will be applied to. Different materials such as metal, plastic, wood, or concrete require different chemical formulations to ensure proper adhesion and optimal performance. Another critical factor is the application's specific requirements, which include environmental conditions, substrate material, desired finish, and performance characteristics. Variations in temperature, humidity, and exposure to UV light and chemicals can significantly impact the coating's effectiveness. Manufacturers must consider these factors carefully to ensure that the chosen product performs well in the intended environment [42].

In addition to the items above, it is crucial to assess factors such as UV exposure, chemical resistance, abrasion resistance, temperature variations, and corrosion protection. The selected coating must meet specific performance requirements related to durability, scratch resistance, chemical resistance, corrosion protection, and esthetics. This ensures that the product will not only look good but also withstand harsh conditions over time. Manufacturers should also factor in maintenance considerations. The product's longevity and ease of maintenance are essential to minimize operational disruptions and costs. Regular maintenance requirements and the potential for reapplication or repair should be considered during the selection process. Equally important are regulatory requirements. Compliance with legal standards and environmental regulations is necessary to avoid fines and ensure responsible production practices. One critical aspect is VOC regulations, which state that carbon-based chemicals known as volatile organic compounds (VOCs) can evaporate into the air, leading to pollution. Many regions have rules limiting the amount of VOCs in paints, coatings, sealants, and adhesives [42]. Manufacturers should ensure that their products meet these VOC limits to reduce environmental impact and comply with environmental laws. In addition to VOC limits, environmental regulations govern waste management, emissions control, and pollution prevention. Manufacturers must comply with regulations on waste disposal, recycling, air emissions, and wastewater treatment. Failure to meet these standards can lead to fines, penalties, and damage to a company's reputation.

Product labeling and safety data sheets (SDS) are another critical regulatory requirement. Manufacturers must provide accurate product labeling and SDS detailing the composition, hazards, handling precautions, and emergency procedures associated with paints, coatings, sealants, and adhesives. Proper labeling ensures that users can safely handle, store, and dispose of these products, minimizing the risk of accidents and exposure to harmful substances. Furthermore, manufacturers may need to comply with industry-specific standards and certifications. These may include specifications related to performance, durability, and environmental impact. Examples of relevant organizations include NACE (National Association of Corrosion Engineers), ISO (International Organization for Standardization), ASTM (American Society for Testing and Materials), and others. Adherence to these standards helps manufacturers mitigate legal and environmental risks, protect human health and safety, and demonstrate their commitment to sustainable manufacturing practices [42]. Hence, regulatory compliance should be integrated into the material selection process. Manufacturers must stay informed about changes to relevant regulations to ensure ongoing compliance and responsible production practices.

6.3.3 Installation and Performance of Anti-Corrosion Coatings and Sealant Systems

In most cases, the concrete substrate must be prepared before a sealer or coating system is applied. Generally speaking, fresh concrete needs to be at least 28 days old, and any impurities such as oils and curing agents that could hinder the adhesion or penetration of coatings and sealers must be eliminated [35)]. For identical reasons as previously stated, oil, grease, rubber, and other impurities should also be removed from old concrete using efficient surface preparation processes. Additionally, to maximize the chemical reactions that demand high alkalinity, it is advised, when applicable, to remove the carbonated surface layer of old concrete before applying specific kinds of penetrating sealers [41]. Most producers of sealers and coatings advise surface preparation on concrete substrates using blast-track (shot-blasting), abrasive blasting (sand or grit blasting), or high-pressure water blasting methods. This ensures that the coating or sealer material and the concrete surface adhere well. After surface preparation is completed, manufacturers may require the substrate to be either moist, saturated, surface-dry, or fully dry, depending on the specifications of their sealer or coating systems.

Concrete pore-walls can be rendered hydrophobic, capillary pores can be blocked, a homogeneous physical barrier can be formed on the surface of the concrete, or the concrete can react with the portlandite that is already present in the concrete to reduce significantly the amount of water that enters the concrete [37, 39]. Specific systems might function by two mechanisms, each consisting of two components. Concrete is more resistant to alkali-aggregate reactions and freeze–thaw cycles when less water infiltration occurs [40, 43]. According to Porter [44], sealers or coatings applied to concrete surfaces help prevent surface degradation caused by freeze–thaw cycles. Surface coatings have been found to improve concrete's chemical resistance and freeze–thaw endurance, according to studies by Almusallam *et al.* [45] and Moon *et al.* [46]. According to Seneviratne *et al.* [47], Al-Dulaijan *et al.* [48], Ibrahim *et al.* [49], and Oshiro and Tanigawa [50], coatings applied to the surface greatly minimize steel corrosion and the permeability of chlorides in reinforced concrete.

By avoiding chloride penetration, a suitable sealer or coating system shields reinforced concrete from corrosion and chemical attacks used to de-ice it, as shown by Palle and Hopwood II [33].

According to Moon *et al.* [46], silicate-based inorganic coatings dramatically lower carbon dioxide infiltration, leading to steel reinforcement corrosion by lowering pore-water pH in concrete. Roughened concrete surfaces and thicker films improve any coating's ability to protect, according to Ho and Harrison [41]. McCarter [51] reported that some surface coatings prevent the liquid absorption of water into concrete while allowing moisture vapor to escape through the coating. These occurrences aid in maintaining the surface of the concrete dry, which gives it considerable resistance against various types of concrete damage. Moreover, concrete's resistivity can be significantly increased by certain surface coatings, according to McCarter [51], enhancing the material's defense against chemical scaling. The results of the literature review, as a whole, show that coating or sealer systems increase concrete's resilience, which significantly increases the service life of concrete structures under field settings.

On the other hand, physical and chemical attacks can also cause surface coatings applied on concrete substrates to deteriorate over time. The bonding strength between a surface sealer and the concrete substrate is a critical factor in the performance of treated (sealed or coated) concrete [48, 50]. This is especially important to prevent or minimize the delamination or blistering of sealers and coatings applied to concrete components or structures. To guarantee that treated concrete performs effectively against harsh conditions, sealers and coatings should bond effectively to the concrete substrate, be applied at an adequate thickness, and provide proper resistance to the infiltration of chloride-rich water [32, 52, 53]. They must withstand the pressures that push water through coating and sealer systems. These stresses include the following:

(a) the force of gravity pushing water through horizontal concrete surfaces, like building roofs and bridge decks;
(b) the hydrostatic pressure acting on one side of horizontal or vertical concrete elements, like building foundation walls, basement walls, and slab-on-grade; and
(c) a difference in air pressure acting on horizontal or vertical concrete elements, like roof slabs and exterior walls.

The sealer or coating system will delaminate if it is unable to withstand the stresses stated above and has insufficient adhesion. A thick coating and a rough surface profile help to increase adhesion and resistance to the previously indicated disruptive forces or stresses. The ability of coating and sealer technologies to transfer moisture vapor from the inside of concrete is another important factor in their performance.

When moisture transfer from the concrete *via* the treated surface is desired, sealers and coatings should have sufficient "breathability" [54]. If not, there could be scorching or delamination on the coating and sealer systems. Several performance standards have been established for the application of sealer and coating systems, as Safiuddin's paper demonstrates. These standards are used to assess how well concrete sealers and coatings perform in terms of their durability and attributes linked to durability under field conditions. Thus, standardized tests are essential in evaluating the performance of coatings and sealants under various conditions. These tests help assess a material's performance in real-world environments by simulating specific environmental exposures such as corrosion, UV radiation, humidity, and mechanical stress. Widely recognized testing organizations such as ASTM and ISO have developed standardized methods to ensure consistency and reliability in evaluating material performance. Some of the most commonly used tests include salt spray testing, accelerated weathering tests, adhesion strength testing, and abrasion resistance testing [55–58].

Salt spray testing (often referred to as *corrosion testing*) is among the most common techniques to assess the coating's corrosion resistance, particularly for metal substrates exposed to saline environments, such as in coastal areas or marine applications. Salt spray testing accelerates corrosion by exposing the material to a controlled saline mist, typically using sodium chloride (NaCl) solution. The standard test method is ASTM B117, "Standard Practice for Operating Salt Spray (Fog) Apparatus." In this test, the specimen is placed in a chamber exposed to a fine mist of salt solution under specified temperature and humidity conditions. This simulates the corrosive effects of saltwater exposure over a much shorter period. The degree of corrosion is assessed by examining the appearance of rust, blistering, or delamination of the coating. Salt spray testing is critical for materials used in automotive, aerospace, and marine industries, where corrosion resistance is paramount. However, while salt spray testing can simulate corrosion, it is important to note that it does not always correlate directly with real-world outdoor performance, particularly in non-saline environments [55].

Accelerated weathering tests are used to simulate the long-term effects of outdoor exposure, such as UV radiation, temperature variations, moisture, and ozone exposure, on coatings and sealants. These tests aim to predict how a material will degrade over time when exposed to environmental elements. ASTM G154, "Standard Practice for Operating Fluorescent Light Apparatus for Exposure of Nonmetallic Materials," and ASTM G155, "Standard Practice for Operating Xenon Arc Light Apparatus for Exposure of Nonmetallic Materials," are widely used methods for accelerated weathering testing. These tests use equipment like fluorescent light or xenon arc lamps that mimic the sun's UV radiation and temperature and humidity cycles to accelerate aging. The accelerated exposure helps predict how coatings will perform over months or years of outdoor use. The results can indicate issues like fading, cracking, chalking, and degradation of material properties. These are crucial for selecting coatings for exterior applications like automotive finishes, architectural coatings, and outdoor furniture. The accelerated weathering tests are invaluable in evaluating the resilience of coatings to UV radiation and other environmental factors that cause aging [56].

Adhesion strength testing measures the bond strength between a coating or sealant and the substrate to ensure that it will remain intact and functional over time. The adhesion strength is fundamental in coatings that are exposed to high mechanical stresses or harsh environmental conditions. ASTM D3359, "Standard Test Methods for Measuring Adhesion by Tape Test," is a widely used method to evaluate the adhesion of coatings to substrates. In this test, an adhesive tape is applied to the coated surface and removed quickly to check whether any coating is detached. The adhesion is graded based on the percentage of coating removed. ASTM D4541, "Standard Test Method for Pull-Off Strength of Coatings Using Portable Adhesion Testers," can be used for more precise measurements. This test involves applying a unique adhesive to the coating surface and pulling on it using a calibrated instrument to measure the force required to remove the coating. These tests ensure that coatings will not peel or delaminate under thermal expansion, mechanical wear, or environmental stress [57].

Abrasion resistance testing evaluates a material's ability to resist mechanical wear and tear, which is particularly important for coatings used in high-traffic areas or where frictional forces are common. ASTM D4060, "Standard Test Method for Abrasion Resistance of Organic Coatings by the Taber Abraser," is one of this test's most commonly used methods. The Taber Abraser uses a rotating abrasive wheel to wear down the coating, and the loss of material is measured in terms of weight loss or the number of revolutions required to cause visible damage. This test helps in evaluating the suitability of coatings for flooring, automotive finishes, and other applications where resistance to wear is critical. The results of abrasion tests can provide insights into the expected service life of coatings in real-world use [58].

In addition to these specific tests, other ASTM and ISO standards evaluate additional performance characteristics such as impact resistance, chemical resistance, and thermal stability. For example, ASTM D2794, "Standard Test Method for Resistance of Organic Coatings to the Effects of Rapid Deformation (Impact)," measures how well a coating resists cracking or delamination under impact forces, which is crucial for applications like automotive and industrial equipment. Similarly, ISO 2812 provides guidance on testing the chemical resistance of coatings, determining how well a material resists degradation when exposed to various chemicals [55–58].

Overall, standardized tests like salt spray testing, accelerated weathering, adhesion strength testing, and abrasion resistance testing are integral to evaluating and selecting coatings and sealants for various applications. These tests allow manufacturers and designers to predict materials' long-term performance and durability in specific environments, ensuring that coatings and sealants provide optimal protection and maintain their functional properties over time. Adherence to ASTM and ISO standards provides a consistent framework for these evaluations, ensuring that the performance characteristics are reliable and can be compared across different materials and products.

6.4 Role of Anti-Corrosion Coatings and Sealants in Concrete Corrosion

The literature has written a great deal about anti-corrosion sealants and coatings, highlighting how well they work to reduce concrete corrosion. Studies show that different coatings offer varying degrees of protection against moisture and chemical infiltration. Examples of these coatings are epoxy, polyurethane, and silane-based sealants as shown in Figure 6.4.

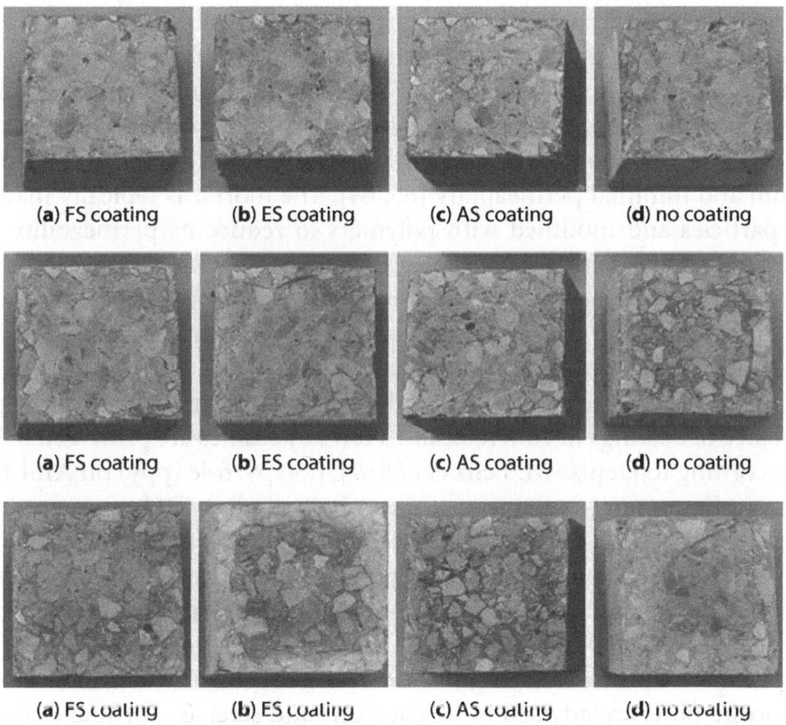

Figure 6.4 Discoloration demarcation on the protective coating of the concrete test block section for (a) 90 days, (b) 180 days, and (c) 270 days by using fluorocarbon coatings, epoxy coatings, and acrylic coatings, respectively [60].

Concrete or steel reinforcement might have a surface treatment called anti-corrosion coating applied to it. Coating the concrete surface is a workable solution for preventing corrosion in reinforced concrete structures [60]. In addition to providing anti-corrosion protection, it has aesthetic benefits and prevents ultraviolet radiation from deteriorating concrete [60]. Generally speaking, the anti-corrosion mechanism of concrete surface coatings works by either lowering the conductivity and corrosion rate of the concrete or preventing the entrance of aggressive substances (such as carbon dioxide, chloride ions, oxygen, and water) [61].

The organic coating can continuously build a film on the surface of concrete and exhibit good chemical resistance against the infiltration of aggressive chemicals. Water within concrete pores can react with silanes, siloxy silanes, and other related compounds to generate a hydrophobic (water-repellent) coating. Since water is the medium *via* which dangerous ions are transported, the organic coating's capacity to repel water is a crucial indicator. Numerous polymer kinds, such as epoxy, polyurethane, chlorinated rubber, and acrylate, are utilized as the organic coating's matrix. The main drawback of organic coating is that it gradually loses its ability to defend against both the high alkalinity of the concrete within and the eroding environment outside [62]. Because inorganic coating lasts longer than organic coating, it is increasingly becoming more popular.

Cement-based inorganic coatings have a millimetric thickness of less than 10 mm and minimal permeability [63, 64]. The mortar is typically made of fine particles and modified with polymers to reduce its permeability and enhance its bond with concrete. High chemical attack resistance geopolymer coatings have been proposed and studied extensively recently. While this study reports on the mechanical strength of cementitious materials, the strengthening effect may be insignificant due to the coating's thinness [65, 66]. Extensive research has been conducted to assess the anticorrosive performance of coatings in concrete, and a few key studies are presented below.

According to Denise M. Lenz *et al.* [67], polypyrrole (ppy) on AISI 1010 steel can be protected against corrosion by including TiO_2 pigment in its matrix. X-ray photoelectron spectroscopy (XPS) was used to examine the pigment's distribution inside the polymeric matrix, and SEM was used to examine the morphology of the composite film. According to them, weight loss and salt spray tests revealed that the anti-corrosion capabilities of the PPy/TiO_2 composite had significantly improved. The aforementioned composite films are advised to be used on mild steel as a primary coating in place of the phosphatized layers.

Studies on the assessment of the effectiveness of organic coatings used for the prevention of steel bar corrosion in concrete structures were conducted by Selvaraj R. *et al.* in 2009 [68]. "Various resins, including acrylic polyol-aromatic isocyanate, polyester polyol-aromatic isocyanate, acrylic resin, and epoxy-silicon-polyamide, with flyash or ordinary Portland cement serving as extenders and titanium dioxide and zinc phosphate serving as the primary pigments on rebar, were subjected to mechanical and accelerated testing to evaluate their performance [68]". Four of the 16 coating compositions that they examined worked well as long-lasting and efficient coatings. Out of all the coatings, the two-pack system, which consists of epoxy, silicone, and polyamide resin combined with flyash and zinc phosphate pigment, fared the best.

In a study published in 2012, Dhoke, S.K. *et al.* investigated the use of nano-Zno pigments (0.1% and 1.0% by wt.) in a water-borne polyurethane dispersion (WPUD) on rebar [69]. The coated mild steel substrate was subjected to humidity, salt spray, and rapid UV weathering. Additionally, mechanical characteristics such pencil hardness, abrasion resistance, and scratch resistance were studied. The relationship between polymer functionalities and nano-ZnO_2 particles is examined using the FTIR method. Also carried out were SEM and AFM. According to their findings, adding nano-ZnO particles to the coatings had a beneficial impact even at lower concentrations (0.1% by weight) as demonstrated by improvements in the coatings' mechanical, UV, and corrosion properties. Additionally, they said that while nano-Zno can function well as a barrier pigment and UV blocker in WPUD, their performance qualities can still be enhanced by improving their concentration and dispersion in the polymer matrix. Ilkhani Ahamed's (2013) article evaluated the corrosion behavior of zinc-rich coatings made of water and zinc dust applied to steel substrates that were cured for 168 h [70].

The evaluation findings, including corrosion potential measurements, electrochemical impedance spectroscopy (EIS), and salt spray testing, showed that adding more nano-SiO_2 to coating formulations shortened the curing period while producing satisfactory corrosion test results. The addition of zinc and SiO_2 elevated the barrier protection stage. Senthil Nathan K.P. *et al.* conducted an analysis of the use of Nito-zinc primer-coated rebar as an anticorrosive agent to prevent corrosion in RC slabs [71]. Epoxy resin: 3:2 (by weight) was used to cast and coat M20 mix R.C.C. slabs, which measured $600 \times 500 \times 50$ mm.

The study observed that the rebar was losing weight and that the coating proved to be 50% more efficient than uncoated rods. Additionally, the bond strength of coated rebar was shown to be nearly identical to that of untreated bar. The modified pullout test used by Xiaofeipei *et al.* (2015) utilized cement as corrosion shielding coatings to evaluate the bond durability and growth length of steel rebars plated with cementitious capillary crystalline as waterproofing material (CCCW) [72]. To examine the influence of curing on the bond, tests were carried out at 7 days and at 90 days following casting on a self-reacting invested T-shaped beam intended to induce stress conditions of flexural structural components. They concluded that the uncoated reinforcing steel development length was 60% shorter for 15-m bars than the calculated design development length and 65% shorter for 20-m bars.

Additionally, the CCCW-coated bars had a lower bond factor and demonstrated a lower bond strength than uncoated bars. Moreover, Bhaumik M. Patel and Hasmukh S. Patel [73] talked about the advancements in the use of renewable resources in the formulation of various coating types. The polyurethane prepolymers were created by adjusting the NCO/OH mole ratio of alkydresin derived from coconut oil and rosinified phenolic resin as polyol containing aromatic and aliphatic diisocyanates. Xylene was employed as the solvent and catalyst, respectively, in place of dibutyltin dilaurate (DBTDL). The pencil hardness, adhesion test, impact resistance, scratch hardness, drying time, flexibility test, and chemical resistance of coated panels were all measured using standard procedures. Compared to PU films made from isophorone diisocyanate (IPDI), the results of scratch hardness, pencil hardness, and chemical resistance are higher in the case of PU films prepared from toluene dissocyanate (TDI).

Every study to date has shown that these coatings greatly limit chloride transport and obstruct the electrochemical reactions that cause rebar corrosion. Furthermore, literature evaluations frequently explore how these materials function in various environmental settings, considering variables like mechanical stress, temperature swings, and UV exposure [74]. Important papers in the field highlight the significance of these protective measures in maintaining concrete infrastructure by offering insights into their long-term effectiveness and durability.

6.5 Conclusion

For concrete structures to remain long-lasting and structurally sound, anti-corrosion sealants and coatings are vital in preventing concrete deterioration. The many kinds of anti-corrosion coatings and sealant systems, as well as their installation procedures and performance indicators, have been highlighted in this chapter, which also offers a thorough understanding of the mechanisms underlying concrete corrosion. The choice of appropriate materials, application methods, and continuous maintenance are all necessary for these protective measures to be effective. Modern, high-performance anti-corrosion technologies are becoming increasingly essential as we deal with more harsh climatic conditions.

Several things will probably influence how concrete's anti-corrosion technologies grow. Research will focus on eco-friendly coatings that offer outstanding protection with low environmental impact. Furthermore, improvements in nanotechnology might result in the development of more resilient coatings with improved barrier and self-healing qualities. It will also be essential to include smart monitoring devices, which allow for real-time evaluations of corrosion activity and coating performance. By facilitating proactive maintenance plans, these technologies can save life-cycle costs and increase the longevity of concrete structures. As the building industry changes, the significance of good anti-corrosion coatings and sealants will remain crucial. Future concrete corrosion concerns will be successfully addressed by long-lasting and sustainable solutions made possible by ongoing research and technical breakthroughs.

AI disclosure statement: The author confirms that no artificial intelligence (AI)-generated content or AI-assisted writing tools were used in the preparation of this chapter (including summaries and the abstract). All text, analysis, and conclusions are the result of the author's original work.

References

1. Mirsayapov, I., Yakupov, S., Hassoun, M., About concrete and reinforced concrete corrosion, in: *IOP Conference Series: Materials Science and Engineering*, vol. 890, IOP Publishing, p. 012061, 2020 Jul 1.
2. Assad, H. and Kumar, A., Understanding functional group effect on corrosion inhibition efficiency of selected organic compounds. *J. Mol. Liq.*, 344, 117755, 2021 Dec 15.
3. Assad, H., Kumar, S., SK, S., Kang, N., Fatma, I., Dahiya, H., PK, S., Thakur, A., Sharma, S., Ganjoo, R., Kumar, A., Evaluating the adsorption and corrosion inhibition capabilities of Pyridinium-P-Toluene Sulphonate on MS in 1 M HCl medium: An experimental and theoretical study. *Inorg. Chem. Commun.*, 153, 110817, 2023 Jul 1.
4. Assad, H., Kumar, S., SK, S., Kang, N., Dahiya, H., Thakur, A., Sharma, S., Ganjoo, R., Kumar, A., A research combined experimental and computational approaches of succinylsulfathiazole hydrate as potent corrosion inhibitor for mild steel in acidic medium. *J. Mol. Liq.*, 388, 122739, 2023 Oct 15.
5. Assad, H., SK, S., Kang, N., Kumar, S., PK, S., Dahiya, H., Thakur, A., Sharma, S., Ganjoo, R., Kumar, A., Electrochemical and computational insights into the

utilization of 2, 2-dithio bisbenzothiazole as a sustainable corrosion inhibitor for mild steel in low pH medium. *Environ. Res.*, 242, 117640, 2024 Feb 1.

6. Assad, H., SK, S., Kang, N., Kumar, S., Dahiya, H., Banerjee, P., Thakur, A., Sharma, S., Ganjoo, R., Kumar, A., Assessment of the inhibitory efficacy of a thiazole derivative as an efficient corrosion inhibitor for augmenting the resistance of MS in acidic environments. *Langmuir*, 40, 31, 16367–83, 2024 Jul 26.
7. Assad, H., Fatma, I., Thakur, A., Kumar, A., 14 Functionalized thin film coatings for reinforced concrete engineering, in: *Corrosion Mitigation Coatings: Functionalized Thin Film Fundamentals and Applications*, vol. 6, p. 347, 2023 Nov.
8. Assad, H., Fatma, I., Kumar, A., An overview of the application of graphene-based materials in anticorrosive coatings. *Mater. Lett.*, 330, 133287, 2023 Jan 1.
9. Assad, H., IA, L., Sihmar, A., Kumar, A., Kumar, A., An overview of contemporary developments and the application of graphene-based materials in anticorrosive coatings. *Environ. Sci. Pollut. Res.*, 24, 1–20, 2023 Nov.
10. Peng, H., Chen, Z., Liu, M., Zhao, Y., Fu, W., Liu, J., Tan, X., Study on the Effect of Additives on the Performance of Cement-Based Composite Anti-Corrosion Coatings for Steel Bars in Prefabricated Construction. *Materials*, 17, 9, 1996, 2024 Apr 25.
11. Assad, H., Sharma, S., Kaya, S., PK, S., Kumar, A., Overview and fundamentals of polymer nanocomposites, in: *Nanocomposites-Advanced Materials for Energy and Environmental Aspects*, pp. 41–66, Woodhead Publishing, 2023 Jan 1.
12. Ganjoo, R., Assad, H., PK, S., Kaya, S., Kumar, A., Fullerenes as anticorrosive coating materials, in: *Smart Anticorrosive Materials*, pp. 91–107, Elsevier, 2023 Jan 1.
13. Dacuan, C.N., VY, A., Canseco, H., Cañete, M.Q., Hydrophobic polyurethane coating against corrosion of reinforced concrete structures exposed to marine environment. *Civ. Eng. Archit.*, 9, 721–36, 2021.
14. Safiuddin, M., Concrete damage in field conditions and protective sealer and coating systems. *Coatings*, 7, 7, 90, 2017 Jun 29.
15. Saraswathy, V. and Karthick, S.P., Effect of ecofriendly sealing coat against corrosion protection of steel rebars in concrete. *Corros. Eng. Sci. Technol.*, 49, 5, 327–34, 2014.
16. Żemajtis, J. and Weyers, R.E., Concrete bridge service life extension using sealers in chloride-laden environments. *Transp. Res. Rec.*, 1561, 1, 1–5, 1996 Jan.
17. Zafar, S., Kahraman, R., Shakoor, R.A., Recent developments and future prospective of polyurethane coatings for corrosion protection– a focused review. *Eur. Polym. J.*, 31, 113421, 2024 Aug.
18. Selvaraj, R., Selvaraj, M., Iyer, S.V., Studies on the Evaluation of the Performance of two Types of Polyurethane Coatings on Steel Rebars for the

Prevention of Corrosion in Concrete Structures. *Port. Electrochim. Acta*, 407, 416, 2008 May 26.

19. Mobin, M., AU, M., Al-Muaili, F., Al-Hajri, M., Performance evaluation of a commercial polyurethane coating in marine environment. *J. Mater. Eng. Perform.*, 21, 1292–9, 2012 Jul.
20. Gaidis, J.M., Chemistry of corrosion inhibitors. *Cem. Concr. Compos.*, 26, 3, 181–9, 2004 Apr 1.
21. Neff, D., Harnisch, J., Beck, M., L'Hostis, V., Goebbels, J., Meinel, D., Morphology of corrosion products of steel in concrete under macro-cell and self-corrosion conditions. *Mater. Corros.*, 62, 9, 861–71, 2011 Sep.
22. Wu, H., Tong, T., Wang, X., Huang, J., Comparative Analysis of Concrete Cracking and Reinforcement Corrosion in Concrete and Ultra-High-Performance Concrete Short Columns after Accelerated Corrosion. *Appl. Sci.*, 14, 17, 7969, 2024 Sep 6.
23. Bertolini, L., Elsener, B., Pedeferri, P., Redaelli, E., Polder, R.B., *Corrosion of steel in concrete: prevention, diagnosis, repair*, John Wiley & Sons, 2013 Feb 26.
24. Pruckner, F. and Gjørv, O.E., Effect of CaCl2 and NaCl additions on concrete corrosivity. *Cem. Concr. Res.*, 34, 7, 1209–17, 2004 Jul 1.
25. Hartt, W.H., Rodney, G.P., Virginie, L., Lysogorski, D.K., *FHWA-HRT-04-093 Final Report*, Florida Atlantic University, USA, 2004.
26. European Standard EN 206-1, Concrete-Part 1: Specification, Performance, Production and Conformity, European Committee for Standardisation, 2000.
27. Shi, X., Fay, L., Yang, Z., TA, N., Liu, Y., Corrosion of deicers to metals in transportation infrastructure: Introduction and recent developments. *Corros. Rev.*, 27, 1-2, 23–52, 2009 Apr.
28. Basheer, L., Kropp, J., Cleland, D.J., *Assessment of the durability of concrete from its permeation properties: a review*, vol. 15, pp. 93–103, Construction and building materials, 2001 Mar 1.
29. Küter, A., Technical University of Denmark, Denmark, 2009.
30. Moreno, M., Morris, W., MG, A., Duffó, G.S., Corrosion of reinforcing steel in simulated concrete pore solutions: Effect of carbonation and chloride content. *Corros. Sci.*, 46, 11, 2681–99, 2004 Nov 1.
31. Tuutti, *Corrosion of Steel in Concrete*, Swedish Foundation for Concrete Research (CBI, Stockholm, 1982.
32. Soudki, K.A., Jeffs, P., Safiuddin, M.D., Concrete Coating Systems on Region of Waterloo Bridges; Final Research Report Submitted to the Region of Waterloo; Department of Civil and Environmental Engineering, Faculty of Engineering, University of Waterloo: Waterloo, ON, Canada, 15 December 2011.
33. Palle, S. and Hopwood, T., Jr., Coatings, Sealants and Fillers to Address Bridge Concrete Deterioration and Aesthetics—Phase 1; Research Report No. KTC06-36/SPR 291-04-1F; Kentucky Transportation Center, College of Engineering, University of Kentucky: Lexington, KY, USA, p. 26, 2006.

34. Wenzlick, J.D., Bridge Deck Concrete Sealers; Final Research Report RI 04-051; Missouri Department of Transportation: Jefferson City, MO, USA, 2007.
35. Attanayaka, A.M.U.B., Duyar, O., Liang, X., Aktan, H.M., Ng, K.Y.S., Fundamentals of use of penetrating sealants for concrete bridge deck protection, in: *Proceedings of the TRB 2003 Annual Meeting*, 12–16 January 2003.
36. Cady, P.D., Sealers for Portland Cement Concrete Highway Facilities; NCHRP Synthesis of Highway Practice 209; Transportation Research Board (TRB), National Research Council: Washington, DC, USA, 1994.
37. Safiuddin, M. and Soudki, K.A., Sealer and coating systems for the protection of concrete bridge structures. *Int. J. Phys. Sciences*, 6, 37, 8188–99, 2011 Dec 31.
38. Thompson, J.L., MR, S., PM, G., Scheetz, B.E., Characterization of silicate sealers on concrete. *Cem. Concr. Res.*, 27, 10, 1561–7, 1997 Oct 1.
39. Nair, P.S. and Gettu, R., Commercially available waterproofing agents in India: A review. *Indian Concr. J.*, 90, 36–53, 2016.
40. Filice, J. and Wong, J., *Best Practice Guidelines for Selecting Concrete Bridge Deck Sealers*, Alberta Transportation and Utilities, Edmonton, AB, Canada, 2001.
41. Ho, D.W. and Harrison, R.S., Influence of surface coatings on carbonation of concrete. *J. Mater. Civ. Eng.*, 2, 1, 35–44, 1990 Feb.
42. https://blog.austinhardware.com/guidelines-for-selecting-paints-coatings-sealants-and-adhesives-in-manufacturing-processes.
43. Litvan, G.G., The effect of sealers on the freeze-thaw resistance of mortar. Cement and concrete research, 1141–7, 1992 Nov 1.
44. Porter, L.C., *Some surface treatments increase concrete durability*, vol. 47, pp. 211–32, Special Publication, 1975 Jan 1.
45. Almusallam, A.A., FM, K., SU, D., Al-Amoudi, O.S., Effectiveness of surface coatings in improving concrete durability. *Cem. Concr. Compos.*, 25, 4-5, 473–81, 2003 May 1.
46. Moon, H.Y., DG, S., Choi, D.S., Evaluation of the durability of mortar and concrete applied with inorganic coating material and surface treatment system. *Constr. Build. Mater.*, 21, 2, 362–9, 2007 Feb 1.
47. Seneviratne, A.M., Sergi, G., Page, C.L., Performance characteristics of surface coatings applied to concrete for control of reinforcement corrosion. *Constr. Build. Mater.*, 14, 1, 55–9, 2000 Feb 28.
48. Al-Dulaijan, S.U., Maslehuddin, M., Al-Zahrani, M.M., Al-Juraifani, E.A., Al-Idi, S.H., Al-Mehthel, M., Performance evaluation of cement-based surface coatings. In Repair, Rehabilitation, and Maintenance of Concrete Structures, and Innovations in Design and Construction. *Proceedings of the ACI Fourth International Conference (ACI SP-193), Seoul, Korea, 19–22 September 2000*, American Concrete Institute, Detroit, MI, USA, pp. 321–335, 2000.
49. Ibrahim, M., AS, A.-G., Maslehuddin, M., Dakhil, F.H., Use of surface treatment materials to improve concrete durability. *J. Mater. Civ. Eng.*, 11, 1, 36–40, 1999 Feb.

50. Oshiro, T. and Tanigawa, S., *Effect of surface coatings on the durability of concrete exposed to marine environment*, vol. 109, pp. 179–98, Special Publication, 1988 Aug 1.
51. McCarter, W.J., Assessing the protective qualities of treated and untreated concrete surfaces under cyclic wetting and drying. *Build. Environ.*, 31, 6, 551–6, 1996 Nov 1.
52. Soudki, K.A., Jeffs, P., Safiuddin, M., Chloride penetration resistance of concrete coating systems on region of waterloo bridges, in: *Final Research Report. Department of Civil and Environmental Engineering*, Faculty of Engineering, University of Waterloo, Waterloo, Ontario, Canada, 2012.
53. Garbacz, A., Courard, L., Bissonnette, B., pp. 73–84, Bulletin of the Polish Academy of Sciences Technical Sciences, 2013.
54. Safiuddin, M. and Soudki, K.A., Water vapor transmission and waterproofing performance of concrete sealer and coating systems. *J. Civ. Eng. Manage.*, 21, 7, 837–44, 2015 Oct 3.
55. Haynes, G., Review of laboratory corrosion tests and standards, in: *Corrosion Testing and Evaluation: Silver Anniversary*, ASTM International, 1990 Jan.
56. Klemann, B.M. and Landrum, P.C., ISO 18930–A New Standard Test Method for Accelerated Weathering, in: *NIP & Digital Fabrication Conference*, vol. 30, Society for Imaging Science and Technology, pp. 73–78, 2014 Jan 1.
57. Francis, A., Tobin, B., Fowler, S., Accelerated Weathering Test Standards for Sealants and Adhesives: Replicating Aqueous Outdoor Environments for Improved Service Life Prediction, in: *Sixth Symposium on the Durability of Building and Construction Sealants and Adhesives*, ASTM International, pp. 94–104, 2018 Aug 1.
58. Jiang, H., Browning, R., Liu, P., TA, C., Sue, H.J., Determination of epoxy coating wet-adhesive strength using a standardized ASTM/ISO scratch test. *J. Coat. Technol. Res.*, 8, 255–63, 2011 Mar.
59 Fan, T., Wu, Y., Yang, M., Xu, P., Li, Y., Wang, L., Chen, H., Attenuation Law of Performance of Concrete Anti-Corrosion Coating under Long-Term Salt Corrosion. *Coatings*, 14, 10, 1249, 2024 Sep 30.
60. Broomfield, J.P., *Corrosion of steel in concrete: understanding, investigation and repair*, Crc Press, 2023 Jan 31.
61. Brenna, A., Bolzoni, F., Beretta, S., Ormellese, M., Long-term chloride-induced corrosion monitoring of reinforced concrete coated with commercial polymer-modified mortar and polymeric coatings. *Constr. Build. Mater.*, 48, 734–44, 2013 Nov 1.
62. Zafeiropoulou, T., Rakanta, E., Batis, G., Performance evaluation of organic coatings against corrosion in reinforced cement mortars. *Prog. Org. Coat.*, 72, 1-2, 175–80, 2011 Sep 1.
63. Al-Zahrani, M.M., SU, A.-D., Ibrahim, M., Saricimen, H., Sharif, F.M., Effect of waterproofing coatings on steel reinforcement corrosion and physical properties of concrete. *Cem. Concr. Compos.*, 24, 1, 127–37, 2002 Feb 1.

64. Goyal, A., HS, P., Ganjian, E., Claisse, P., A review of corrosion and protection of steel in concrete. *Arabian J. Sci. Engineering*, 43, 5035–55, 2018.
65. Aguirre-Guerrero, A.M., RA, R.-S., de Gutiérrez, R.M., A novel geopolymer application: Coatings to protect reinforced concrete against corrosion. *Appl. Clay Sci.*, 135, 437–46, 2017 Jan 1.
66. Zhang, Z., Yao, X., Zhu, H., Potential application of geopolymers as protection coatings for marine concrete: II. Microstructure and anticorrosion mechanism. *Appl. Clay Sci.*, 49, 1-2, 7–12, 2010 Jun 1.
67. Lenz, D.M., Delamar, M., Ferreira, C.A., Application of polypyrrole/TiO2 composite films as corrosion protection of mild steel. *J. Electroanal. Chem.*, 540, 35–44, 2003 Jan 2.
68. Selvaraj, R., Selvaraj, M., Iyer, S.V., Studies on the evaluation of the performance of organic coatings used for the prevention of corrosion of steel rebars in concrete structures. *Prog. Org. Coat.*, 64, 4, 454–9, 2009 Mar 1.
69. Dhoke, S.K., *et al.*, Effect of Nano-Zinc oxide Particles on the performance behaviour of water borne polyurethane composite coatings. *Int. J. Material Sci.*, 2, 47–55, 2012.
70. Ilkhani, A., Corrosion protection of carbon steel by using Zinc-Rich inorganic water-based silicate coating comprising different amounts of nano-silica. *Int. J. Chem.*, 75–82, 2013.
71. Senthil Nathan, K.P., *et al.*, Experimental study on Corrosion Prevention in R.C. slabs. *Int. J. Eng. Res. Technol.*, 3, 1008–1013, 2014.
72. Pei, X., Noël, M., Fam, A., Green, M., Development length of steel reinforcement with corrosion protection cementitious coatings. *Cem. Concr. Compos.*, 60, 34–43, 2015 Jul 1.
73. Patel, B.M. and Patel, H.S., Polyurethane surface coatings derived from (rosinified phenolic resin-coconut alkyd resin) blends with aliphatic and aromatic diisocyanates. *Int. J. Chem. Stud.*, 3, 116–9, 2015.
74. Dai, J.G., Akira, Y., FH, W., Yokota, H., Zhang, P., Water repellent surface impregnation for extension of service life of reinforced concrete structures in marine environments: The role of cracks. *Cem. Concr. Compos.*, 32, 2, 101–9, 2010 Feb 1.

7

Role of Cathodic Protection Systems in Preventing Concrete Corrosion

Samir H. Shafek[1], Ahmed H. Elged[1], Mohamed A. Shenashen[2], Emad A. Badr[1] and Hassan H. H. Hefni[2*]

[1]*Surfactant Lab, Petrochemicals Department, Egyptian Petroleum Research Institute (EPRI), Cairo, Egypt*
[2]*Polymers Lab, Petrochemicals Department, Egyptian Petroleum Research Institute (EPRI), Cairo, Egypt*

Abstract

Reinforced concrete structures play an important role in global infrastructure networks, and their firmness and serviceability depend on the reinforcing steel's quality. Cathodic protection (CP) is a practical approach for stopping or preventing corrosion caused by chloride in steel reinforced concrete structures. CP is split into sacrificial anode cathodic protection (SACP) and impressed current cathodic protection (ICCP). SACP consists of an extrinsic sacrificial anode, a metal that precedes iron in the electrochemical series, electrically coupled to another metal that serves as the anode. At the same time, ICCP includes passing an electrical current from outside the anode across the concrete to the rebar. The current induces steel polarization, electrochemical processes, and ion transport. Generally, the anode is spread across rather extensive surface regions, even where the steel is inert. Corrosion area varies according to anode size. Among the benefits is the rise in pH at the steel's exterior due to oxygen reduction. As the pH rises, the corrosion rate and current demand fall. Electrochemical studies use chloride-free and chloride-contaminated pore mixtures under completely hydrated and deaerated conditions. Linear and dynamic polarization and electrochemical impedance spectroscopy tests are used to evaluate (CP). This chapter aims to comprehend the latest studies on the cathodic protection of steel in building materials, which can be valuable for designers working on bridge construction, bridge supervisors, mechanical repairs, healing, and protection. This chapter covers both CP

**Corresponding author*: hassanhefni@epri.sci.eg; hassanhefni@yahoo.com

Ruby Aslam, Qihui Wang, Zhitao Yan and Afroz Aslam (eds.) Corrosion in Concrete Structures: Integrating Advanced Technologies and Sustainable Practices, (187–212) © 2026 Scrivener Publishing LLC

techniques (ICCP and SACP), anode resources, forecasting anode lifespans and evaluating CP execution, observing approaches, thermally coated zinc anode assembly and substitution, and current advances in CP innovations.

***Keywords*:** Corrosion of reinforcement concrete, cathodic protection, sacrificial anode cathodic protection (CP), sacrificial anode cathodic protection (SACP), impressed current cathodic protection (ICCP)

7.1 Introduction

Steel plays a crucial part in modern infrastructure constructions worldwide. The steel sheet in concrete could preserve form dissolution by forming a passive layer on the outer substrate. Also, the concrete protects the metal by acting as a coating layer. Two methods can degrade concrete's environmental protection and the rods' corresponding passivity impact. The first way is lowering the alkalinity by migrating alkaline through water or reacting with carbon dioxide. The second way is to dissolve the protecting layer on the steel through electrochemical reaction with chloride and oxygen. Chloride can produce rebar corrosion after critical concentration on the rebar surface. The corrosion process generates rusting, which may consume more volume of the underlying steel, leading to strain challenges and fractures in concrete, facilitating the entry of dangerous species into the embedded steel. The degree of corrosion directly impacts the remaining service life of a concrete structure, causing structural deformity and early structural failure [1–4].

Cathodic protection (CP) is an electrochemical method used to prevent and dissolve steel in reinforced concrete [5, 6]. The broad utilization of CP was documented; specific advances in the area have been described previously. Fortunately, using CP has been accompanied by adverse effects [e.g., alkali-silica reaction (ASR), bond strength deterioration], etc. As a result, any advancements to the CP method that reduce the range of side effects are a key feature in corrosion control and corrosion protection of reinforced concrete structures.

Cathodic protection works by transferring a tiny electrical current from the anode to the corroded rebar, thus shielding it from additional corrosion by increasing the hydroxyl ions nearby. At Mouchel, cathodic safeguarding is primarily employed in concrete restorations to recover damaged roadway concrete structures with varying chloride contaminants, corroding, and breakage levels. Nevertheless, cathodic protection is frequently used on reinforced concrete structures, including parking lots, tunnels, ports and harbor services (jetties/wharves), commercial and residential developments, and ships. Cathodic protection systems are typically classified into impressed current cathodic protection (ICCP) and sacrificial anode cathodic protection (SACP). The choice of which of the two cathodic protection (CP) systems to use is frequently impacted by a variety of criteria, which include but are not restricted to the current state of the construction, the financial capacity of the customer, and probable life anticipation of the building that follows the repair process [7].

Both CP systems are suitable for mending RC structures, but the ICCP approach provides extra versatility and permanence [8]. Because of the current output, this may be modified to give a vital current in circumstances of high resistivity of concrete constructions. Choosing anode equipment is also crucial, as it helps the system's duration by removing damaging products from the rebar to the placed anode. Fortunately, using ICCP requires a significant amount of energy from the steady direct current (DC). Unlike the SACP method, this depends on the mean current drawn from anodes with greater electrode potentials than iron. The ICCP technique's anode components comprise conducting coatings, titanium-based meshes in cementitious covering, conducting overlaying with carbon fibers, flame-sprayed zinc, and discrete anode systems [7]. However, the current anodes utilized have not been demonstrated to be efficient, and the choice of the anode components depends mainly on the duration of the building types of corrosion and building, anode placing operations, processes, a desire for periodic repairs, and the overall expense of the CP framework [9]. Zinc is frequently employed as a reactive protecting layer; hence, it is less noble compared to the bulk of metals [9] and can preserve the reinforcement from corrosion. When steel in concrete corrodes, the binding force with the steel inserted into it falls as it loses its quality.

The ambient temperature affects the cathodic protection potential setting. As a result, the appropriate cathodic protection potential must be determined. On the reverse side, the pores' design of concrete provides a protective impact on the CP. As a result, this work examines the best cathodic protection potential at different temperatures *via* electrochemical and submerged tests. It explores the electrochemical reaction that happens to the alloy in the passageway when the metal in the most significant crack attains the best cathodic protection potential [10].

7.2 Cathodic Protection

Cathodic protection (CP) was first applied to concrete structures in 1973 when Stratfull utilized it for bridge decks contaminated by de-icing salts in

California. Since then, CP has become a widely adopted technique, especially in scenarios where traditional rehabilitation methods are costly or ineffective, particularly in high chloride environments. In the 1980s, titanium mesh anodes coated with noble metal oxides (like ruthenium and iridium) were developed, allowing the technique to be used not only on bridge decks but also on substructures, marine structures, parking garages, and various buildings [11].

In the late 1990s, sacrificial cathodic protection (SCP) emerged as an alternative to impressed current systems. SCP uses zinc or aluminum to provide security through their natural dissolution, eliminating the need for an external power source. Millions of square meters of concrete have since been treated with impressed current and sacrificial systems.

Initially, CP was not recommended for pre-stressed concrete structures due to concerns about hydrogen embrittlement in high-strength steel. However, advancements in European standards have led to safe applications with stringent monitoring protocols. More recently, CP has been applied to underground and submerged structures, often utilizing sacrificial anodes, and these developments are now included in international standards [12, 13].

Additionally, CP has been shown to repassivate steel in carbonated concrete, a method referred to as "continuous realkalization." Although more economical alternatives are usually preferred, CP has been used in specific cases where other methods failed to provide lasting results, such as in Italy and the Netherlands during the 1960s and 1970s [13, 14].

7.3 Cathodic Protection (CP) Monitoring Techniques in Concrete Structures

Cathodic protection (CP) monitoring techniques in concrete structures, particularly reinforced concrete, are crucial to mitigate the risk of corrosion in embedded steel reinforcements. Concrete is often a barrier against corrosion, but factors like chloride ion ingress (from saltwater exposure or de-icing salts) and carbonation can compromise this protection.

7.3.1 Embedded Reference Electrodes

Embedded reference electrodes play a critical role in assessing the effectiveness of cathodic protection (CP) systems within concrete structures by monitoring the potential of reinforcing steel relative to a stable reference point. These electrodes are placed near the steel reinforcement and continuously measure the electrical potential of the steel, offering real-time data on the system's performance. The potential difference between the steel and the electrode can be monitored by embedding the electrodes into the concrete, which helps detect early signs of corrosion [15]. The steel is adequately protected if the measured potential exceeds a specific threshold. This continuous monitoring technique enables timely adjustments to the CP system, minimizing corrosion risks and ensuring long-term maintenance and durability. Particularly in aggressive environments, such as marine or chloride-rich conditions, this method is essential for evaluating the performance of CP systems in real time and tracking corrosion risks to safeguard the structural integrity of concrete [16].

7.3.2 Linear Polarization Resistance (LPR)

Linear polarization resistance (LPR) is a highly effective technique used to measure the corrosion rate of steel reinforcement in concrete structures, playing a critical role in evaluating the performance of cathodic protection (CP) systems. This method involves applying a small perturbation to the reinforcement's potential and recording the resulting current [17]. The ratio of the applied voltage change to the current response yields the polarization resistance, which is inversely proportional to the corrosion rate. LPR can quantify the corrosion rate, offering more direct insights into the ongoing corrosion process than straightforward potential measurements. This technique is particularly valuable in assessing the effectiveness of CP in reducing the corrosion rate, ensuring that the corrosion remains below acceptable levels. One of the strengths of LPR is its ability to provide real-time, quantitative data on the CP system's performance, allowing for precise adjustments to maintain the structure's integrity. Unlike other monitoring techniques, LPR gives detailed feedback on the active corrosion mechanisms, making it indispensable for long-term durability assessments and maintenance of concrete structures exposed to harsh environments [18].

7.3.3 Concrete Resistivity Monitoring

Concrete resistivity plays a crucial role in the effectiveness of cathodic protection (CP) systems, as it directly influences the ability of protective currents to reach steel reinforcement within concrete structures. Concrete resistivity monitoring is essential for evaluating whether the material is conductive enough for the CP system to function correctly. Devices like resistivity meters are used to measure the electrical resistance of the concrete, providing insights into its ability to support the protective currents,

Table 7.1 Empirical relationship between resistivity and corrosion rate [19].

Resistivity (Ωm)	Corrosion rate
<100	High
100–500	Moderate
500–1,000	Low
>1,000	Negligible

as shown in Table 7.1. This monitoring helps identify zones within the concrete that may have become carbonated or contaminated with chlorides, both of which can increase resistivity and, in turn, reduce the efficiency of CP systems [16, 19]. For instance, chloride contamination can set up macrocell corrosion, while carbonation can lead to uniform corrosion, which is influenced by concrete resistivity. Regular resistivity measurements allow timely adjustments to the CP system to ensure its effectiveness, particularly in environments with high moisture or chloride exposure. When concrete resistivity becomes too high, the protective current may not penetrate sufficiently to protect the steel, leading to localized areas of corrosion [20]. Thus, understanding resistivity fluctuations due to environmental factors like moisture content is critical for maintaining optimal CP performance. This monitoring technique tracks the concrete's condition and predicts changes in corrosion rates, which is essential for long-term durability and maintenance planning [21].

7.3.4 Galvanostatic Pulse Measurements

Galvanostatic pulse measurements are an essential technique to assess the real-time performance of cathodic protection (CP) systems in reinforced concrete structures. This method involves applying a small, controlled current to the steel reinforcement and measuring the response to evaluate the polarization of the steel. By analyzing the rate of depolarization, engineers can gauge the effectiveness of the CP system. A slow depolarization indicates that the steel remains well-protected, while a fast depolarization suggests that the steel is more susceptible to corrosion. This technique provides valuable insights into the current performance of the CP system and the potential corrosion risk if the system is turned off. When CP is temporarily deactivated, the galvanostatic pulse method allows for monitoring how quickly the steel reverts to a corrosion-prone state, offering critical data on the long-term stability of the reinforcement. The ability to assess the immediate protection and future risks makes this technique particularly useful for maintaining the durability of concrete structures exposed to harsh environments. Additionally, this method helps optimize CP system performance by providing real-time feedback, allowing for timely adjustments to ensure the reinforcement remains safeguarded against corrosion. It is especially important in environments with fluctuating conditions, such as marine or industrial settings, where constant monitoring and adjustments are necessary to protect the steel from corrosion [22] effectively.

7.3.5 Wireless and Remote Monitoring Systems

Recent technological advancements have revolutionized cathodic protection (CP) monitoring systems in concrete structures, mainly by developing wireless and remote monitoring solutions. These systems use embedded sensors to collect data on key parameters such as potential, current density, resistivity, and even temperature, transmitting it remotely to a central database for real-time analysis. This enables more efficient, continuous monitoring and adjustment of CP systems, especially in large or difficult-to-access structures like bridges, tunnels, and marine platforms. Wireless systems eliminate the need for frequent manual inspections, reducing labor costs and allowing quicker responses to any issues. The real-time feedback provided by these systems ensures that the performance of the CP system can be optimized continuously, helping to extend the lifespan of critical infrastructure.

In addition to monitoring CP performance, tracking temperature is essential for understanding processes like concrete hardening, carbonation, and the degradation of embedded steel. Different sensor technologies, such as micro-electrochemical systems (MEMS), fiber optic sensors, infrared imaging, RFID sensors, and thermocouples, are used to measure temperature variations within the concrete. Temperature monitoring is crucial because extreme or fluctuating temperatures can compromise concrete strength and lead to cracking, accelerating steel corrosion. For example, the heat generated during hydration can be distributed unevenly, increasing the risk of structural failure. Sensors are often placed near the steel reinforcement to capture accurate temperature data. Environmental factors like humidity also affect these readings, making continuous monitoring vital. While wireless temperature sensors have been introduced to address these needs, challenges remain regarding signal stability and sensor durability in harsh environmental conditions. Despite these obstacles, integrating wireless monitoring for both CP performance and temperature enhances reinforced concrete structures' long-term maintenance and durability.

These monitoring techniques provide comprehensive insights into corrosion protection in reinforced concrete structures, which is crucial in ensuring long-term durability while minimizing maintenance costs. By employing methods such as embedded reference electrodes, linear polarization resistance (LPR), concrete resistivity monitoring, and galvanostatic pulse measurements, engineers can effectively assess the ongoing performance of cathodic protection (CP) systems in real time. Each technique offers a unique perspective on different aspects of corrosion prevention, from measuring potential differences and corrosion rates to evaluating concrete resistivity and the overall health of the steel reinforcement. This multifaceted approach allows for the early detection of issues and timely adjustments, reducing the likelihood of costly repairs and extending the life of the infrastructure. Collectively, these monitoring techniques serve as a robust defense against the progressive risk of steel corrosion, ensuring the longevity and effectiveness of CP systems in concrete. The integration of advanced technologies, such as wireless and remote monitoring systems, has further enhanced the efficiency of these processes, allowing for continuous data collection and real-time analysis even in large or inaccessible structures. This improves the ability to respond quickly to emerging corrosion risks and reduces the need for labor-intensive manual inspections. By maintaining optimal performance of CP systems, these monitoring techniques contribute to safeguarding critical infrastructure in aggressive environments, such as marine and industrial settings, where exposure to chlorides, moisture, and other corrosive elements is constant. The result is improved structural resilience, reduced maintenance costs, and extended service life for concrete structures, ensuring their functionality and safety for years [23].

7.4 Anode Installation and Replacement

Anodes are crucial components in cathodic protection (CP) systems for reinforced concrete structures, as they provide the necessary current to protect the steel reinforcement from corrosion. Installing and replacing anodes requires careful planning and execution to ensure long-term effectiveness.

7.4.1 Anode Installation

7.4.1.1 Site Assessment and Design

Before installation, a thorough site assessment is essential to ensure the long-term effectiveness of the cathodic protection system. Factors such as the concrete's condition, environmental exposure, soil resistivity, moisture levels, and the existing level of corrosion must all be carefully evaluated. In addition, the location of embedded steel reinforcement, chloride content, and the presence of other contaminants play a significant role in determining the protection requirements. Design considerations include selecting the appropriate type and configuration of anodes (sacrificial or impressed current) most suitable for the specific application, ensuring efficiency and durability.

7.4.1.2 Preparation of the Concrete Surface

Proper surface preparation is critical for successful anode installation, as it directly impacts the effectiveness of the cathodic protection system. This may involve meticulously cleaning the concrete surface to remove contaminants, such as dust, oil, grease, or existing coatings, which could hinder adhesion and electrical contact. Techniques such as sandblasting, high-pressure water cleaning, or chemical cleaning agents are often employed to ensure a clean substrate. Additionally, the surface may need to be roughened to enhance mechanical interlocking between the anode and the concrete. Any surface defects, such as cracks or spalls, should be repaired before anode installation to promote optimal performance.

7.4.1.3 Anode Placement

Anodes should be installed as close as possible to the areas with the highest corrosion activity, typically where the concrete is most damaged or chloride concentration is highest. Properly identifying these critical zones is essential for maximizing protection. The anodes can be embedded directly in the concrete for impressed current systems, while sacrificial anodes can be attached to the surface or embedded within it. Ensuring the placement allows for even current distribution across the reinforcement is vital to avoid localized corrosion. Additionally, the orientation and spacing of the anodes should be carefully considered to facilitate effective electrochemical reactions and enhance the overall efficiency of the cathodic protection system. Proper documentation of the anode placement is also necessary for future monitoring and maintenance [24].

7.4.1.4 Electrical Connections

Secure and reliable electrical connections are essential for adequate cathodic protection (CP). Connections should be made using corrosion-resistant materials compatible with the environment to ensure longevity. Additionally, these connections must be insulated to prevent moisture ingress, which can lead to deterioration over time. Proper insulation techniques should be employed to maintain system integrity. Electrical testing should be performed to confirm appropriate connections before covering the anodes, including checks for continuity and resistance. This testing phase is critical, as it identifies potential issues early on, ensuring optimal performance and reliability of the entire cathodic protection system in the long term [24].

7.4.2 Anode Replacement

7.4.2.1 Monitoring and Evaluation

Anodes do not last indefinitely, and their performance must be regularly monitored. This involves measuring the potential and current output to assess their condition. When a significant drop in performance is noted or if the anodes are nearing the end of their expected life, replacement should be planned [25, 26].

7.4.2.2 Removal of Old Anodes

The removal process requires careful handling to minimize damage to the surrounding concrete. This may involve chiseling or cutting away the concrete to access the anodes. The old anodes should be disposed of according to environmental regulations, especially if they contain hazardous materials [26, 27].

7.4.2.3 Installation of New Anodes

Once the old anodes are removed, the new anodes should be installed following the same procedures as the initial installation. Proper alignment, electrical connections, and surface preparation are critical to ensure the longevity and effectiveness of the new anodes [27, 28].

7.4.2.4 Post-Installation Testing

After installation, the new anodes should undergo thorough testing to ensure they are functioning correctly. This may include checking the electrical connections, current output, and overall effectiveness of the CP system. Continuous monitoring is recommended to ensure that the system remains operational and effective over time [28].

7.5 Anode Materials

Anode materials of cathodic protection in reinforcement concrete as has been mentioned above is classified to two strategies as discussed below.

7.5.1 Sacrificial Anode Cathodic Protection (SACP)

A sacrificial anode cathodic protection (SACP) system includes an external sacrificial anode, typically made of a metal with a higher electrode potential than steel reinforcement. This anode is electrically connected to the steel, causing it to act as the anode [8], as illustrated in Figure 7.1.

A comparative experiment was conducted by Wang *et al.* [29] to evaluate the cathodic protection performance of zinc alloy, magnesium alloy, and aluminum alloy as sacrificial anodes in chloride-contaminated reinforced concrete, as can be seen in Figure 7.2. The findings reveal that the aluminum alloy anode fails to deliver adequate cathodic protection over the long term due to its loss of activity. Conversely, both zinc and magnesium alloys exhibit strong electrochemical activity and stability as sacrificial anodes. Among them, the magnesium alloy anode demonstrates superior cathodic protection performance for reinforcing steel compared to the zinc alloy anode.

A market study by Krishnan *et al.* [30] indicates that a patch repair (PR) strategy without the use of galvanic anodes often results in continued corrosion due to factors such as the halo effect and residual chlorides, necessitating another significant repair within approximately 5 years. Repeated patch repairs can further accelerate corrosion, leading to structural replacement

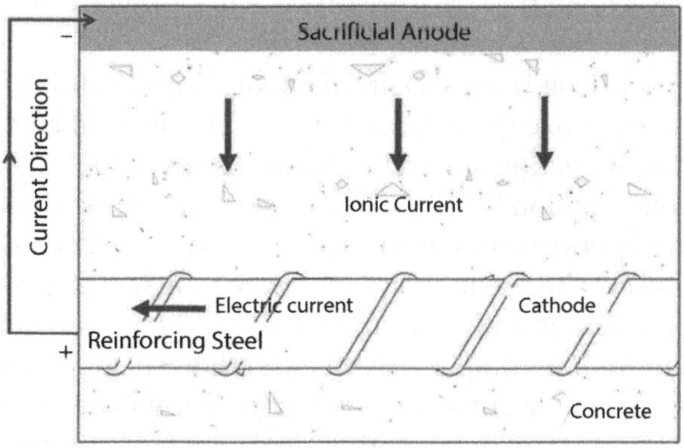

Figure 7.1 Circuit diagram for the SACP system [8].

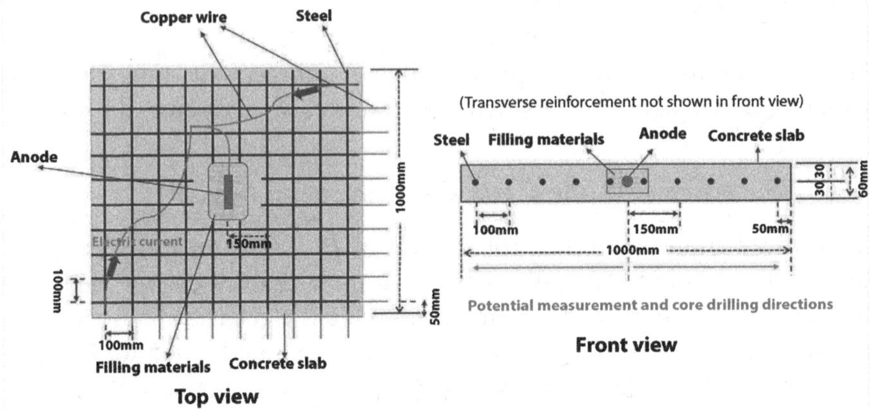

Figure 7.2 Schematic illustration of the experimental setups [29].

and substantial life cycle costs (LCC). In contrast, employing cathodic protection (CP) with galvanic anodes has significantly extended service life and reduced LCC. Long-term data demonstrate that galvanic anodes effectively controlled chloride-induced corrosion for up to 14 and 10 years.

Moreover, the additional cost of incorporating galvanic anodes represents only about 4% of the total repair cost for a jetty structure, challenging the misconception that CP involves prohibitively high capital expenses. A framework developed to estimate the LCC of PR and CP strategies revealed that CP and cathodic prevention (CPrev) approaches are significantly more cost-effective than PR. An analysis of 30 repair projects confirmed that adopting CP can result in LCC savings of up to 90% over 30 years following the initial repair. Furthermore, CP and CPrev strategies can indefinitely extend the structure's service life, provided that anodes are replaced periodically at a minimal cost.

According to Cheng *et al.*, sacrificial anode cathodic protection (SACP) systems aim to optimize the balance between efficacy and efficiency by (1) maximizing the reduction of steel corrosion current and (2) minimizing anode mass consumption [31]. The findings indicate that halving the anode spacing improves corrosion current suppression more than doubling the anode size. However, the reduction in corrosion current is accompanied by a substantial increase in cathodic current, resulting in over 50% additional anode consumption. A comprehensive evaluation suggests that in a SACP system, when the anode corrosion current is maintained at 5.0–7.5 times the steel corrosion current, an optimal balance between system efficacy and efficiency can be achieved.

7.5.2 Impressed Current Cathodic Protection (ICCP)

In the case of impressed current cathodic protection (ICCP), an external power source is utilized to deliver a controlled current to the reinforcing steel, counteracting the metal ion flow that causes corrosion. Typically, a platinum-based anode is employed for its efficiency in reducing the corrosion rate, as illustrated in Figure 7.3. The external anode supplies electrons to the steel reinforcement within reinforced concrete (RC) structures, ensuring that the steel remains cathodic and preventing further corrosion [8]. A real system typically comprises a DC power supply, wiring, and monitoring equipment in practical applications.

The study of Carmona *et al.* [32] outlines the use of cathodic protection (CP) and cathodic prevention (CPrev), sometimes preceded by electrochemical chloride extraction (ECE), on representative reinforced concrete specimens. The anodic system employs a graphite–cement paste as a surface coating. The results demonstrate the anode's efficiency and ability to facilitate a combined ECE process followed by CP application.

The carbon nanotube-carbon fiber (CNT-CF)/cement-based composite served as a novel auxiliary anode [33] in the form of overlay for the cathodic protection of reinforced concrete [27] was proposed by Li *et al.* It showed that the resistivity of CNT-CF/cement-based composite monotonously decreased with increasing content of CNT up to 2 wt.%. When the CNT content was 0.5 wt.%, the resistivity of the composite reached the minimum value of 22.3 Ω·cm. The resistivity for all the groups decreased

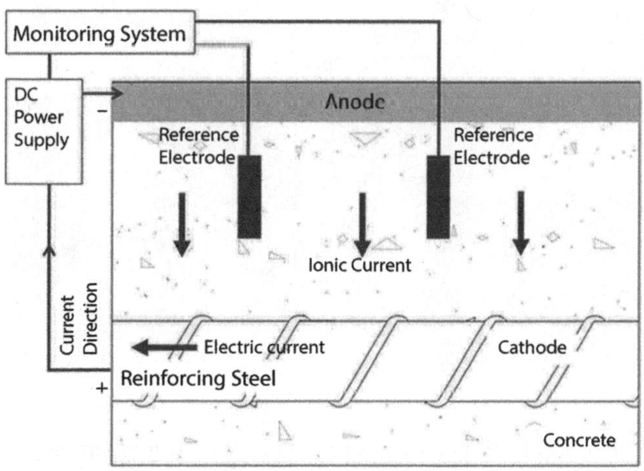

Figure 7.3 Circuit diagram for the ICCP system [8].

with increasing age in the first 7 days. Both the experimental and simulated results showed that the CNT-CF/cement-based auxiliary anode improves the cathodic protection effectiveness for RC.

Anwar *et al.* conducted an evaluation study using lightweight conductive cementitious mortar as an anode material for impressed current cathodic protection (ICCP) in concrete structures. The research highlighted that incorporating lightweight aggregates, such as pumice, into the mortar significantly improved its electrical conductivity, making it a

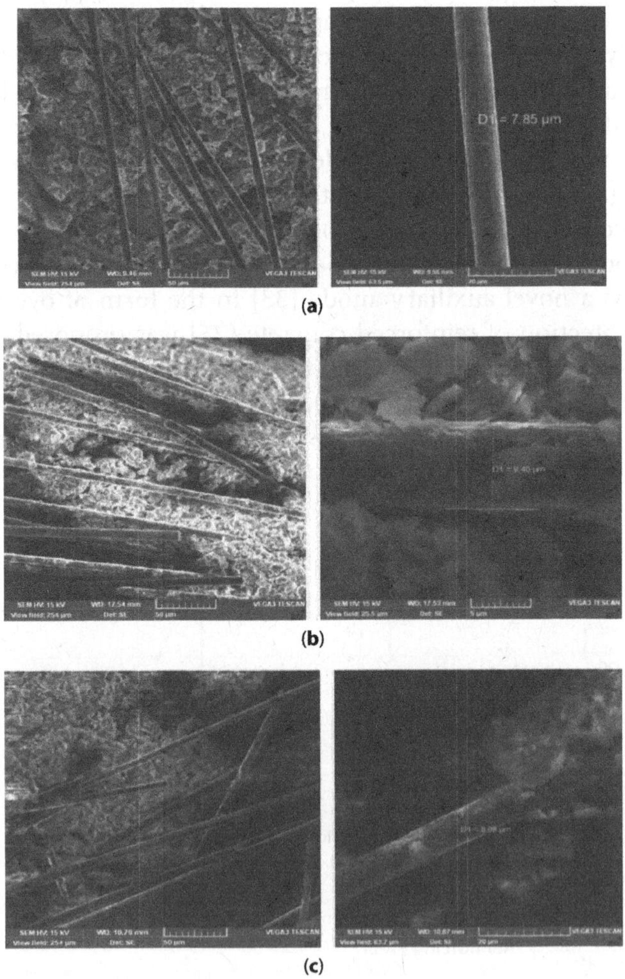

Figure 7.4 SEM image of 10% pumice mortar with 1.1% CF before galvanostatic test (a), after 14 days of galvanostatic test in SPS (b), and after 14 days of galvanostatic test in SPS + 3% NaCl (c), with magnification of 500_ (left) and 2,000_ (right), respectively [34].

viable option for ICCP applications. This method presents a promising approach to enhancing the durability and service life of reinforced concrete structures in corrosive environments [34]. The anode consists of a mixture incorporating fine pumice aggregate, carbon fibers, and cement, with MMO-Ti serving as the primary anode, as illustrated in the SEM images in Figure 7.4. The findings indicate that replacing 20%–30% of the pumice aggregate and including 1.1% carbon fiber in the mortar represent the percolation threshold, achieving a maximum conductivity of 0.2 S/cm. Additionally, the anode's weight is reduced by approximately 50% compared to traditional anodes. Electrochemical analysis confirmed the suitability of this material as an anode for cathodic protection in reinforced concrete structures.

Zuo *et al.* investigated the use of carbon nanotube-carbon fiber/cement-based composites as auxiliary anodes for the cathodic protection of reinforced concrete [35]. The study found that cement composites containing 0.4 wt.% carbon fiber and 0.5 wt.% carbon nanotubes exhibited the best combination of electrical and mechanical properties. Compared to a

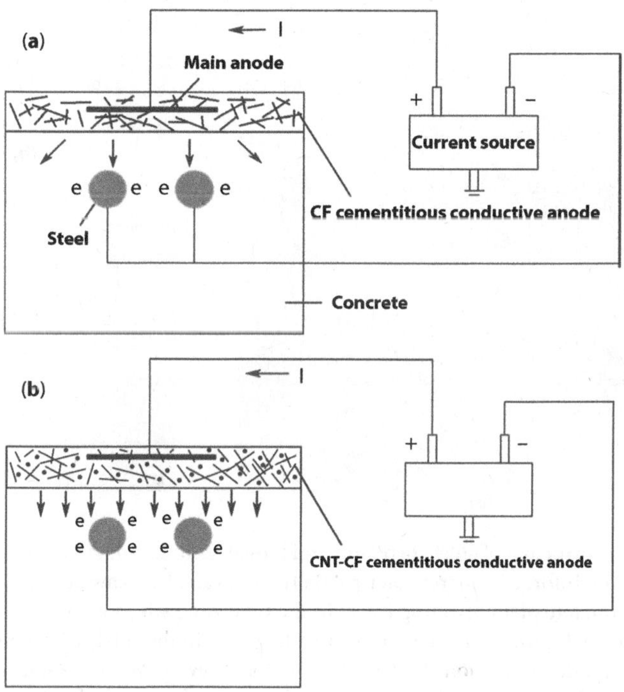

Figure 7.5 Protection regime diagrams of cementitious conductive anode for cathodic protection: (a) CF cementitious conductive anode and (b) CNT-CF cementitious conductive anode [35].

carbon fiber cement-based conductive anode system, the carbon nanotube–carbon fiber system reduced the corrosion current of steel rebars. Additionally, a thorough analysis of the mechanism behind the efficiency of cathodic protection using this composite anode is presented, as illustrated in Figure 7.5.

A novel biological cathodic protection (CP) technology for reinforced concrete structures in marine environments was explored by Dubuit *et al.* [36], where a bioanode is embedded in marine sediments, and electrons are generated through the oxidation of various substrates by electro-active microorganisms, as shown in Figure 7.6. Both numerical and experimental

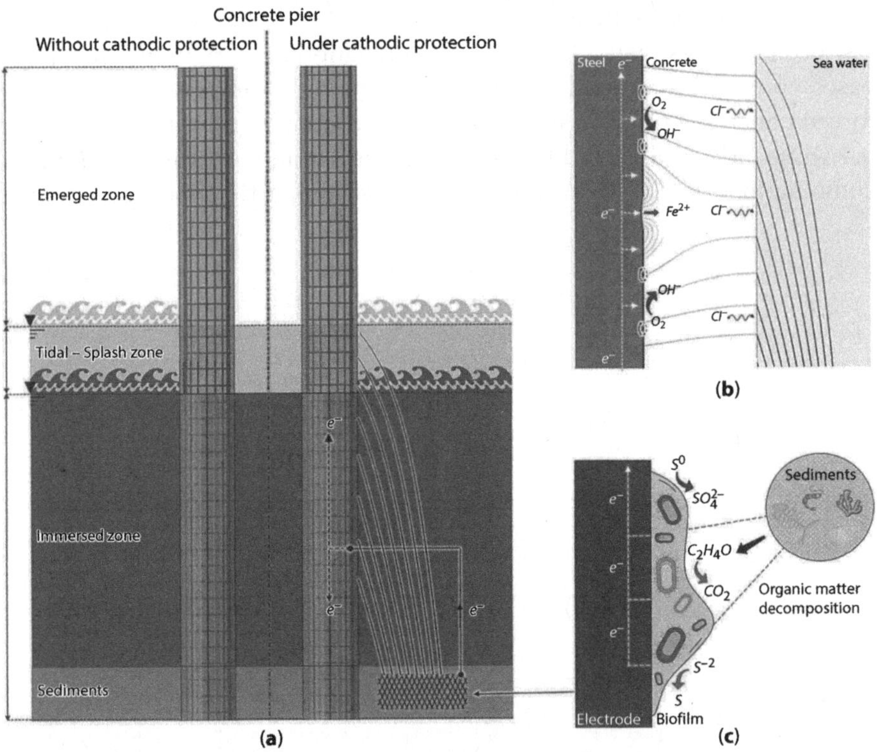

Figure 7.6 The principle of steel corrosion and biogalvanic cathodic protection is illustrated on a reinforced concrete pier partially submerged in seawater. (a) Left: a reinforced concrete pier corroding freely in the tidal zone without any cathodic protection. (a) Right: the same pier has a corroding area in the tidal zone but is under biogalvanic cathodic protection (BGCP). (c) Detailed view of the corrosion system under BGCP, showing the reduction of corrosion current streamlines (red) by the cathodic protection current streamlines (blue) generated by the biological anode system, as calculated using comsol multiphysics®. (d) A schematic of the biological anodic process occurring in sediments, where electrons are supplied to the system [36].

approaches were employed, including a pioneering experimental verification on a laboratory specimen. This innovative solution offers three key advantages: autonomous operation, an environmentally friendly approach, and a free current supply.

7.6 Factors Affecting Cathodic Protection Design in Concrete Corrosion

Designing an adequate cathodic protection (CP) system for concrete structures requires a comprehensive understanding of several interrelated factors influencing the system's performance. The factors included are discussed below.

7.6.1 Chloride Content

Chloride ions are one of the primary causes of corrosion in concrete caused by steel reinforcement. High chloride levels, often due to exposure to de-icing salts or marine environments, disrupt the passive oxide layer on steel, accelerating corrosion [16]. CP systems must be designed to counteract the elevated corrosion rates by supplying adequate protective current. The threshold chloride content that initiates corrosion typically ranges from 0.4% to 1% by weight of cement, depending on environmental conditions and concrete quality.

7.6.2 Concrete Resistivity

Concrete resistivity determines how effectively the CP current can flow through the concrete matrix to the steel reinforcement. Moisture content, chloride concentration, and concrete composition influence resistivity [37]. Lower resistivity, often seen in chloride-contaminated or wet concrete, facilitates current flow, while higher resistivity in dry or dense concrete may necessitate higher driving voltages or specialized anode configurations. Designing CP systems in high-resistivity concrete environments often involves carefully selecting anode materials and placement to ensure uniform current distribution.

7.6.3 Steel Reinforcement Geometry and Spacing

The arrangement, diameter, and spacing of steel reinforcement impact current distribution. Closely spaced or intricate reinforcement geometries can create "shielding effects," where certain areas receive insufficient current due to obstruction by other steel elements [38]. This necessitates detailed modeling during the design phase to identify and mitigate potential shielding zones. Additional considerations may include applying supplementary anodes or modifying system voltage in areas with complex reinforcement layouts [39].

7.6.4 Concrete Cover Depth

The depth of the concrete cover over the steel reinforcement affects the ease of current flow. Thicker covers increase resistive pathways, requiring higher energy input for CP systems. Conversely, thin covers may be more susceptible to chloride ingress, increasing demand.

7.6.5 Environmental Exposure Conditions

The external environment plays a significant role in determining the performance of CP systems. For instance, structures in marine environments are exposed to constant chloride ingress, necessitating robust CP designs. Conversely, inland structures in arid climates may experience slower chloride ingress but require considerations for higher concrete resistivity due to low moisture content. Seasonal variations in temperature and humidity can also influence system performance, necessitating periodic monitoring and adjustments.

7.6.6 Compatibility of Materials

The interaction between the anode materials, steel, and concrete must be carefully considered to avoid adverse reactions. For example, some anode materials may produce byproducts that could influence the alkalinity of the surrounding concrete, potentially affecting its protective properties [40].

By addressing these factors, CP designs can be tailored to the structure's specific needs, ensuring long-term protection and minimizing maintenance requirements.

7.7 New Trends of Cathodic Protection in Concrete Corrosion

These advanced CP methods contribute to more sustainable and durable concrete structures, making them better suited for applications in challenging environments like coastal and marine infrastructure.

Recent trends in cathodic protection (CP) techniques for reinforced concrete structures focus on enhancing durability and reducing environmental impacts. Here are some notable advancements.

A hybrid approach combining ICCP and galvanic anodes, fusion anodes provide an initial burst of current to passivate existing corrosion while offering ongoing protection through embedded galvanic anodes. This two-phase protection system is designed for long-term maintenance-free performance, making it suitable for large-scale applications.

Khalilah Kamarulzaman *et al.* investigated the efficacy of impressed current cathodic protection (ICCP) and sacrificed anode cathodic protection (SACP) in preventing steel corrosion in concrete with reinforcement [27]. First, the SACP approach is used on chloride-contaminated samples that have been air-cured for 150 days before being switched to a dry/wet sodium chloride cycling. The results demonstrate that the initially observed shielding current was extraordinarily high but gradually declined dramatically until it balanced. The depolarization results are higher in dry/wet conditions than in air conditions. Additionally, the ICCP procedure is used with various protected current densities, and the dissolution of the steel rods speeds up earlier. The results show that greater current density indicates an elevated depolarization level. Even with a lower current density, crossing the 100-mV threshold remained possible after 56 days of testing. Depending on the two outcomes, ICCP and SACP were integrated into a "hybrid system" to extend SACP's service life. The initial current demand was managed in the ICCP step before moving on to the SACP step. Data demonstrates that the potential value raises steadily in the ICCP step and becomes stable after moving to the SACP step, where the depolarization amount in the SACP stage displays a more substantial stable magnitude than the in SACP. After 56 days of evaluation, additionally, a lower current density was able to reach the 100-mV requirement. To lengthen the service lifetime of SACP, ICCP and SACP were assembled to create a "hybrid framework" according to both outcomes. The starting current use was controlled in the ICCP step before switching to the SACP stage. The results indicate that the potential magnitude steadily rises in the ICCP step and regulates after switching to the SACP step. In contrast, the depolarization magnitude in the SACP step is more stable [41].

Hamid Fatemia *et al.* introduced a specific sustainable cementitious matrix with strong acid resistance designed for use as anode reinforcement in ICCP systems. The suggested concrete substantially reduces the consequences of acidity near the anode, significantly improving concrete resilience and, as a result, bridge service life with ICCP units. The mortar's mechanical and resilience profitability in an acidic media is assessed using a variety of evaluations involving gravimetric evaluation, dimensional alterations, diffusing depths, compressive strength, direct tension, slant shear, impact resistance, and interface integrity assessments and another spectroscopic test. The findings show that the suggested mortar efficiently overcomes acidification issues while improving the profitability and longevity of ICCP frameworks in hostile maritime settings [42].

A suitable electrical source is needed for ICCP systems to mitigate corrosion. Autonomous networks require power for concrete constructions positioned in isolated and challenging-to-reach areas. Solar-powered equipment could polarize CP networks throughout the day and replenish storage batteries to deliver current at night and on overcast days in remote and hilly places without access to external current.

One of these techniques is the cathodic protection system with the galvanic anode, in which the galvanic anodes supply the necessary shielding current. However, because they provide electric current to the metal framework, these anodes gradually lose mass despite protecting water or underground, and eventually, they stop working. The life of the anodes should be prolonged because replacing galvanic anodes is expensive and time-consuming. This led to the study of a technique that prolongs the life of the anodes in cathodic protection systems that use galvanic anodes. Solar energy is one of the renewable energy sources that is employed in this arrangement. By offering cathodic protection, this technology increased the anode's lifespan. Lastly, a comparison between the created system and the standard systems was used to examine the results [43].

Abdulrahman *et al.* introduced an innovative method for dealing with electrochemical corrosion locally in precise, constrained areas instead of using complete cathodic protection (CP) treatment regions. An induced current cathodic protection cell (ICCPC) was made and securely positioned in the center of a steel structure surface to prevent localized corrosion in stationary or mobile steel structures. The designed ICCPC has three crucial components: an anode, a cathode, and an artificial electrolyte. The last one was established to replicate the role of the natural electrolyte in CP. A proportional-integrated-derivative (PID) is intended to stabilize this potential beneath the ICCPC at a cathodic potential of −850 mV, which is essential for adequate protection. The protective technique's operator was made for automatic turning on the ICCPC as the alloy's structure's surroundings are exposed to moisture. The protective potential, current, and power percentages increased by 411%, 688.74%, and 2842.3%, correspondingly, at temperatures between 27°C and 35°C as the relative humidity increased from 10% to 100%. Excellent preventive corrosion, noticeable increases in electric conductivity, and a decrease in anode–cathode resistance were all demonstrated by the anticipated application of sodium chloride (NaCl) mixture at varying doses on the structure, ranging from 1.5% to 5%. The energy dispersive spectroscopy and scanning electron microscopy (SEM-EDS) data verify that ICCPC coating effectively inhibits corrosion compared to exposed regions. In contrast with areas outside the ICCPC's coverage, these tests showed no signs of oxidation below it. Correspondingly, the iron and carbon levels in regions not covered by ICCPC decreased from 57.1% to 47.0% and 41.4% to 22.4%, while oxygen levels increased from 1.6% to 30.6% [44].

Xuanbo He *et al.* utilized (ICCP) technology to conduct anti-corrosion trials at the steel wire platform to the cable scale, studied the deterioration prevention features of steel wire ICCP, determined the corrosion cause ICCP diffusing approach, and disclosed the spatial ICCP procedure of the original harmed cable. The investigation demonstrates that ICCP may successfully suppress the degrading impact of steel wires and cables while reducing mechanical property loss. Its preventive efficacy rises as an increase in the total amount of its preventive potential and falls with a rise in dead load. ICCP inhibits the propagation of corrosion variables throughout the cable space, causing the corrosion location span, corrosion component level, and diffusion coefficient of the original impaired cable to decrease substantially as the total amount of the protection potential increases. As the corrosion process progresses, the open circuit potential increases in a favorable orientation, and the hindering impact of corrosive element diffusion shifts from joint involvement by sacrificial anode cathodic protection and ICCP to the dominance of ICCP [45].

Galvanic batteries are an additional power supply for an ICCP framework, allowing cathodic protection of restricted networks in remote places while avoiding the costly expense of installing a voltage regulator. When necessary, more batteries can be linked to provide increased current needs.

Yongjing Wang *et al.* determined the impact of the form of the auxiliary anode on the protective potential dispersion in an impressed current cathodic protection (ICCP) framework, and an electrochemical and level-set optimizing approach were coupled. To achieve this, the square of the deviation over the structure's and protective potential has been employed as the objective function, and the anode shape was developed using level set theory. After a certain amount of iterations, the optimal anode form was determined. Utilizing the optimized anode, a current was introduced into the anode surface, and the potential distribution was computed. Testing outcomes revealed that the potential dispersion of the treated surface was unified, which may adequately protect the surface of the protected structures. This allowed preventing excessive protection or underprotection [46].

Karina Vilela Rodrigues *et al.* evaluated the effect of Al anodes on the weight loss of galvanized steel utilized for the fertigation process with white potassium chloride solution at 10 g L^{-1}. They also verified the effect of the cathode/anode area ratio on galvanized steel protection and anode degradation. To evaluate the total weight loss of the galvanized steel and Al anodes, they ran immersion tests that simulated 10 years of fertigation. The findings revealed that Al anodes substantially decrease mass loss in galvanized steel after contact with KCl solution; however, there is a slight variation in mass loss with increasing cathode/anode area ratio. In the case of Al anodes, there was no substantial variation in mass loss as the cathode/anode area ratio increased [47].

7.8 Challenges in Implementing Cathodic Protection in Concrete and Addressing Environmental and Sustainability Considerations

Implementing cathodic protection (CP) in concrete structures presents several challenges. One major issue is the variability in concrete resistivity, which can significantly impact the uniform distribution of protective current. High resistivity areas, often due to dry conditions or dense concrete, may require increased system voltage or specialized anode configurations, complicating installation and maintenance [16]. Complex reinforcement geometries can lead to shielding effects, where certain steel areas receive insufficient protection, necessitating meticulous design and monitoring. Additionally, sustainability considerations arise from using nonrenewable materials, such as titanium or zinc anodes, and the energy required to operate impressed current CP systems [48]. Environmental factors, including exposure to aggressive chemicals or fluctuating climates, further strain CP systems, potentially increasing the frequency of maintenance or replacement. Addressing these challenges calls for innovative approaches, such as developing sustainable anode materials, leveraging renewable energy sources, and integrating advanced monitoring technologies to reduce long-term environmental impacts [49]. Emphasizing life-cycle assessments and environmentally friendly solutions can enhance the sustainability of CP systems in concrete [50].

7.9 Conclusion

Cathodic protection of reinforced concrete has developed as a successful method for mitigating chloride-induced deterioration, prolonging the useful life of reinforced concrete constructions. Technological breakthroughs have rendered CP more appealing to builders by providing new options, as evidenced by upgrading modern electronic machinery that allows for efficient evaluation and oversight of the technique's functioning.

This overview discusses key aspects of cathodic protection for concrete construction, such as anode supplies, anode evaluation, anode lifespan estimation, CP tracking, thermally sprayed zinc anode assembly and substitution, and numerical approaches. Such experience can assist in establishing optimal procedures for deciding, installing, maintaining, and replacing CP technologies.

Cathodic protection has grown to the point where revolutionary developments have occurred. However, ongoing and gradual improvements in current components, machinery, and evaluation approaches are expected. Innovative conductive paint anodes with the ability to heal themselves could be created to reduce the danger of mechanical degradation and hence improve system reliability.

Several present CP systems that use chemically applied Zn anodes may approach or surpass the designs shortly, causing them to work poorly or inadequately, requiring the replacement of the old anodes. In this scenario, there is an immediate need to investigate the most economically viable technique of replacing old Zn anodes and create a concrete substrate strategy for modern anodes.

Basic study must be focused on improving the expertise essential to more accurate CP efficacy and dependability involving the investigation of electrochemical aged effects on thermally-sprayed anodes and its disconnect that underlies concrete. Such a flexible procedure seems complicated within the polarized circumstances; however, the outcomes were an improved foundation for figuring out anode service life and a greater understanding of when dropping and terminal end cathodic protection.

AI disclosure statement: The author confirms that no artificial intelligence (AI)-generated content or AI-assisted writing tools were used in the preparation of this chapter (including summaries and the abstract). All text, analysis, and conclusions are the result of the author's original work.

References

1. Enevoldsen, J.N., Hansson, C.M., Hope, B.B., Binding of chloride in mortar containing admixed or penetrated chlorides. *Cem. Concr. Res.*, 24, 8, 1525–33, 1994.
2. Gjørv, O.E. and Vennesland, O., Diffusion of chloride ions from seawater into concrete. *Cem. Concr. Res.*, 9, 2, 229–38, 1979.
3. Glass, G.K. and Buenfeld, N.R., The presentation of the chloride threshold level for corrosion of steel in concrete. *Corros. Sci.*, 39, 5, 1001–13, 1997.

4. Yajun, L. and Xianming, S., Cathodic Protection Technologies for Reinforced Concrete: Introduction and Recent Developments. *Rev. Chem. Eng.*, 25, 5-6, 339–88, 2009.
5. Koleva, D.A., Guo, Z., van Breugel, K., de Wit, J.H.W., Conventional and pulse cathodic protection of reinforced concrete: Electrochemical behavior of the steel reinforcement after corrosion and protection. *Mater. Corros.*, 60, 5, 344–54, 2009.
6. Page, C.L. and Sergi, G., Developments in Cathodic Protection Applied to Reinforced Concrete. *J. Mater. Civ. Eng.*, 12, 1, 8–15, 2000.
7. Wilson, K., Jawed, M., Ngala, V., The selection and use of cathodic protection systems for the repair of reinforced concrete structures. *Constr. Build. Mater.*, 39, 19–25, 2013.
8. Mohd Nadzri, N.I. and Mohd Amin, N., A Review of Cathodic Protection in Repairing Reinforced Concrete Structures. *J. Mech. Eng. (JMechE)*, 2, 183–98%V 16, 2021.
9. Naik, Y.A., Venkatesha, T.V., Nayak, P.V., Electrodeposition of zinc from chloride solution. *Turk. J. Chem.*, 26, 5, 725–34, 2002.
10. Chen, J., Ji, L., Song, J., Study of Crevice Corrosion Behavior and Cathodic Protection of Carbon Steel Reinforcement in Concrete. *Int. J. Electrochem. Sci.*, 17, 1, 220140, 2022.
11. Bertolini, L., Pedeferri, P., Redaelli, E., Pastore, T., Repassivation of steel in carbonated concrete induced by cathodic protection. *Mater. Corros.*, 54, 3, 163–75, 2003.
12. Sergi, G. and Page, C., Sacrificial anodes for cathodic prevention of reinforcing steel around patch repairs applied to chloride-contaminated concrete. *Eur. Fed. Corros. Publications(UK)*, 31, 93–100, 2000.
13. Mietz, J., Polder, R., Elsener, B., *Corrosion of Reinforcement in Concrete: Corrosion Mechanisms and Corrosion Protection: Papers from EUROCORR'99*, Woodhead Publishing Limited, Germany, 2000.
14. Grefstad, K., Cathodic protection applied on Norwegian concrete bridges. Experience and recommendations. *Eurocorr05 Lisboa*, 74, 2005.
15. Araujo, A., Panossian, Z., Lourenço, Z., Cathodic protection for concrete structures. *Rev. IBRACON de Estruturas e Materiais*, 6, 178–93, 2013.
16. Bertolini, L., Elsener, B., Pedeferri, P., Redaelli, E., Polder, R.B., *Corrosion of steel in concrete: prevention, diagnosis, repair*, John Wiley & Sons, (Wiley-VCH Verlag GmbH & Co. KGaA, Weinheim) 2013. DOI:10.1002/3527603379.
17. Pollet, V. and Dieryck, V., Re-alkalisation: specification for the treatment application and acceptance criteria. Annual progress report 1999, 2000.
18. Redaelli, E. and Bertolini, L., Electrochemical repair techniques in carbonated concrete. Part I: electrochemical realkalisation. *J. Appl. Electrochem.*, 41, 7, 817–27, 2011.
19. Pedeferri, P., *Corrosion Science and Engineering*, P. Pedeferri (ed.), Springer International Publishing, Cham, pp. 509–548, 2018, DOI: 10.1007/978-3-319-97625-9_23.

20. Polder, R. and Peelen, W., Cathodic protection of steel in concrete–experience and overview of 30 years application. *MATEC Web of Conferences. 199*, EDP Sciences, p. 01002, 2018.
21. Du, Y., Gu, F., Ke, T., Jiang, Z., The Latest Developments on Cathodic Protection Technologies in China. *AMPP Middle East Corrosion Conference and Exhibition*, AMPP, 2023, AMPP-MECC-2023-19994.
22. Sathiyanarayanan, S., Natarajan, P., Saravanan, K., Srinivasan, S., Venkatachari, G., Corrosion monitoring of steel in concrete by galvanostatic pulse technique. *Cem. Concr. Compos.*, 28, 7, 630–7, 2006.
23. Sohawon, H., Service life extension of reinforced concrete structures using hydrophobic impregnation, 2018.
24. Guo, Z., Xiao, Z., Chen, H., Zhou, X., Wang, P., Luo, J., *et al.*, Review of Cathodic Protection Technology for Steel Rebars in Concrete Structures in Marine Environments. *Appl. Sci.*, 14, 19, 9062, 2024.
25. Daniyal, M. and Akhtar, S., Corrosion assessment and control techniques for reinforced concrete structures: a review. *J. Build. Pathol. Rehabil.*, 5, 1, 1, 2019.
26. Polder, R., Cathodic protection of reinforced concrete structures in The Netherlands-experience and developments. *Heron*, 43, 1, 3–14, 1998.
27. Garcés, P., Andión L, G., De la Varga, I., Catalá, G., Zornoza, E., Corrosion of steel reinforcement in structural concrete with carbon material addition. *Corros. Sci.*, 49, 6, 2557–66, 2007.
28. Hornbostel, K., Angst, U., Elsener, B., Larsen, C., Geiker, M., On the limitations of predicting the ohmic resistance in a macro-cell in mortar from bulk resistivity measurements. *Cem. Concr. Res.*, 76, 147–58, 2015.
29. Wang, F., Xu, J., Xu, Y., Jiang, L., Ma, G., A comparative investigation on cathodic protections of three sacrificial anodes on chloride-contaminated reinforced concrete. *Constr. Build. Mater.*, 246, 118476, 2020.
30. Krishnan, N., Kamde, D.K., Doosa Veedu, Z., Pillai, R.G., Shah, D., Velayudham, R., Long-term performance and life-cycle-cost benefits of cathodic protection of concrete structures using galvanic anodes. *J. Build. Eng.*, 42, 102467, 2021.
31. Cheng, X., Xia, J., Wu, R.-., Jin, W.-., Pan, C.-., Optimisation of sacrificial anode cathodic protection system in chloride-contaminated reinforced concrete structure. *J. Build. Eng.*, 45, 103515, 2022.
32. Carmona, J., Garcés, P., Climent, M.A., Efficiency of a conductive cement-based anodic system for the application of cathodic protection, cathodic prevention and electrochemical chloride extraction to control corrosion in reinforced concrete structures. *Corros. Sci.*, 96, 102–11, 2015.
33. Li, W., Zuo, J., Xu, J., Li, Y., Zheng, H., Yang, H., Cathodic protection of reinforced concrete facilitated by using CNT-CF/cement-based auxiliary anode. *Fullerenes Nanotubes Carbon Nanostruct.*, 32, 6, 595–602, 2024.
34. Anwar, M.S., Sujitha, B., Vedalakshmi, R., Light-weight cementitious conductive anode for impressed current cathodic protection of steel reinforced concrete application. *Constr. Build. Mater.*, 71, 167–80, 2014.

35. Zuo, J., Yao, W., Xu, J., Chen, Y., Liu, X., Effects of Carbon Nanotube-Carbon Fiber Cementitious Conductive Anode for Cathodic Protection of Reinforced Concrete. *J. Test. Eval.*, 45, 5, 1777–86, 2017.
36. Dubuit, J., Laurens, S., Bertron, A., Garcia, D., Etcheverry, L., Deby, F., *et al.*, Biogalvanic cathodic protection of reinforced concrete structures in marine environments. *Constr. Build. Mater.*, 403, 133180, 2023.
37. Andrade, C. and González, J., Quantitative measurements of corrosion rate of reinforcing steels embedded in concrete using polarization resistance measurements. *Mater. Corros.*, 29, 8, 515–9, 1978.
38. Pedeferri, P., Cathodic protection and cathodic prevention. *Constr. Build. Mater.*, 10, 5, 391–402, 1996.
39. Bertolini, L., Bolzoni, F., Pastore, T., Pedeferri, P., Behaviour of stainless steel in simulated concrete pore solution. *Br. Corros. J.*, 31, 3, 218–22, 1996.
40. Harriott, D., Cathodic Protection of Reinforced Concrete Bridge Elements: A Stateof-the-Art Report. SHRP-S-337 report, 1993.
41. Kamarulzaman, K., Astuti, P., Caronge, M.A., Nguyen, L., Sagawa, Y., Combination of impressed current and sacrificial anode cathodic prevention to improve corrosion prevention system in reinforced concrete. *Advances in Construction Materials Proceedings of the ConMat'20*, 2020.
42. Fatemi, H., Hadigheh, S.A., Tao, Y., Adam, G., Development of a novel and specialised cementitious matrix overlay for anode embedment in impressed current cathodic protection (ICCP) systems for reinforced concrete bridges. *Case Stud. Constr. Mater.*, 20, e02908, 2024.
43. Ma, L. and Zhang, D., *Smart protective coatings for corrosion control*, John Wiley & Sons, (Wiley-VCH GmbH, Boschstraße, Weinheim, Germany) 2024, DOI:10.1002/9783527847150.
44. Khudhair, A.B. and Hussein, F.I., Implementation of a control and monitoring system for a cathodic protection cell to mitigate localized corrosion in fixed and mobile steel structures. *Results Eng.*, 23, 102553, 2024.
45. He, X., Yao, G., Long, H., Ying, L., Guo, Z., Zhao, L., *et al.*, Corrosion protection failure test analysis of the initial damaged cable ICCP mechanism. *Case Stud. Constr. Mater.*, 20, e03227, 2024.
46. Wang, Y., Liu, Y., Dai, L., 2-D shape optimization of auxiliary anode for ICCP. *Mater. Chem. Phys.*, 323, 129600, 2024.
47. Rodrigues, K.V., Lima, L.A., Thebaldi, M.S., Corrêa, F.V., Cathode/Anode Area Ratio on the Sacrificial Cathodic Protection Against Mass Loss of Galvanized Steel Used in Potassium Chloride Fertigation. *Arabian J. Sci. Eng.*, 49, 10, 13491–9, 2024.
48. Angst, U.M., A critical review of the science and engineering of cathodic protection of steel in soil and concrete. *Corrosion*, 75, 12, 1420–33, 2019.
49. Rodrigues, R., Gaboreau, S., Gance, J., Ignatiadis, I., Betelu, S., Reinforced concrete structures: A review of corrosion mechanisms and advances in electrical methods for corrosion monitoring. *Constr. Build. Mater.*, 269, 121240, 2021.
50. Cherry, B. and Green, W., *Corrosion and protection of reinforced concrete (1st ed.)*, CRC Press, London, 2021. https://doi.org/10.1201/9781003081302.

8

Role of Corrosion Inhibitors in Preventing Concrete Corrosion

Shweta Goyal[1,2*], Vijay Luxami[2,3] and Sonia Rani[2,3]

[1]*Department of Civil Engineering, Thapar Institute of Engineering and Technology, Patiala, Punjab, India*
[2]*Centre of Excellence in Emerging Materials, Thapar Institute of Engineering & Technology, Patiala, Punjab, India*
[3]*Department of Chemistry and Biochemistry, Thapar Institute of Engineering and Technology, Patiala, Punjab, India*

Abstract

Corrosion of reinforced concrete (RC) is the major reason for the deterioration of concrete structures, significantly impacting economic development and creating serious societal issues. According to the survey, the global cost of recently maintaining public utilities from corrosion is around \$2.5 trillion or 3.4% of the world's global GDP. Generally, the compact passive film on the concrete structure protects it from corrosion. The film sustain itself at pH of around 12.5-13.5. The damage to the passive film happens when pH of the pore solution reduces and/or chloride ion concentration reaches the critical value. Porous concrete allows the corrosive ions to enter the steel bar, leading to the localized corrosion of reinforcing steel. Many new advancements have been made to decrease/prevent the corrosion process of RC and increase the service life of the structure. There are various physical and electrochem ical methods for corrosion protection/prevention. Nowadays, corrosion inhibitors are employed as a cost-effective measure for protecting concrete structures. These substances have been used in very low concentrations to reduce corrosion rate and can serve dual purposes, inhibition and self-healing. Corrosion inhibitors can be classified as organic and inorganic. The inhibitors can also be classified according to their application methods, protection mechanism, and chemical composition. In this chapter, the role of inhibitors has been studied based on the mechanism of action in different aggressive environments. The effect of corrosion inhibitors on

**Corresponding author*: shweta@thapar.edu

Ruby Aslam, Qihui Wang, Zhitao Yan and Afroz Aslam (eds.) Corrosion in Concrete Structures: Integrating Advanced Technologies and Sustainable Practices, (213–256) © 2026 Scrivener Publishing LLC

other properties of concrete has also been explored. The new class of green corrosion inhibitors that are non-toxic and non-carcinogenic are also discussed.

***Keywords*:** Organic, inorganic, heterocyclic, migratory, admixed, smart

8.1 Introduction

The two constituents of reinforced concrete (RC) structures are steel and concrete, which are simultaneously used; the reinforced steel procured higher tensile strength while concrete protects the steel from the aggressive outer environment. The steel in the concrete is protected from corrosion through thin oxide layer of $\gamma\ Fe_2O_3.H_2O$ formed by its highly alkaline environment. The two major factors that lead to the breakdown of the passive film are either the ingress of chloride ions or carbonation. Above a critical threshold, chloride ions near steel causes localized film breakdown, leading to pitting corrosion. On the other hand, the pH of the alkaline pore solution is reduced by the reaction of concrete components with atmospheric carbon dioxide that diffuses through pores, which destroys the overall passive film around the rebar, thus making it vulnerable to corrosion [1]. The corrosion of rebar ultimately decreases the strength and ductility of concrete structures. Further, the corrosion products (namely γ-Fe_2O_3, γ-FeOOH) are more voluminous than the parent metal. Therefore, the expansive corrosion products formed causes tensile stresses in the concrete environment, leading to the development of cracks and ultimate delamination of the concrete cover, thereby compromising the structure's overall integrity [2].

Therefore, corrosion protection is the most important need for the building industry. The corrosion process can be reduced by limiting the access of aggressive species from the outside environment to reach the steel surface. This can be done by reducing the permeability of concrete using a low water: cement ratio of the mix, use of supplementary cementitious materials in the mix, proper curing, and higher thickness of concrete cover [3]. Apart from this, various other prevention methods like corrosion-resistant rebars, cathodic protection, external protective coatings, anodization, and corrosion inhibitors can be utilised to enhance service life of the structure. All these techniques are broadly categorized as electrochemical or non-electrochemical, and are listed in Figure 8.1.

Electrochemical techniques depend solely on the equipment and facilities; therefore, they are costly and require trained operators for installation and monitoring. Moreover, their maintenance causes an interruption in their functioning, which can have detrimental consequences for the structure. To work in versatile conditions, corrosion inhibitors (CoIs) as a non-electrochemical technique can be successfully applied to prevent

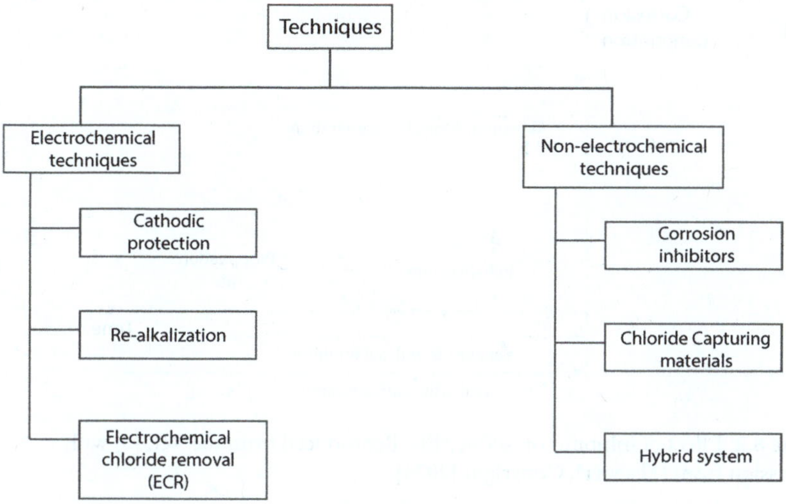

Figure 8.1 Classification of corrosion prevention techniques.

corrosion in almost all industries [4]. Moreover, they are cost-effective, easy to handle, and generally required in low concentrations to produce the desired results.

The present chapter revolves around the significant role of corrosion inhibitors as a corrosion prevention strategy in RC structures.

8.2 Definition of Corrosion Inhibitors

According to ISO 8044:2020 terminology, corrosion inhibitors can be defined as chemical substances used in corrosive systems in the required concentration. They can decrease the corrosion process without affecting the overall working of the system [5]. The use of CoIs is known since the beginning of the 19th century, with the initial applications targeting the protection of pipelines and storage tanks for oil and gas exploration. Their usage in RC applications began in the 1950s [6]. In RC, they are generally used in minimal concentration and are assumed to have no detrimental effect on other concrete properties. They work either by refining the cover concrete's pore structure, thereby decreasing diffusion of corrosive species, or by forming the barrier layer on the rebar surface and strengthening its passivity. Because of this dual effect, they efficiently increase the corrosion initiation and propagation period, which is schematically represented by modifying the Tuutti's general service life model (Figure 8.2) [7].

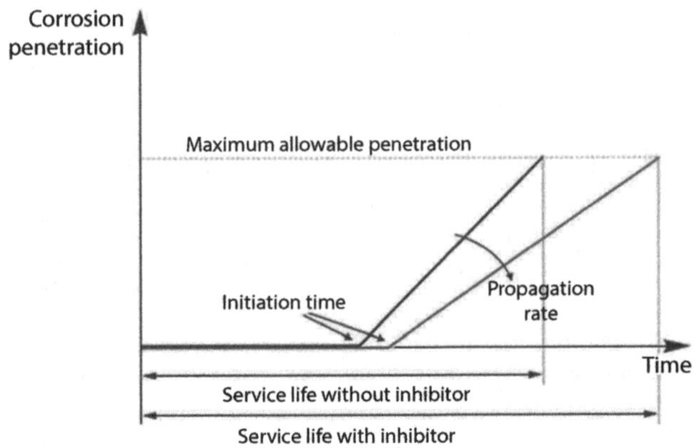

Figure 8.2 Effect of inhibitor on service life. Reproduced from the Ref. [3] with permission from [Elsevier], Copyright [2024].

8.3 Mechanism of Action of Inhibitor

Various mechanisms are followed by corrosion inhibitors to resist or delay the corrosion process. The mechanisms are divided into four main groups: Passivation, Adsorption, Film-forming, Smart release.

8.3.1 Passivation

Passivation means the formation of the protective passive layer, a thin film of oxide, hydroxide or any other corrosion-resistant material on the metal surface, that prevents the metal from further oxidation by the environment. Various inorganic compounds such as molybdates, chromates, and phosphates inhibit metal through the passivation process. Organic molecules like amino acids, alcohols, carboxylic acids, and polyphenols are used to passivate the metal surface.

8.3.2 Adsorption

Inhibitors containing different functional groups (heteroatoms), such as amine, acid, thiols, and hydroxyl groups, interact with the metal surface through the lone pair of electrons. Also, the compound containing aromatic rings has π electrons to adsorb physically on the rebar surface. The mechanism is termed physio-chemisorption. The inhibitors used in this category, being environmentally friendly and non-toxic provide adequate corrosion protection.

8.3.3 Film-Forming

The inhibitors form a film on steel surface, either by physical interaction (adsorption) on the surface or by chemical reaction of metal with the inhibitor. Different organic compounds, synthetic polymers, and green inhibitors form durable films on the rebar surface and provide long-term prevention.

8.3.4 Smart Release

The above mechanisms are based on adsorption and passivation to enhance the corrosion resistance. Smart inhibitors use smart technology, the responsive strategy of the core-shell of the inhibitor, to react differently in corrosive environments. These inhibitors are prone to act in changing environments like pH, temperature, and humidity changes. The inhibitor itself modifies its structures and properties in response to the change.

The type of mechanism adopted by different inhibitors is discussed in detail in the subsequent sections.

8.4 Classification of Corrosion Inhibitors

The classification of the corrosion inhibitors depends on various factors such as mechanism of action, chemical composition, method of their introduction in concrete and the process of synthesis. The flow chart classifying corrosion inhibitors is shown in Figure 8.3. Depending on the mechanism of action on the target metal, the inhibitors are classified as anodic inhibitor, cathodic inhibitors, and mixed inhibitors. In the structures, they can be either admixed in concrete at the time of casting or can be introduced at the later stage either by electrochemical injection in the concrete or by surface application of the concrete to act as migratory corrosion inhibitors [8]. Further, depending on the fundamental composition, the CoIs are classified as organic and inorganic compounds. These compounds can be either commercially available generic compounds or can be synthesised in the laboratory using simple synthetic procedures. Some biological compounds (green compounds) are also being explored for their use as efficient corrosion inhibitors. These classifications have been discussed in detail in the following sections.

8.5 Classification Based on Mechanism of Action

Corrosion process of steel in RC is an electrochemical reaction where anodes and cathodes are simultaneously formed on the rebar and are surrounded

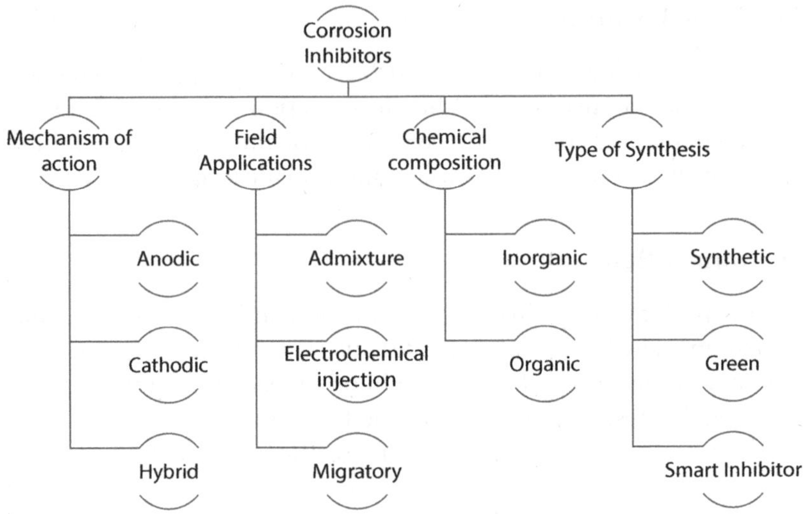

Figure 8.3 Classification of corrosion inhibitors.

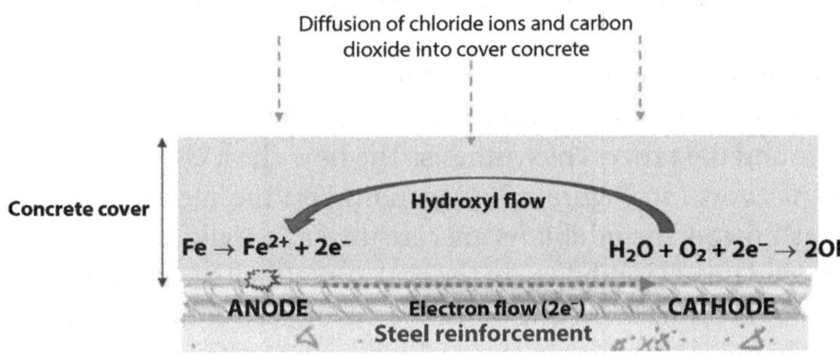

Figure 8.4 Schematic representation of corrosion of steel in concrete.

by concrete pore solution (electrolyte) that allows the movement of ions between the two electrodes. The schematic representation of the redox reaction on steel rebar is depicted in Figure 8.4. This process involves the following reactions:

(i) Iron metal atom losses electron at anode and become ferrous (Fe^{2+}) ions. This process is called half-cell oxidation or anodic reaction, as shown in equation 8.1.

(ii) The free electrons in the electrolytic solution move to the other site called cathode, binding with water and oxygen in

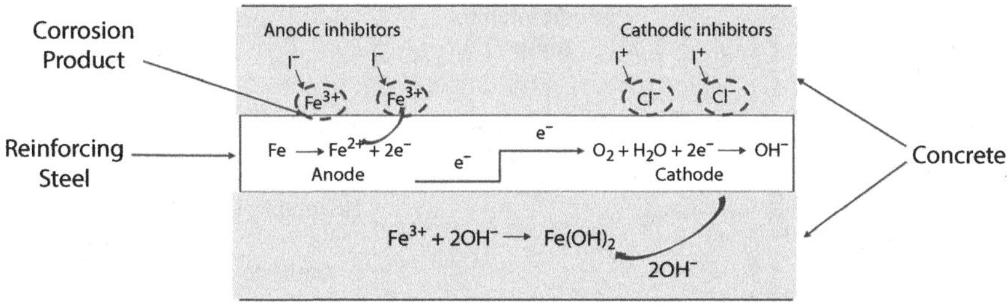

Figure 8.5 Mechanism of action of anodic and cathodic inhibitors.

concrete. The respective cathodic reaction is shown in eq. 8.2, which is called as reduction half-cell reaction.

$$2Fe \rightarrow 2Fe^{2+} + 4e^{-} \tag{8.1}$$

$$2H_2O + O_2 + 4e^{-} \rightarrow 4OH^{-} \tag{8.2}$$

Depending on this mechanism of action, the corrosion inhibitors are categorized as anodic, cathodic, and hybrid (or mixed) inhibitors. The overall mechanism of anodic and cathodic inhibitors is depicted in Figure 8.5.

8.5.1 Anodic Inhibitors

The inhibitors that control the anodic reaction of metal and restrict the metal dissolution process are called anodic inhibitors. Various inhibitors, like chromates, nitrates, and phosphates, act as anodic corrosion inhibitors [9]. These inhibitors create a barrier passive layer on the metal surface, thereby shifting metal's anodic polarisation curve potential to a more positive value (to the passive zone).

8.5.2 Cathodic Inhibitors

The other class of inhibitors that restrict the cathodic reaction of the corrosion process are called cathodic inhibitors. Inhibitors precipitate the salts present in water, and the reduction reaction of the corrosion process altogether ceased. The precipitated salt forms the barrier layer on the steel surface to reduce oxygen transfer to the cathodic compartment. Hence, oxygen reduction leads to decreased corrosion potential and corrosion rate. Cathodic inhibitors shift the cathodic reaction potential to a more negative value and reduce the cathodic reaction rate. Examples of cathodic

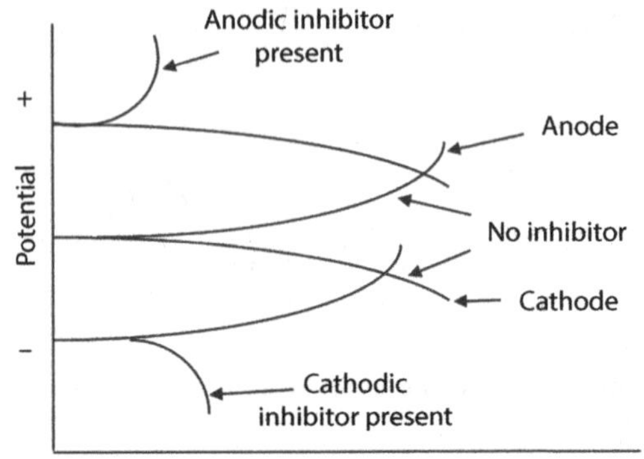

Figure 8.6 Schematic diagram shows the effect of inhibitors on the polarisation curve.

inhibitors are hydroxides of zinc, cadmium, aluminium and magnesium [10]. The inhibition effect of cathodic inhibitors is less than that of anodic inhibitors.

The polarisation curves are shown in Figure 8.6, indicating the effect of anodic and cathodic inhibitors on potential and current.

8.5.3 Hybrid Inhibitors

The inhibitors that simultaneously restrict the cathodic reaction and anodic metal dissolution process are termed hybrid inhibitors. These inhibitors get uniformly adsorbed on the metal surface, shifting the potentials of both cathodic and anodic reactions [11]. Mainly, organic compounds fall into the hybrid or mixed-type inhibitors category.

8.6 Classification Based on Field Application

Depending on the mode of introduction into concrete, the inhibitors can be classified as admixed or migratory inhibitors.

8.6.1 Admixed Inhibitors

The inhibitors admixed during casting at the time of construction of any structure are called admixed inhibitors. Most inorganic inhibitors introduced in the early 1950s were primarily used in admixed form. They protect by forming a barrier layer on the rebar by physio-chemisorption [12], which delays the duration of corrosion and reduces the corrosion rate.

However, using inorganic admixed inhibitors led to a decrease in concrete's compressive strength by about 10-20%. Some compounds caused increased air voids in concrete, lowering the overall quality of concrete. Apart from this, some inorganic compounds are known to be carcinogenic and are very harmful to human life, and some inhibitors pollute soil and water bodies. In the later decades, some organic corrosion inhibitors were introduced in the admixed form, which does not have any detrimental effect on the resultant properties of concrete [13].

8.6.2 Migratory Inhibitors

Due to the drawbacks of admixed inhibitors on the new and some hardened concrete properties, a range of migratory type inhibitors were delved. These compounds are applied on the concrete surface once it hardens and migrate through the concrete cover to reach the rebar level to protect it. Since they are used later, they can effectively repair existing RC structures [14]. These inhibitors are applied on the concrete surface and, *via* capillary action, migrate through concrete pores very easily. The compounds also reach the rebar level through electrostatic interaction between metal and inhibitor, as the inhibitor has excess electrons and steel has positive charge [11]. Some researchers observed that these surface-applied inhibitors form insoluble compounds with the concrete mix, block the pores of concrete, and termed them pore blockers. The inhibitors protect the rebar by creating a passive hydrophobic barrier when reaching the rebar level. Because of the combined action of pore blocking and passive film formation, the corrosion rate reduces, and hence, the service life of RC increases. Examples of migratory inhibitors are- amino acids, salicyldehyde, 2-amino benzoic acid, amino pyridine etc. Although these inhibitors efficiently attain corrosion inhibition, they have some drawbacks. The amount of inhibitor applied on the surface is much higher than the amount reaching the rebar level for inhibition [15], which leads to the loss of resources and wastage of chemicals. Therefore, to enhance the migration and approachability of inhibitors, the bidirectional electro-migration rehabilitation (BIEM) technique was used by some researchers.

8.6.3 Bi-Directional Electro-Migration (BIEM)

In this technique, the inhibitor is pushed into concrete by electric current, while at the same time, the aggresive ions like chlorides are forced out of the concrete cover. An electric field is applied to inject the inhibitor into the non-carbonated concrete to extract chloride ions from its cover is studied by Xu *et al.* Triethylenetetramine (TETA) was used for this electro-migration rehabilitation. The amount of inhibitor obtained at the rebar level was increased using BIEM, that led to greater inhibition efficacy [16]. Rajput *et al.* injected 2-aminopyridine by BIEM. The migration of the inhibitor and chloride extraction was found to increase with the increase in treatment duration. Also, with the application of BIEM, at 30 mm depth, the inhibitor concentration increased by 800% compared to the inhibitor concentration at 15 mm depth without BIEM [17]. Therefore, applying BIEM creates a win-win situation in attaining higher inhibitor concentration even at higher depths.

8.7 Classification Based on Chemical Composition

Based on the chemical composition of the primary compound, the inhibitors are classified into inorganic, organic and heterocyclic inhibitors. The earliest used CoIs were inorganic inhibitors. However, the research interest shifted towards sustainable organic corrosion inhibitors because of environmental and health concerns related to inorganic inhibitors.

8.7.1 Inorganic Inhibitors

These inhibitors contain inorganic parts, mainly nitrates, phosphates, chromates, molybdates, silicates and nitrites. Some examples are sodium nitrate, calcium nitrate, magnesium hydroxide, calcium phosphate, etc. Different inorganic ions and their mechanism of inhibition is shown in Table 8.1. Depending on the mechanism of action, they worked as anodic or cathodic inhibitors; although most inorganic inhibitors are anodic. Different compounds possess different mechanisms to restrict the anodic reaction. Nitrates or chromates form a protective film on the steel rebar without oxygen, while phosphates or silicates hinder the anodic reaction in the presence of oxygen [18]. Their mechanism is discussed in detail in the following sections.

8.7.1.1 Nitrate and Nitrite Based Inhibitor

The use of nitrite-based inhibitors was started in the 1950s. Inorganic nitrite like calcium nitrite was investigated as a corrosion inhibiting molecule and used as a protection strategy on different structures over 1980s and 90s [19]. Most commonly used nitrite-based inhibitors are calcium nitrite and sodium nitrite [20, 21]. These nitrite compounds are anodic

Table 8.1 Different inorganic functional groups as a corrosion inhibitor.

Functional group	Molecular structure	Protection type	Mechanism	Examples
Nitrites	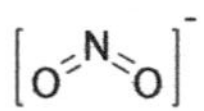	Anodic	Nitrite ions compete with chloride ions to bind with Fe^{3+} ions to form barrier film	Sodium nitrite, calcium nitrite
Molybdate		Anodic	These ions promote the passivation of active site of steel by forming an effective layer and displacing the corrosion current to more anodic values.	Zinc molybdate, sodium molybdate
Chromates	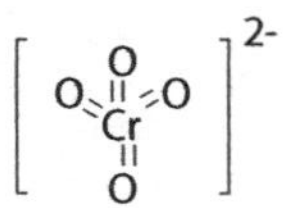	Anodic	These ions help rebar to maintain its passive film by preventing disruption	Chromium chromate, chromic acid
Phosphates	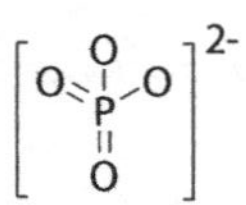	Anodic or Cathodic	As anodic inhibitor, they form passive layer and as cathodic inhibitor, the $[PO_4^{3-}]/[Cl^-]$ <0.6	Calcium monofluorophosphates, zinc monofluorophosphates

passivating inhibitors since they work on anodic part of the electrochemical reaction. The generalized redox reactions using sodium nitrite (nitrite ions) are shown as follows:

$$2Fe^{2+} + 2OH^- + 2NO_2^- \rightarrow 2NO + \gamma\text{-}Fe_2O_3 + H_2O \tag{8.3}$$

$$2Fe^{2+} + 2OH^- + 2NO_2^- \rightarrow 2NO + \gamma\text{-}FeOOH \tag{8.4}$$

As can be seen in the equations, the ferrous ions react with hydroxide ions to form a stable passive layer of either FeOOH and Fe_2O_3, by the interference of nitrite ions in the metal (anode) dissolution process [22].

Major work on the use of nitrite-based inhibitors is reported in chloride induced exposure. In chloride-based environment, the corrosion inhibition effectiveness is proportional to the amount of free chlorides present, which is expressed in the terms of the molar ratio of $[NO_2^-]/[Cl^-]$. It is observed that a molar ratio of 0.5-0.6 was adequate to protect the chloride-induced corrosion. As the molar ratio is more than 0.6, the rebar corrosion was negligible (Figure 8.7). Further, the critical chloride threshold can increase to 0.4% by the weight of cement with the introduction of calcium nitrite [23].

The major drawback of using nitrite-based inhibitors is the cost and concerns related to their toxicity. These are not usually recommended for working in an open system (exposure to air). Also, sodium nitrite admixed with chloride in concrete reduces concrete's compressive strength [24].

Another class of inhibitors as an alternative to nitrite-based inhibitors is nitrate-containing inhibitors, like calcium nitrates, sodium nitrates. They offer

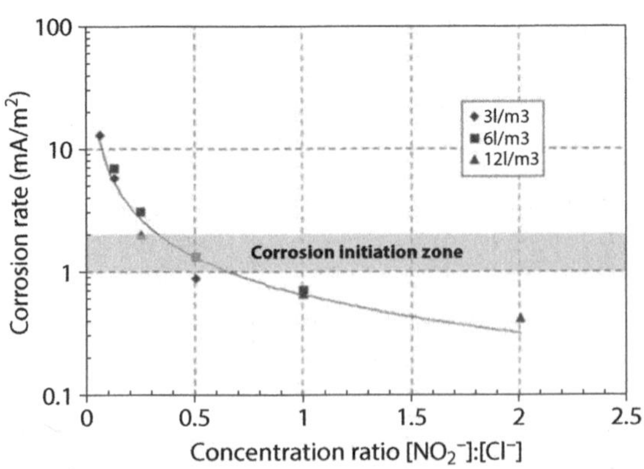

Figure 8.7 Corrosion rate in reference to $[NO_2^-]/[Cl^-]$. Reproduced from Ref. [3] with permission from [Elsevier], Copyright [2024].

two major advantages in comparison to nitrites, Firstly, they are cheaper and secondly, they are known for their high solubility and biodegradable nature (environmentally friendly). Therefore, they are used as fertilizers and food preservatives [25]. It has also been found that an anodic inhibitor can resist the anodic dissolution process and works on oxidising Fe(II) to Fe(III) and reducing NO_3^- to NO_2^-. This mechanism is suggested by Saura *et al.*, and the mechanistic process is as follows:

$$2Fe^{2+} + 2OH^- + NO_3^- \rightarrow \gamma\text{-}Fe_2O_3 + NO_2^- + 3H^+ \tag{8.5}$$

This mechanism provides information about the stable, protective layer of γ-Fe_2O_3 in an alkaline environment.

Most commonly used nitrate-based inorganic corrosion inhibitor is calcium nitrate. This compound is already used in concrete production as a set accelerator. The compound affects the chloride diffusion into concrete as nitrate competes with chloride ions to bind with Fe^{2+} ions [26]. Justnes and Nygaard used 3.86% calcium nitrate by cement weight in concrete samples with a 0.55 W/C ratio. To induce the corrosive environment for concrete samples, 5% NaCl solution was used, and after 14 months, the samples were tested, and it was observed that no significant corrosion rate of rebar was obtained [27]. However, no intimation about the critical chloride threshold was provided. Similarly, Al-Amoudi *et al.* [28] used more than 3% of the cement weight of calcium nitrate to protect rebar and reduce the corrosion rate.

Additionally, a 0.8% chloride content threshold was detected; however, in an alkaline concrete medium 1% chloride content by cement weight is already present just by casting and curing the concrete. Therefore, the results are not reliable. The above literature findings showed that nitrates as good corrosion inhibitors to attain long term efficiency by reducing and delaying the rebar corrosion rate, but the long-term efficacy of corrosion inhibition using calcium nitrate in admixture with concrete specimens was not achieved and confirmed the low inhibition efficiency by EIS (electrochemical impedance spectroscopy), as reported by Dhouibi *et al.* [29] therefore, efficiency of nitrates is a controversial topic, as mentioned in the literature. Majority of work with nitrate ions has been done in a chloride-contaminated concrete environment. Stefanoni *et al.* reported the inefficiency of using nitrates in a carbonated environment. High pH of concrete lead to the stable iron oxide passive film. In this condition, chlorides are the only threat to the passive layer, as they rupture the layer. Nitrates quickly re-passivate the layer in alkaline pH. However, the pH reduces in carbonated concrete, it destabilises passive layer and oxidation of metal becomes faster in the presence of nitrates, which restricts the re-passivation process. The corrosion rate increased with increasing concentration, as reported by Stefanoni and co-workers [30].

8.7.1.2 Phosphate Based Inhibitor

Phosphate-based corrosion inhibitors contain phosphate ions, either with inorganic or organic base. The commonly reported phosphate compounds with inorganic bases are sodium phosphate, disodium phosphate, and monofluoro phosphate. Phosphate ions are considered to act as mixed corrosion inhibitors, reported by Yohai *et al.* [31]. They can be used either as admixed form or as migratory-type inhibitors. By admixing Na_3PO_4 in concrete mortar simulating a chloride environment, it was found that the localized corrosion was reduced to a certain extent. Phosphate based inhibitors are sometimes coupled with different organic compounds like BTA, tolyltriazole etc. and are reported to be effective in carbonated environment [32]. The major limitation of these inhibitors is that they are toxic to the environment, as they contribute to eutrophication and foster algal growth that pollutes water bodies. Higher concentrations of these compounds can cause acidification that hinders the formation of a protective film; therefore, the highest inhibition efficiency cannot be achieved.

8.7.1.3 Chromates and Molybdates

There are various other efficient inorganic corrosion inhibitors, including chromates and molybdates. Chromates are efficient corrosion inhibitors used to inhibit alloys in the aerospace industry. The utilisation of these inhibitors created sustainable drawbacks accompanying issues such as environmental degradation, serious health issues, and toxicity. Despite being efficient corrosion inhibitors [33], chromates are considered carcinogenic.

8.7.2 Organic Inhibitors

Organic compounds are the chemical compounds of covalently bonded C and H atoms and hence are generally represented as C-H. These compounds may contain other electronegative elements, such as oxygen, nitrogen, and sulfur called heteroatoms. These compounds are found in almost all living organisms. Organic compounds as corrosion inhibitors gained importance in the late 1990s, but their use started in the early 80s [12]. Initially, commercially available organic compounds were used for this purpose. Different novel organic compounds have been synthesised to obtain cost-effective, sustainable, and eco-friendly corrosion inhibitors. Their contribution towards sustainable construction is increasing due to their low production cost and the requirement of lower concentration (as compared to inorganic inhibitors) for attaining inhibition efficiency [34].

The organic inhibitors work by physisorption and chemisorption. Physisorption is the adsorption method in which the inhibitor molecules attach to the steel surface due to the electrostatic interaction between the organic species and the surface. Chemisorption is mode of adsorption in which the inhibitor molecule interacts through chemical interaction with the metal surface. These chemical interactions can occur due to the presence of extra pair of electrons (called lone pair or pi-electrons) that facilitates the chemical bonding with positive metal surface (Figure 8.8). The Organic inhibitors contain various heteroatoms like N, O or S, which have lone pairs of electrons. With these lone pairs of electrons (electron rich centres) the inhibitors bind chemically with the metal ions through chemical adsorption. Further, different functional groups like carboxylates, amines and alcohols, thiols, etc. are attached to a carbon chain length that acts as a physical barrier to restrict the movement of chloride ions towards steel surface. Some compounds have the adsorption ability by forming chelate rings with metal/steel surfaces, indicating chemical interactions of the compound with the rebar [35].

Concerning the type of protection provided by the organic compounds, they act mainly as mixed-type inhibitors, effectively reducing redox reactions, i.e. anodic and cathodic corrosion processes. Due to this, they tend to cover the whole surface of the corroding metal with a thick film composed of several monolayers and change the double-layer structure at the metal interface, thereby reducing the depolarization rate. They can act as a barrier layer by obstructing anodic and cathodic active sites or reducing

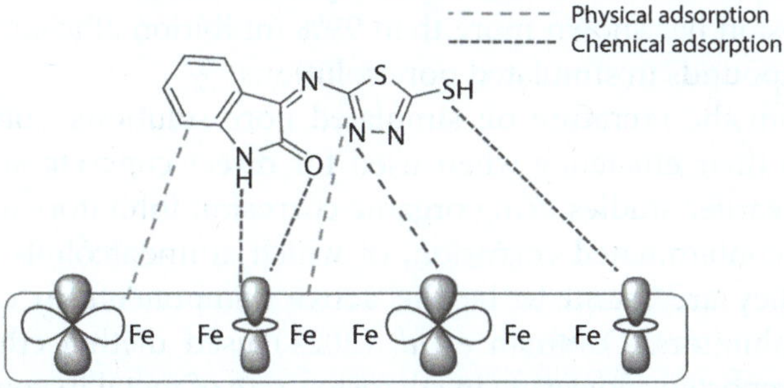

Figure 8.8 Mode of adsorption of molecule on metal surface.

electro active species to and fro transport rate to the metal surface. Morris *et al.* 2004 state that organic inhibitors provide protection by adsorption and form a protective film on passive steel surface (Moreno *et al.*, 2004) [36].

The most critical parameters that dictate a compound's effectiveness in acting as corrosion inhibitors is its chemical structure, the presence of the functional group. The most reported functional groups are amines, carboxylic acid, aldehyde, and alcohols. Further, the compounds can be either monofunctional or multifunctional depending on the number of functional groups present. For instance, alkanolamines have a combination of alcohol and amine group; amino acids are a combination of amine and carboxyl groups. A description of all these functional groups and their corresponding mechanisms is provided in Table 8.2.

Apart from the primary benefit of corrosion prevention, another benefit of using organic compounds is their positive effect on concrete's fresh and hardened properties. For instance, Owamah *et al.* (2021) reported an increase in the compressive and tensile strength of concrete after adding diethanolamine (DEA) up to 4% by weight of cement. The researcher further reported that the inhibitor acted as a plasticizer for self-compacting concrete (SCC) [37]. Similarly, Afshar *et al.* reported a 20% enhancement in compressive strength with the addition of DMEA (dimethyl-ethanolamine) [38], Due to these benefits, these compounds are recommended to be used in either admixed form or migratory form.

8.7.2.1 Performance of Organic Corrosion Inhibitors Depending on the Functional Groups

As already stated, the performance of corrosion inhibitors is directly related to the type of functional groups present. These compounds can be used to prevent chloride and carbonation-induced corrosion. The compounds with respective functional groups in the simulated pore solution are tabulated in Table 8.3. Most of the reported literature on chloride-induced corrosion has shown more than 99% inhibition efficiency by using various compounds in simulated pore solutions.

Apart from the literature on simulated pore solutions, some work is available on their efficiency when used for direct concrete applications. Most of the earlier studies using organic corrosion inhibitors were related to chloride-contaminated corrosion, of which aminoalcohols are mostly reported. They are found to be the active components in commercial corrosion inhibitors. Owamah *et al.* (2021) used diethyl ethanolamine (DEA) and reported a lowering in corrosion rate of metal in concrete from 7.456 to 4.305 mm/year [39]. Similarly, Bolzoni *et al.* admixed dimethyl

Table 8.2 Different functional groups used as corrosion inhibitors are tabulated below.

	Functional group	Molecular structure	Protection type	Mechanism	Examples
Monofunctional	Amines	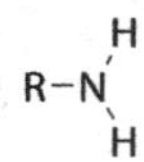	Mixed	Amines being polar act as reaction centers and adsorption occurs through a lone pair of nitrogen	Cyclohexylamines, propyl amine [37], 4-aminobenzoic acid,
	Carboxylic Acid	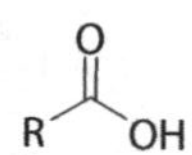	Mixed	Acids form a chelating complex with metal on a steel surface through a lone pair of oxygen	Oxalic acid, Glutaric acid [37], ascorbic acid, tartaric acid, citric acid, acetic acid, 4-amino benzoic acid
	Aldehyde	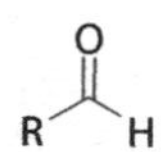	Mixed	Aldehyde participate in physical adsorption through carbonyl oxygen	Salicyldehyde, benzaldehyde

(Continued)

Table 8.2 Different functional groups used as corrosion inhibitors are tabulated below. (*Continued*)

	Functional group	Molecular structure	Protection type	Mechanism	Examples
Multifunctional	Amino-alcohols	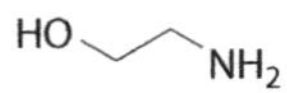	Mixed	Alkanolamines adsorb through chelation on the steel surface *via* -O and -N atoms of inhibitor	Triethyl amine, diethanol amine, D-glucosamine,
	Amino acids	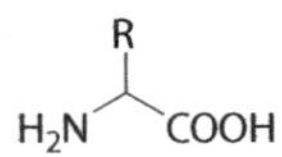	Mixed or anodic	Amino-acids adsorb through physiochemisorption on steel surface in chloride as well as carbonated environment	Arginine, Serine, cysteine, glutamic acid, asparagine, tryptophan
	Amino-ester	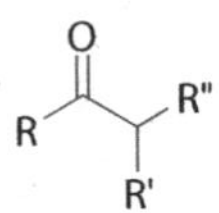	Mixed	Amino-ester blocks the concrete pores, restricting the further movement of corrosive ions into the concrete.	L-methionine methyl ester, α-amino acid ester

Table 8.3 Organic and heterocyclic corrosion inhibitors in simulated concrete pore solution.

Inhibitor	Effective functional group	Exposure conditions	IE (%)	References
2-Aminopyridine	$-NH_2$	Chloride, carbonation	84.5	[40]
4-Aminobenzoic acid	$-NH_2$, -COOH	Chloride, carbonation	99.9	[40]
Salicyaldehyde	-CHO, -OH	Chloride	96.5	[41]
Triethylphosphate	$-PO_4^{3-}$	Chloride	83.2	[41]
Acetic Acid	-COOH	Chloride	81.0	[42]
Citric acid	-COOH, -OH	Chloride	91.0	[42]
Ethylenediaminetetraacetic acid	-COOH, $-NH_2$	Chloride	77.7	[42]
Tartaric acid	-COOH, -OH	Chloride	87.2	[42]
Ascorbic acid	$-COO^-$, -OH	Chloride	96.4	[43]
Ascorbic acid	$-COO^-$, -OH	Chloride	88.9	[44]
Caffeine	-C=O, =N	Chloride	59	[45]
L-Arginine	$-NH_2$, -COOH	Chloride	70.0	[45]
Diethanolamine	$-NH_2$, -OH	Chloride	42.3	[46]

(Continued)

Table 8.3 Organic and heterocyclic corrosion inhibitors in simulated concrete pore solution. (*Continued*)

Inhibitor	Effective functional group	Exposure conditions	IE (%)	References
D-glucosamine	-O-, -OH, $-NH_2$	Chloride	73.4	[46]
L-Arginine	$-NH_2$, -COOH	Chloride	96.0	[47]
Sodium dodecylbenzene sulfonate	$-SO_4^{2-}$, Na^+	Chloride	85.7	[48]
Benzimidazole, imidazoline derivative SMIF (9:1)	=N, N-H	Chloride	88.4	[32]
Benzotriazole (BTA), disodium hydrogen phosphate (1:1)	N=N, N-H	Chloride	97.0	[49]
Diisoctyl sebacate, sodium D-gluconate, $ZnSO_4$	-OH, COO^-	Chloride	96.8	[50]
Polyacrylamide (PA), tetradecylpyridinium bromide (TDPB) (1:4)	-CONH, -C=N	Chloride	97.7	[50]
Sodium D-gluconate and sodium nitrite composite	-OH, COO^-	Chloride	89.9	[51]

(*Continued*)

Table 8.3 Organic and heterocyclic corrosion inhibitors in simulated concrete pore solution. (*Continued*)

Inhibitor	Effective functional group	Exposure conditions	IE (%)	References
Sodium nitrite, sodium hexametaphosphate, triethylamine, sugar and urea	$-NH_2$	Chloride	97.0	[52]
Sodium potassium tartrate, sodium nitrite (2:1)	$-COOH$, $-NO_2^-$	Chloride	91.2	[53]
Trisodium citrate, sodium nitrate (3:1)	$-COOH$, $-NO_3$	Chloride	91.7	[53]
Triethanclamine (TEA), tris-isopropylamine (TIPA), sodium monofluorophosphate, sodium molybdate (2.3:1:2.2:2.2)	$-OH$,$-NH_2$	Chloride	95.3	[54]

ethanolamine (DMEA) and reported a delay in corrosion initiation by nearly 10 months with the addition of the compound. Cement weight increased the critical chloride concentration to 1.2-1.3% [7].

Regarding their mechanism of action, most of the studies claimed that these compounds act as mixed inhibitors. These compounds often get adsorbed on the rebar surface, forming a protective layer of thickness 2-10 nm that reduces steel rebar's corrosion rate and mass loss. However, some amino-alcohols (i.e. ethanolamine and diethanolamine) acted as cathodic inhibitors and decreased corrosion by blocking cathodic sites only where O_2 gains electrons and is reduced to OH^- ions. On the other hand, Jamil *et al.* concluded that the inhibitor's film was formed on the steel surface, which hinders the anodic sites, and they acted as anodic inhibitors [55]. Agrawal *et al.* [56] studied the performance of different alkanol-amine compounds (viz. monoethanolamine (MEA), methyl ethanolamine (MeEA), dimethyl ethanolamine (DMEA) and ethyl monoethanolamine (EMEA)) as a corrosion inhibitor in RC structures and assessed their mechanism of action. These inhibitors worked as vapour phase corrosion inhibitors as they have vapour transmission capability that allowed them to percolate through concrete pores (spaces) to the rebar surface. These inhibitors act as mixed-type corrosion inhibitors in direct contact mode. They formed film on the steel surface and showed more excellent resistance towards corrosion. Further, they authors stated that the compounds work by chelation with metal ions accompanying a dense protective layer. The structure of the protective layer formed by the inhibitors on the rebar surface by chelation is shown in Figure 8.9.

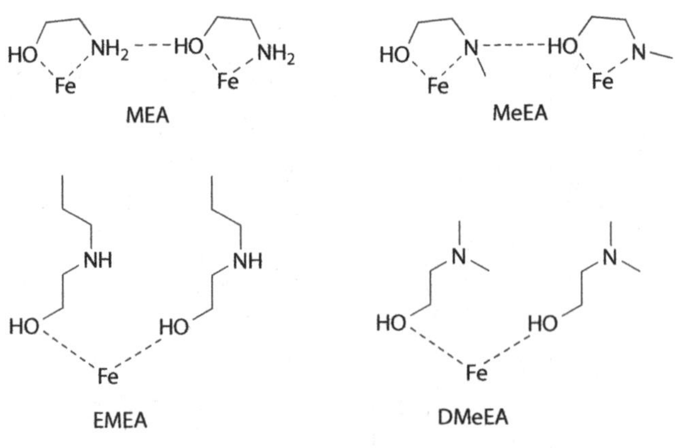

Figure 8.9 Structure of protective film formed by inhibitor by chelation.

Other than amino alcohol, amines, amino-acids, and carboxylates are also explored for their inhibition efficiency. Ormellese *et al.* analysed the inhibition efficiency of several compounds, including amines, amino acids, and carboxylates, as functional groups [57]. Carboxylates (mono- and polycarboxylates) was concluded to be the best among all in terms of inhibition capacity. It was found that despite the carboxyl group's electrostatic character and steric effect, the carboxylates compounds can also form complexes through chelation process. Ormellese *et al.* 2011 reported that amino-acids perform better than amines. Of the amino acids, sodium glutamate and sodium tartrate were among the best inhibiting compounds [58]. Cabrini *et al.* (2015) investigated the inhibitive properties of aspartic acid and compared them with nitrite in admixed concrete with critical chloride threshold concentration. The investigation discloses that the aspartate ions probably adsorb on Fe^{2+} oxide surface due to its chelating property. It inhibits corrosion through negative charge, creating repulsion by its un-adsorbed carboxylate group [59]. Ormellese *et al.* [60] tested and compared almost 80 pure organic compounds' inhibition efficiency in chloride-induced corrosion. Out of these, sodium benzoate (ester based), sodium glutamine and asparagine (amino acids) were tested in concrete. All these compounds increased critical chloride content up to 1.5-2% by weight cement. These inhibitors were poured into concrete specimens in 1% by cement weight so that the workability and mechanical properties of concrete remained unaffected.

Shen *et al.*, [61] used various amine-based compounds like guanidine, 1,6-hexamethylenediamine (HMDA), and 3-aminopropyltriethoxysilane (APS). Among these compounds, APS and HMDA were very effective as these compounds served dual purpose of pore blockers and water Repeller that helped reduce the permeation of active corrosion-causing agents. APS was found to increase the hydrophobic nature of concrete by forming Si-O-Si bonds, and their alkyl groups restricted the water molecule to avoid wetting of pores. At the same time, HMDA decreased the water absorption ability of concrete.

In another report, the author worked on long-term corrosion testing of organic compounds such as amines, amino acids, and carboxylate-based compounds in concrete and solution. sodium glutamate, sodium tartrate were the essential and efficient inhibitors that showed the increased critical chloride threshold of 50% concerning the control sample.

As discussed above, diverse work has been done in chloride-contaminated areas, but the carbonation in the concrete environment is still the major issue many researchers have not targeted. Very few researchers have worked on developing corrosion inhibitors that worked simultaneously in both chloride and carbonated concrete or solely in carbonation. Few reports are listed below:

Tong *et al.* studied the efficacy of ethanolamine in the alkalization process to repair carbonated concrete. The inhibitor successfully penetrated through concrete pores to form a stable, passive layer on the metal surface, as suggested by the detection of alcohol and amine functional groups on the steel surface (surface analysis techniques). Carbonation reduces the pH of the system, which causes corrosion to occur. Here, in the process, the OH^- ions were released in massive amounts that enhanced the alkalinity of the concrete mix so that the concrete restored its properties to a certain extent [62]. Kaur *et al.* 2016 studied two commercially available (amino ether and amino alcohol based) and two generic compounds (ethanol amine and anthranilic acid) in the concrete mix as migratory corrosion inhibitors. Of these inhibitors, amino-alcohol based and anthranilic acids successfully inhibit corrosion induced through carbonation [63, 64].

Purnima *et al.* used commercially available amino acids such as serine (Ser), cysteine (Cys), asparagine (Asn), glutamine (Gln), phenylalanine (Phe) and tryptophan (Trp) (Figure 8.10) to check their inhibition efficiency for steel in the carbonated concrete environment. All the inhibitor achieved an inhibition efficiency of more than 70%. The order of inhibition efficiency was Glu>Asn>Ser>Phe>Cys>Trp. The author also tested their migration ability in concrete to reach the rebar. These inhibitors were applied on the surface to act as migrating corrosion inhibitors. Four of these amino acids percolated through concrete pores within 15 days of application. The order of percolation was Ser>Asn>Glu=Phe. These amino acids, through carbon groups, either long chains or aromatic rings, form a

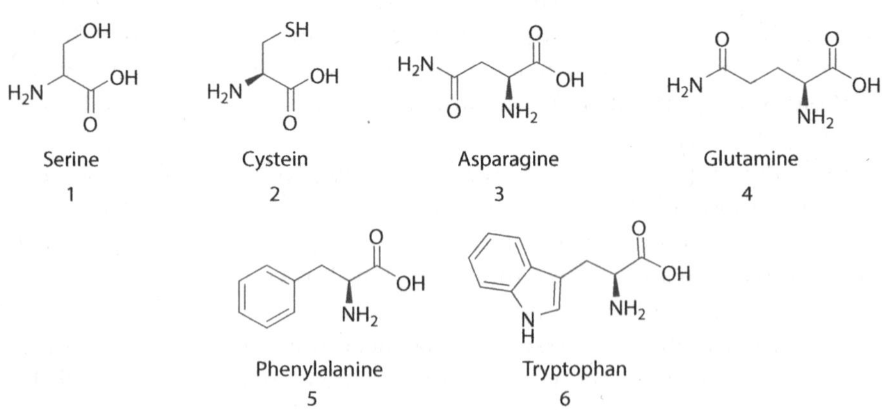
Serine 1, Cystein 2, Asparagine 3, Glutamine 4, Phenylalanine 5, Tryptophan 6

Figure 8.10 Structures of amino acids.

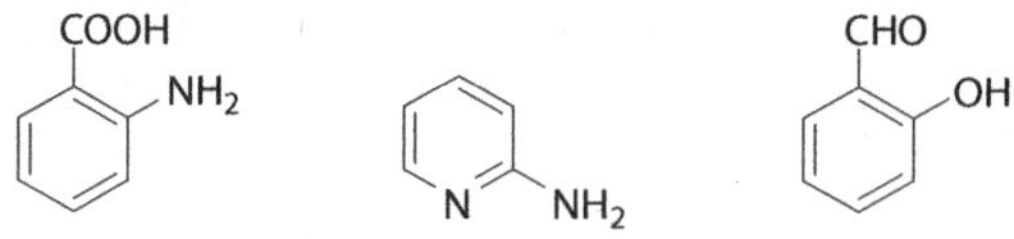

7 8 9

Figure 8.11 Structure of ABA (7), AP (8), SA (9).

passive layer by adsorbing physically and form chemical interactions with metal ions through heteroatoms, providing a physio-chemisorption mechanism [15, 65, 66].

Tiwari *et al.* studied the performance of three compounds (2 aminobenzoic acid (ABA), 2-aminopyridine (AP) and Salicyldehyde (SA)) (Figure 8.11) in the most aggressive environment with simultaneous presence of both chlorides and carbonation [67]. All these compounds were found to be efficient in aggressive exposure, and the variation of inhibition efficiency was in the order ABA>AP>SA. Further, the authors checked the migration ability of the compounds and found that ABA has a maximum percolation ability compared to AP and SA in different concretes. The inhibition efficiency of compounds varies in the order ABA>AP>SA, which is directly proportional to the amounts of certain compounds reaching the rebar surface. These compounds had no side effects on the compressive strength of the concrete.

8.7.2.2 Heterocyclic Compounds as Inhibitors

Heterocyclic compounds are the new range of corrosion inhibitors that have been recently explored for their corrosion inhibition potential. Heterocyclic inhibitors are organic compounds in which one or more carbon of the ring structure is replaced with heteroatoms like N, O, and S. Heteroatoms in the ring can improve the interactions between the metal surface and the inhibitor by better adsorption also, they help to work these compounds in different corrosive environments. Organic chemistry has a vast library of heterocyclic compounds. Commonly used heterocyclic compounds have the following categories such as imidazolines (10), thiazoles (11), triazoles (12), benzotriazoles (13), pyrazoles (14), and oxazoles (15) [68]. There structure is shown in Figure 8.12. These inhibitors have been used in oil, gas, pipeline, and industries, and they can reduce steel corrosion in neutral, acidic, and alkaline mediums. Some of these compounds have been tested for their efficiency in concrete simulated pore solution; however, a few researchers have only studied their applicability

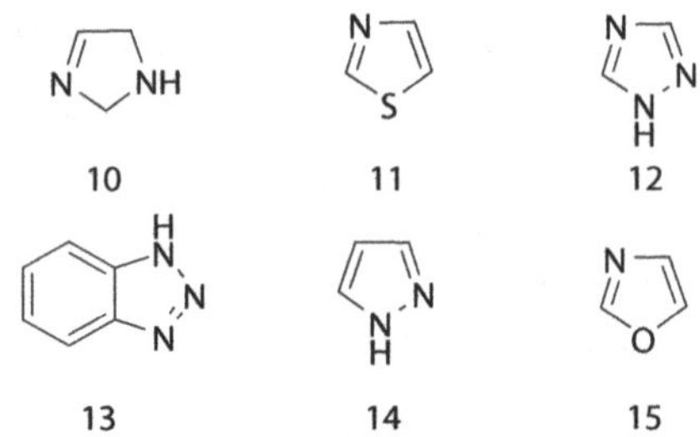

Figure 8.12 Structures of heterocyclic compounds.

in reinforced concrete. As Tiwari *et al.* reported, 2-aminopyridine is an efficient corrosion inhibitor in combined aggressive environments such as chlorides and carbonation. The authors further enhanced its migration ability by using BIEM in which the inhibitors were pushed into concrete by applying small amount of current for a limited period (Rajput *et al.*) [17].

8.8 Green Inhibitors

Though inorganic and organic inhibitors efficiently protect reinforced concrete from corrosion, some have negative impacts, such as environmental toxicity and human beings. Their hazardous effects can be experienced during synthesis and application in the field. They can cause health problems to various organs. Therefore, the focus of researchers has shifted to develop novel natural green compounds that work as corrosion inhibitors in concrete. These green compounds are extracted from natural plants and their biodegradable and vegetable waste, available freely worldwide. They are easy to synthesise without using toxic solvents through green synthesis procedures, and they are cheap, nontoxic, and biocompatible. They contain multiple bonds (C=C, C=O, C=N) and aromatic rings, which can form physical and chemical adsorption with the rebar. Further, some compounds contain heteroatoms such as N, O, and S., which have lone pairs of electrons that can form coordinated bonds with metal ions to form protective films on the metal surface. These inhibitors can form cationic ions and adsorb on the cathodic part of the metal surface, restricting the corrosion reaction. The significant challenge of using these inhibitors is their expensive and complex synthetic procedure and the need for a large amount of plant extract to satisfy the demands of the growing market.

The use of green inhibitors in concrete is the emerging field of corrosion science. The type of green inhibitor used by various authors, the type of aggressive environment, and the inhibitor efficiency, are summarised in Table 8.4. It can be seen that most of the inhibitors are tested in simulated concrete pore solutions contaminated with chlorides, with few reports on the applicability of these inhibitors on concrete specimens, specifically in carbonated environment. Almost all the authors have reported very high inhibitor efficiency. The results of the work performed in reinforced concrete have been discussed in detail here.

Asaad *et al.*, produced green inhibitor by extracting *Elaeis guineesis*, EG from oil palm leaves and doping it with silver nanoparticles (AgNPs), an effective alternative as an eco-friendly and non-toxic corrosion inhibitor (EG/AgNPs). In palm oil production, E. guineas is a waste by-product that is expelled in abundance, and silver nanoparticles are environmentally friendly compared to other metallic nanoparticles, thus making the process highly sustainably. Although the composition of the extract was complex, it accommodates various phytoconstituents that can adsorb on the metal surface to have anticorrosion behaviour. The extract components, namely, saponins, terpenoids, flavonoids, tannins, and alkaloids, mainly have aromatic rings and polar functional groups. The compound was admixed to the reinforced concrete mix, with the optimum percentage of 5% by weight, and the corrosion potential of rebar in chloride contaminated environment was monitored. It was observed that the corrosion rate tends to decrease with the days of immersion in seawater, as shown in Figure 8.13. The inhibitor adsorbs on the metal surface and it forms wider, crystalline calcium silicate hydrate (C-S-H) in concrete, which can work as pore blocker, that would block the penetration of chloride ions through the capillary pores [69].

Molina and co-workers extracted nopal slime (mucilage of Opuntia ficus-indica (OFI)) to employ it to work as corrosion inhibitor for steel rebar in mortar immersing in chloride rich environment. The synthesized compound had amino acids and carbohydrates as its major components, which have already been used as efficient corrosion inhibitors. The optimum concentration of the compound was reported to by 8% by weight of cement. When admixed in concrete, the compound decreased water penetration, ultimately enhancing impermeability of aggressive ions, which led to the reduction in permeation of chlorides [70].

Wang *et al.* [71] compared corrosion inhibition of ginger extract in mortar sample with three normally used CoIs such as $Ca(NO_2)_2$, commercial organic compound and kelp extract. The ginger extract contains 6-gingerol, 8-gingerol, cyclocurcumin, arginine as main components which could easily adsorb on the steel surface and worked with the optimum concentration of 1.5%, which enhanced both the fresh and hardened properties of the mix. The authors proposed three mechanisms of adsorption:

Table 8.4 Different green inhibitors containing various functional groups.

Testing environment	Inhibitor	Functional groups	Exposure	IE (%)	References
Concrete Specimen	Gum Arabic	-OH	Chloride, Carbonation	94.5	[73]
	Azadirachta indica (leaves)	-OH, -OAc, -COOR	Chloride	95.0	[74]
	Cocos nucifera (coconut coir dust)	-OH, -COOH	Chloride	NR	[75]
	Eucalyptus tereticornis (fine grained leaves)	-CO-, -OH	Chloride	NR	[76]
	Lawsonia inermis (powdered leaves)	-OH, -CO-	Chloride	37.0	[77]
	Nicotinic acid (vitamin B3)	=N-, -COOH	Chloride	NR	[78]
	Pyridoxine (vitamin B6)	=N-, -OH	Chloride	NR	[78]

(Continued)

Table 8.4 Different green inhibitors containing various functional groups. (*Continued*)

Testing environment	Inhibitor	Functional groups	Exposure	IE (%)	References
Concrete simulated pore solution	Davidia involucrate (leaves)	-OH, -COOH	Chloride	96.0	[79]
	Eucommia (bark and leaves)	-OH, -CONH, -COOH	Chloride	85.5	[80]
	Glycyrrhiza glabra (root)	-CO-, -COOH, -OH	Chloride	80.0	[81]
	Olea europaea (dried leaves)	-OH, -COOR	Chloride	91.9	[82]
	Pinus resinosa (powdered Conifer cone)	$-NH_2$, CH=N	Chloride	81.2	[83]
	Saccharun officinarum (bagasse ash)	-OH, -CO-	Chloride	NR	[84]
	Arabic gum, guar gum (1:2)	-OH	Chloride	NR	[85]
	Locust beam gum	-OH	Chloride	95.0	[86]
	Stevioside	-OH, -COOR	Chloride	88.6	[45]

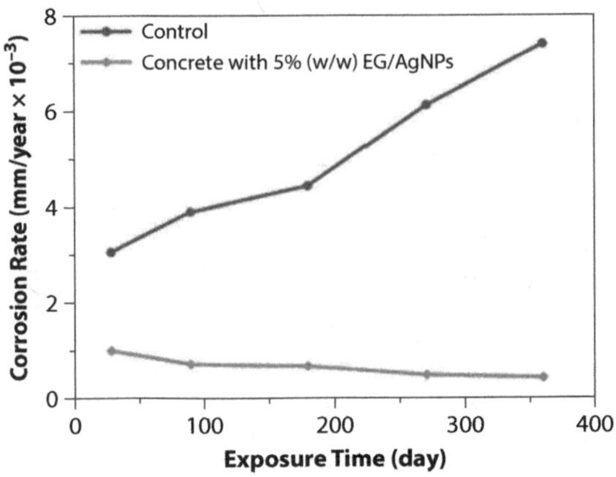

Figure 8.13 Variation of corrosion rate of rebar with days of immersion. Reproduced from Ref. [74] with permission from [Elsevier], Copyright [2024].

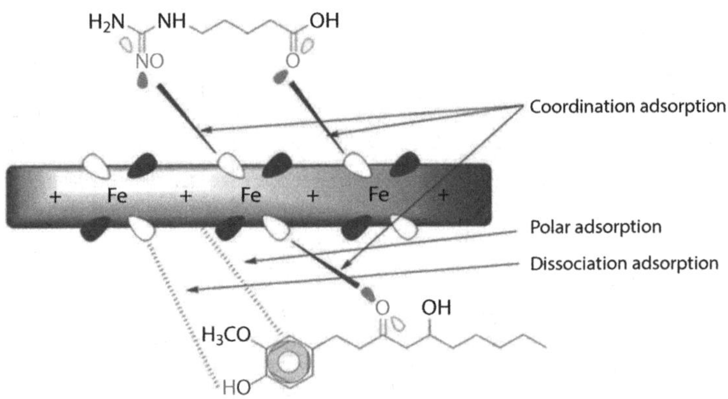

Figure 8.14 Inhibition mechanism of extracts of green inhibitor. Reproduced from Ref. [76] with permission from [Elsevier], Copyright [2024].

(i) presence of C=O, C=N bonds adsorption through electron donor-receptor interaction, (ii) dissociation by active hydrogen of the phenolic group, (iii) adsorption through π-bonds as shown in Figure 8.14.

Pradipta *et al.* [72] compared the anti-corrosion properties of green tea extract with inorganic commercial inhibitor, and observed that the former performed exceptionally better than the later. Green tea extract behaved as mixed type corrosion inhibitor as it formed the protective barrier layer on the steel surface as suggested by other inhibitors in this category.

8.9 Smart Inhibitors

The primary issue related to the use of corrosion inhibitors is their long-term performance. It is observed that the inhibitors get consumed over time and require reapplication for long-term performance. This led to a new exploration of using the smart release strategy of corrosion inhibitors. This strategy allows the release of inhibitors smartly on perturbations in the system, such as temperature changes, pH, humidity, and attack of corrosive ions to protect the surface against corrosion. Their properties and structures are modified in response to changes in the environment. Some examples are different Micro/nano-containers encapsulated corrosion inhibitors (organic/inorganic), polyelectrolyte multilayers, pH-responsive polymers, layered-double hydroxides, and organic-core shell corrosion inhibitors. The core-shell responds to various changes during the corrosion process in these inhibitors. Once the core shell is triggered, it ruptures to release the inhibitor smartly, and the inhibitor's inhibitive performance can be significantly increased.

These inhibitors have been used in simulated pores (model solutions), and their actual application in the concrete structure is underway. The smart corrosion inhibitors used in the simulated pore solution and concrete are tabulated in Table 8.5.

Wang *et al.* synthesised a microcapsule incorporating calcium hydroxide that is being released/ triggered by decreasing pH value. The release of the inhibitor led to the release of hydroxide ions, which helped restore the rebar's alkaline conditions [87]. The schematic representation of microcapsule admixed in the concrete mix is shown in Figure 8.15. Hu *et al.* and Zhu *et al.* synthesised BTA (benzotriazole)-incorporated microcapsule based on poly (ethylene oxide)-b-polystyrene (PEO-b-PS) copolymer [88]. The authors observed that the corrosion rate was reduced significantly, with subsequent adsorption Fe-B TA complex on the steel surface. The inhibitor BTA was released at low pH, which triggered the outer wall's rupture and enhanced the inhibitor's release (Figure 8.16).

While these inhibitors are sustainable corrosion inhibitors, there are still some difficulties related to the scalability and effectiveness in all the aggressive environments. Most of the studies have been carried out in simulated pore solution. Even when studied in concrete mixes, their long-term performance is not reported. Also, their effects on the mechanical properties of concrete have not been studied.

Table 8.5 Smart corrosion inhibitor in simulated concrete pore solution and concrete.

Inhibitor	Solution system	Conditions	IE (%)	References
Core-shell encapsulated benzotriazole (BTA)	SPS (saturated $Ca(OH)_2$, 0.06 mol/L NaOH	Chloride	>93.0	[89]
pH-sensitive $CaCO_3$ microparticles loaded with 5 wt% sodium lignosulfonate (SLS)	Synthetic concrete pore solution	Chloride	90.8	[90]
Zeolitic imidazolate framework (ZIF-8)	Cement extract	Chloride	87.8	[91]
$NaNO_2$ encapsulated in Urea-formaldehyde core-shell	Simulated concrete pore solution	Chloride	97.48	[92]

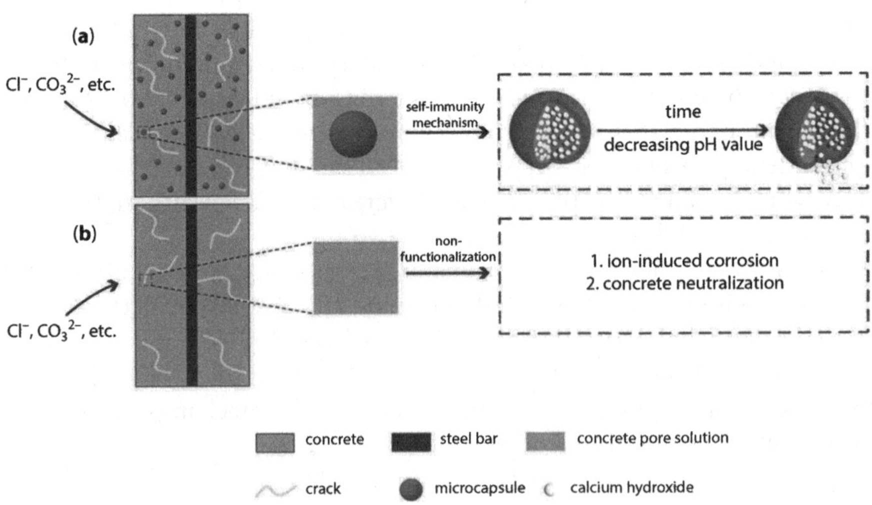

Figure 8.15 The schematic representation of microcapsule admixed in the concrete mix [87].

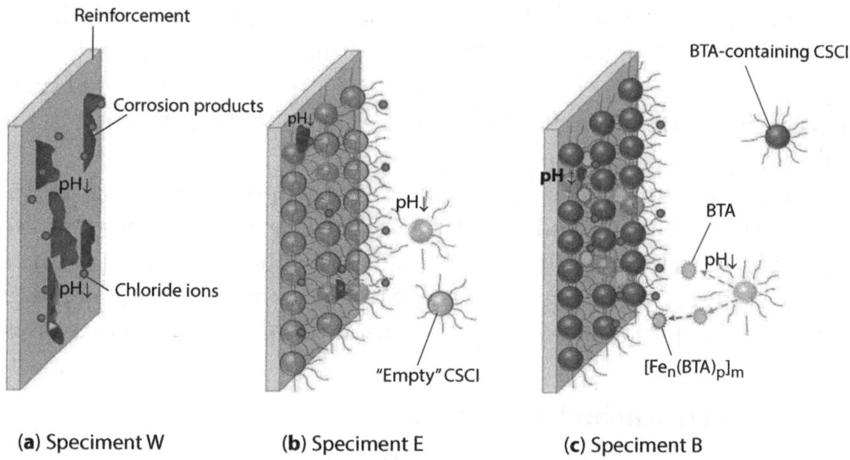

Figure 8.16 Release of inhibitor (BTA) as result of trigger of pH. Reproduced from Ref. [93] with permission from [Elsevier], Copyright [2024].

8.10 Testing and Evaluation Methods of Inhibitors

An assessment of the performance of corrosion inhibitors is necessary. Therefore, various testing and evaluation methods are used to ensure better performance and reproducibility, including mass loss measurement (gravimetrical analysis), electrochemical procedure, solution analysis, surface analysis, etc.

8.10.1 Gravimetric Analysis

This is the preliminary method to examine the corrosion inhibition efficiency. In this method, a steel specimen of known dimension is immersed in a testing solution containing inhibitors with different concentrations for a fixed duration of time and then taken out. Then, the mass loss is measured after removing corrosion products by washing them with running water and cleaning them with a soft cloth.

Mass loss measurements are efficient in the calculation of corrosion rate in mm/year, depending on the time of immersion. Similarly, the inhibition efficiency of an inhibitor can be calculated using the formula:

$$\eta\% = \frac{w_0 - w_i}{w_0} \times 100 \tag{8.6}$$

here, wo and wi are the weights of steel specimens taken out from the AS with or without inhibitor at different concentrations.

If the mass loss of the inhibited sample is less than the uninhibited specimens, it can indicate a reduced rate and greater inhibition efficiency of the inhibitor.

However, this method solely depends on manpower, sample preparation, solution, weighing and calculation. Therefore, any human error can cause variation in the result. That is why this method can only provide a rough overview of the performance of the result. Moreover, a small sample is required, and it is impossible to test mass loss directly on concrete structures.

8.10.2 Electrochemical Analysis

Electrochemical method is the testing technique used to evaluate corrosion resistance of a metal or alloy by analysing its electrochemical behaviour in a corrosive environment. These methods can be applied to liquid and solid materials. Electrochemical techniques such as half cell potential (HCP) and linear polarisation resistance (LPR) are used to test the performance of the inhibitor. HCP is used to assess the probability of corrosion in RC structures. In measuring the reinforcing steel's potential concerning half-cell potential on RC, HCP is defined as the potentail developed at steel relative to reference in an electrochemical cell. In the cell setup, concrete worked as an electrolyte, and corrosion can be measured through the potiential differenc eusing voltmeter. The more negative value of HCP indicates the maximum probability of rebar corrosion.

LPR is the electrochemical method for the measurement of corrosion resistance. By applying electrochemical perturbation to the rebar, the measurement is carried out by putting electrodes on the concrete surface. The slight potential shift (ranging ± 20mV) from rest potential is considered perturbation. The resulting current and resistance are related as follows:

$$R_p = \Delta E/\Delta I \tag{8.7}$$

However, the corrosion current density is calculated by,

$$I_{corr} = B/R_p \tag{8.8}$$

Here, B is proportionality constant

LPR is a Non-destructive technique that evaluate corrosion rate and corrosion current density. Therefore, it can be used for on-site application (RC structures).

Potentiodynamic polarisation (PDP) is limited to laboratory applications as it is adestructive technique and can not be used for direct structure applications.

8.10.3 Surface Analysis

The performance of inhibitors and their action to protect the rebar surface can be evaluated using various surface analysis techniques. These techniques help decide the morphology of the rebar surface. Most common surface analysis techniques used are Optical Microscopy (OM), Field Emission Scanning Electron Microscopy (FE-SEM), Tunneling Electron Microscopy (TEM), Sanning Tunneling Microscopy (STM), Atomic Field Microscopy (AFM). Different techniques are used to study various properties of the surface. For instance, TEM and STM techniques used to evaluate the material's properties such as structure and composition at molecular level like surface thickness and passive layer thickness formed by the corrosion in hibitor on the rebar surface. However, FE-SEM and OM were used to study the topography of the passive layer of inhibitor adsorbed on the rebar surface. Moreover, surface roughness occurred due to the corrosion products formed can be analyzed using AFM. These techniques are also limited to rebar application and are not applicable on direct site application as they are all instrumentation-based testings.

8.10.4 Standard Test Method

Apart from the other method, a standard designated test method (ASTM G109-23) covers the methodology for determining the effect of chemical admixtures on the corrosion of steel embedded in concrete exposed to chloride-induced corrosion [93].

8.11 Challenges and Future Perspectives

Apart from the significant development of corrosion inhibitors, some challenges still need to be sought for smart corrosion inhibitors to be applied in actual working conditions. The production cost of these inhibitor is very high as the raw material is very costly and produces low output, which restricts its application in the practical world. Moreover, these inhibitors (micro-shell capsules) are prepared using toxic resins and polymers. Also, these inhibitors are well studied in concrete pore solutions and scarcely used for their practical applicability, such as in mortar. Therefore, from a future perspective, cost-effective raw materials like industrial waste must be utilized for their synthesis, and simplified production process must be developed. Also, initiatives must be taken to replace toxic substances with more environmentally friendly alternatives to ensure their green applications. Most inhibitors are used for preventive strategies on fresh and new structures; however, they must be developed to repair existing concrete structures.

8.12 Conclusion

This chapter explores corrosion inhibitors' role in enhancing reinforced concrete's durability. Various inorganic, organic, heterocyclic, and green inhibitors have been used to increase concrete structures' durability and life span. Inorganic inhibitor ions such as nitrites, nitrates, molybdates, and chromates were efficient in attaining optimum inhibition efficiency. However, these groups offered some health and environmental hazards, so the introduction of organic corrosion inhibitors came into play. Various organic corrosion inhibitors contain amino acids, alcohol amines, and carboxy groups that inhibit corrosion through passivation and adsorption. These inhibitors are costly, and some harm the environment and human beings. All inhibitors are efficient and demonstrate good corrosion resistance ability. Due to toxicity, other health hazards and cost, researchers have made various attempts to modulate effective, non-toxic corrosion inhibitors. Therefore, the synthesis of green corrosion inhibitors is fascinating in the area of corrosion inhibitors. These inhibitors are prepared using an environmentally friendly green synthetic procedure. Also, innovative corrosion inhibitors are in the emerging field of concrete corrosion to counteract the need for the application of inhibitor repeatedly. These inhibitors are well-known corrosion inhibitors in concrete pore solutions; however, their applicability is tedious and costly. Further, detailed cost analysis and sustainable preparation methods for innovative and green inhibitors are required, enabling them to be used for commercial applications.

AI disclosure statement: The author confirms that no artificial intelligence (AI)-generated content or AI-assisted writing tools were used in the preparation of this chapter (including summaries and the abstract). All text, analysis, and conclusions are the result of the author's original work.

References

1. Rivetti, M.L.S., da Silva Andrade Neto, J., de Amorim Junior, N.S., Ribeiro, D.V., Corrosion inhibitors for reinforced concrete, in: *Corrosion inhibitors, principles, and recent applications*, 2018.
2. Sawada, S., Kubo, J., Page, C.L., Page, M.M., Electrochemical injection of organic corrosion inhibitors into carbonated cementitious materials: Part 1. Effects on pore solution chemistry. *Corros. Sci.*, 49, 3, 1186–1204, 2007.
3. Bolzoni, F., Brenna, A., Ormellese, M., Recent advances in the use of inhibitors to prevent chloride-induced corrosion in reinforced concrete. *Cem. Concr. Res.*, 154, 106719, 2022.
4. Zomorodian, A. and Behnood, A., Review of corrosion inhibitors in reinforced concrete: Conventional and green materials. *Buildings*, 13, 5, 1170, 2023.
5. Alvarez, L.X., Troconis de Rincón, O., Escribano, J., Rincon Troconis, B.C., Organic compounds as corrosion inhibitors for reinforced concrete: a review. *Corros. Rev.*, 41, 6, 617–634, 2023.
6. Gartner, N., Kosec, T., Legat, A., The efficiency of a corrosion inhibitor on steel in a simulated concrete environment. *Mater. Chem. Phys.*, 184, 31–40, 2016.
7. Bolzoni, F.A.B., II, Coppola, L., Goidanich, S., Lazzari, L., Ormellese, M., Pedeferri, M.P., Corrosion inhibitors in reinforced concrete structures Part 1: Preventative technique. *Corros. Eng. Sci. Technol.*, 39, 3, 219–228, 2004.
8. Song, H.W. and Saraswathy, V., Analysis of corrosion resistance behavior of inhibitors in concrete using electrochemical techniques. *Met. Mater. Int.*, 12, 4, 323–329, 2006.
9. Elsener, B. and Angst, U., Corrosion inhibitors for reinforced concrete, in: *Science and Technology of Concrete Admixtures*, Elsevier Ltd, Woodhead Publisher, 2015.
10. de Rincón, T., Perez, O., Paredes, E., Caldera, Y., Urdaneta, C., Sandoval, I., Long-term performance of ZnO as a rebar corrosion inhibitor. *Cem. Concr. Compos.*, 24, 1, 79–87, 2002.
11. Lee, H.S., Saraswathy, V., Kwon, S.J., Karthick, S., Corrosion inhibitors for reinforced concrete: A review. *Corros. Inhibitors Princ. Recent Appl.*, 95–120, 2018.
12. Wang, X., Liu, J., Jin, M., Yan, Y., Tang, J., Jin, Z., A review of organic corrosion inhibitors for resistance under chloride attacks in reinforced concrete: Background, Mechanisms and Evaluation methods. *Constr. Build. Mater.*, 433, 136583, 2024.
13. Rivetti, M.L.S., Netto, J., Junior, M.A., Ribeiro, D.V., Corrosion inhibitors for reinforced concrete. *Corros. Inhibitors Princ. Recent Appl.*, 35–58, 2018.
14. Goyal, A., Ganjian, E., Pouya, H.S., Tyrer, M., Inhibitor efficiency of migratory corrosion inhibitors to reduce corrosion in reinforced concrete exposed to high chloride environment. *Constr. Build. Mater.*, 303, August, 124461, 2021.

15. Tiwari, A.K., Goyal, S., Luxami, V., Developing the inhibition mechanism for amide-based amino acids in carbonated concrete environment and assessing the migration ability in concrete. *J. Build. Eng., 76*, 107048, 2023.
16. Xu, C., Jin, W.L., Wang, H.L., Li, Z.Y., Mao, J.H., Organic corrosion inhibitor of triethylenetetramine into chloride contamination concrete by eletro-injection method, Constr. *Build. Mater.*, 115, 602–617, 2016.
17. Rajput, T.S., Goyal, A., Goyal, S., Effect of 2-amino pyridine inhibitor on the rehabilitation of chloride contaminated reinforced concrete with bidirectional electromigration method. *Constr. Build. Mater.*, 412, 134754, 2024.
18. Soeda, K. and Ichimura, T., Present state of corrosion inhibitors in Japan. *Cem. Concr. Compos.*, 25, 117–122, 2003.
19. Sagoe-Crentsil, K., Yilmaz, V., Glasser, F., Properties of inorganic corrosion inhibition admixtures in steel-containing OPC mortars. Part 1: chemical properties. *Adv. Cem. Res.*, 4, 15, 91–96, 1992.
20. Rosenberg, A.M., Gaidis, J.M., Kossivas, T.G., Previte, R.W., Corrosion Inhibitor Formulated With Calcium Nitrite for Use in Reindorced Concrete, 1977.
21. Andrade, C., Alonso, C., Gonzales, J., Some laboratory experiments on the inhibitor effect of sodium nitrite on reinforcement corrosion. *Cem. Concr. Aggr.*, 8, 110, 1986.
22. Dhouibi, L., Refait, P., Triki, E., Génin, J.-M., Interactions between nitrites and Fe(II)- containing phases during corrosion of iron in concrete-simulating electrolytes. *J. Mater. Sci.*, 41, 4928–4936, 2006.
23. Ann, K.Y., Jung, H.S., Kim, H.S., Kim, S.S., Moon, H.Y., Effect of calcium nitrite-based corrosion inhibitor in preventing corrosion of embedded steel in concrete. *Cem. Concr. Res.*, 36, 3, 530–535, 2006.
24. Okeniyi, J.O., Omoniyi, O.M., Okpala, S.O., Loto, C.A., Popoola, A.P., II, Effect of ethylenediaminetetraacetic disodium dihydrate and sodium nitrite admixtures on steel-rebar corrosion in concrete. *Eur. J. Environ. Civ. Eng., 17*, 5, 398–416, 2013.
25. Justnes, H. and Nygaard, E.C., Technical calcium nitrate as set accelerator for cement at low temperatures. *Cem. Concr. Res.*, 25, 8, 1766–1774, 1995.
26. Justnes, H., Concrete corrosion problems solved by calcium nitrate, in: *25th Conference on Our World in Concrete and Structures*, pp. 349–356, 2000.
27. Justnes, H. and Nygaard, E.C., The influence of technical calcium nitrate additions on the chloride binding capacity of cement and the rate of chloride induced corrosion of steel embedded in mortars, in: *Corrosion protection of steel in concrete*, pp. 491–502, 1994.
28. Al-Amoudi, O.S.B., Maslehuddin, M., Lashari, A.N., Almusallam, A.A., Effectiveness of corrosion inhibitors in contaminated concrete. *Cem. Concr. Compos.*, 25, 4, 439–449, 2003.
29. Dhouibi, L., Triki, E., Raharinaivo, A., The application of electrochemical impedance spectroscopy to determine the long-term effectiveness of corrosion inhibitors for steel in concrete. *Cem. Concr. Compos.*, 24, 1, 35–43, 2002.

30. Stefanoni, M., Angst, U., Elsener, B., Influence of calcium nitrate and sodium hydroxide on carbonation-induced steel corrosion in concrete. *Corrosion*, *75*, 7, 737–744, 2019.
31. Yohai, L., Schreiner, W., Vázquez, M., Valcarce, M.B., Phosphate ions as effective inhibitors for carbon steel in carbonated solutions contaminated with chloride ions. *Electrochim. Acta*, 202, 231–242, 2016.
32. Wang, D., Pan, C., Liu, Z., Chen, K., Chen, N., Liu, S., Synergetic effect of two inhibitors for enhanced corrosion protection on the reinforcing steel in the chloride-contaminated carbonated solutions. *Constr. Build. Mater.*, 286, 122916, 2021.
33. Winkler, D.A., Breedon, M., White, P., Hughes, A.E., Sapper, E.D., Cole, I., Using high throughout experimental data and *in silico* models to discover alternatives to toxic chromate corrosion inhibitors. *Corros. Sci.*, 106, 229–235, 2016.
34. Söylev, T.A. and Richardson, M.G., Corrosion inhibitors for steel in concrete: state-of-the-art report. *Constr. Build. Mater.*, 22, 4, 609–622, 2008.
35. Moreno, M., Morris, W., Alvarez, M.G., Duffó, G.S., Corrosion of reinforcing steel in simulated concrete pore solutions effect of carbonation and chloride content. *Corros. Sci.*, 46, 11, 2681–2699, 2004.
36. Mohamed, A., Visco, D.P., Bastidas, D.M., Significance of p-electrons in the design of corrosion inhibitors for carbon steel in simulated concrete pore solution. *Corrosion*, 77, 976–990, 2021.
37. Afshar, A., Jahandari, S., Rasekh, H., Afshar, A., Shokrgozar, A., Corrosion resistance evaluation of rebars with various primers and coatings in concrete modified with different additives. *Constr. Build. Mater.*, 262, 120034, 2020.
38. Tiwari, A.K., Dogra, P., Goyal, S., Luxami, V., Water-based corrosion inhibitor for prevention and repair of reinforcement in chloride and carbonated environment. *Corros. Eng. Sci. Technol.*, 1478422X241257033, 59, 6, 2024.
39. Owamah, H.I., Ikpeseni, S.C., Dharmaraj, R., Gopikumar, S., Malathy, R., Uma, S., Sada, S.O., Oyebisi, S.O., Influence of diethanolamine on the properties of concrete, corrosion rate of rebar and renewable energy generation. *Arab. J. Sci. Eng.*, 46, 11487–11496, 2021.
40. Tiwari, A., Goyal, S., Luxami, V., Chakraborty, M.K., Prabhakar, G., Assessment of corrosion inhibition efficiency of generic compounds having different functional groups in carbonated pore solution with chlorides and migration ability in concrete. *Constr. Build. Mater.*, 290, 123275, 2021.
41. Teymouri, F., Allahkaram, S.R., Shekarchi, M., Azamian, I., Johari, M., A comprehensive study on the inhibition behaviour of four carboxylate-based corrosion inhibitors focusing on efficiency drop after the optimum concentration for carbon steel in the simulated concrete pore solution. *Constr. Build. Mater.*, 296, 123702, 2021.
42. Kasatkin, V.E., Dorofeeva, V.N., Kasatkina, I.V., Korosteleva, I.G., Kornienko, L.P., Andreev, N.N., Gedvillo, I.A., Zhmakina, A.S., Ascorbic acid as a

corrosion inhibitor of steel in chloride-containing solutions of calcium hydroxide. *Int. J. Corros. Scale Inhib.*, 11, 727–751, 2022.

43. Brixi, N.K., Sail, L., Bezzar, A., Application of ascorbic acid as green corrosion inhibitor of reinforced steel in concrete pore solutions contaminated with chlorides. *J. Adhes. Sci. Technol.*, 36, 1176–1199, 2022.
44. Kurtay, M., Gerengi, H., Chidiebere, A., Yildiz, M., , Corrosion inhibition of reinforcement steel in mixture water by caffeine and. *Adhes. Sci. Technol.*, 36, 134–167, 2022.
45. Ramesh, T., Suji, D., Quraishi, M.A., Evaluation of greener corrosion-inhibiting admixtures for steel reinforcements in concrete. *Arab. J. Sci. Eng.*, 47, 13451–13466, 2022.
46. Singh, J.K., Yang, H.M., Lee, H.S., Mandal, S., Aslam, F., Alyousef, R., Role of L-arginine on the formation and breakdown of passive film onto the steel rebars surface in chloride contaminated concrete pore solution. *J. Mol. Liq.*, 337, 116454, 2021.
47. Tang, R., Liu, G., Shi, C., Zhang, B., Tian, J., Gan, H., Liu, Y., Inhibition and adsorption behavior of sodium dodecyl-benzene sulfonate on Q235 steel in simulated concrete pore fluid. *J. Chin. Soc. Corros. Prot.*, 41, 857–863, 2021.
48. Xiong, L.P., Wang, P.J., He, Z.Y., Chen, Q., Pu, J.B., Zhang, R.H., Corrosion behaviors of Q235 carbon steel under imidazoline derivatives as corrosion inhibitors: experimental and computational investigations. *Arab J. Chem.*, 14, 102952, 2021.
49. Guo, Y., Jin, P.A., Dong, S.G., Du, R.G., Lin, C.J., Effect of an environmentally friendly diisooctyl sebacate-based mixed corrosion inhibitor on reinforcing steel. *Acta Phys. Chim. Sin.*, 38, 2003033, 2022.
50. Shi, R.L., Zhang, J., Shi, S.H., Synergistic inhibition effect of polyacrylamide and tetradecylpyridinium bromide on corrosion behavior of carbon steel rebar in simulated cement-mortar pore solution. *Int. J. Electrochem. Sci.*, 16, 210626, 2021.
51. Xu, P.Z., Zhou, J., Li, G.G., Wang, P.G., Wang, P., Li, F., Zhang, B.H., Chi, H., Corrosion inhibition efficiency of compound nitrite with D-sodium gluconate on carbon steel in simulated concrete pore solution. *Constr. Build. Mater.*, 288, 12, 2021.
52. Park, T.J. and Jang, H., Effects of inhibitors on corrosion resistance of steel in CaCl2 solution based on response surface analysis. *Corros. Sci. Technol. Korea*, 20, 129–142, 2021.
53. Johari, M., Allahkaram, S.R., Teymouri, F., Azamian, I., Shekarchi, M., The effect of carboxylate compounds on controlling nitrite's environmental side effects for carbon steel corrosion protection in the simulated concrete pore solution. *Constr. Build. Mater.*, 308, 125037, 2021.
54. Cui, L., Hang, M.Y., Huang, H.H., Gao, X.J., Experimental study on multi-component corrosion inhibitor for steel bar in chloride environment. *Constr. Build. Mater.*, 313, 125533, 2021.

55. Jamil, H.E., Shriri, A., Boulif, R., Bastos, C., Montemor, M.F., Ferreira, M.G.S., Electrochemical behaviour of amino alcohol-based inhibitors used to control corrosion of reinforcing steel. *Electrochim. Acta*, 49, 17–18, 2753–2760, 2004.
56. Agrawal, H., Patil, A.A., Muhammad, S., Performance assessment of alkoamine corrosion inhibitors for use as an admixture in reinforced concrete structures: Experimental investigation and mechanism of action. *J. Build. Eng.*, 95, 110266, 2024.
57. Ormellese, M., Lazzari, L., Goidanich, S., Fumagalli, G., Brenna, A., A study of organic substances as inhibitors for chloride-induced corrosion in concrete. *Corros. Sci.*, 51, 12, 2959–2968, 2009.
58. Ormellese, M., Bolzoni, F., Lazzari, L., Brenna, A., Pedeferri, M., Organic substances as inhibitors for chloride-induced corrosion in reinforced concrete. *Mater. Corros.*, 62, 2, 170–177, 2011.
59. Cabrini, M., Fontana, F., Lorenzi, S., Pastore, T., Pellegrini, S., Effect of Organic Inhibitors on Chloride Corrosion of Steel Rebars in Alkaline Pore Solution. *J. Chem.*, 2015, 1, 521507–521517, 2015.
60. Ormellese, M., Bolzoni, F., Goidanich, S., Pedeferri, M., Brenna, A., Corrosion inhibitors in reinforced concrete structures. Part 3 – migration of inhibitors into concrete –. *Corros. Eng. Sci. Technol.*, 46, 334–339, 2011.
61. Shen, L., Jiang, H., Cao, J., Zhang, M., Zhang, H., A comparison study of the performance of three electro-migrating corrosion inhibitors in improving the concrete durability and rehabilitating decayed reinforced concrete. *Constr. Build. Mater.*, 238, 117673, 2020.
62. Tong, Y., Wang, J., Ye, J., Li, M., Sun, J., Li, Z., Lin, S., Effect of ethanolamine as a steel corrosion inhibitor in realkalization of carbonated concrete. *J. Asian Archit. Build. Eng.*, 24, 3, 1475–1486, 2024.
63. Kaur, K., Goyal, S., Bhattacharje, B., Oh, T., Kim, J., Lee, C., Park, S., Electrochemical Impedance Spectroscopy to study the carbonation behavior of concrete treated with corrosion inhibitors, 172–182, 2017a.
64. Kaur, K., Goyal, S., Bhattacharje, B., Oh, T., Kim, J., Lee, C., Park, S., Electrochemical Impedance Spectroscopy to study the carbonation behavior of concrete treated with corrosion inhibitors. *J. Adv. Concr. Technol.*, 15, 12, 738–748, 2017b.
65. Purnima, G,.S. and Luxami, V., Exploring the corrosion inhibition mechanism of Serine (Ser) and Cysteine (Cys) in alkaline concrete pore solution simulating carbonated environment. *Constr. Build. Mater.*, *384*, 131433, 2023.
66. Purnima, G,.S. and Luxami, V., Corrosion inhibition mechanism of aromatic amino acids for steel in alkaline pore solution simulating carbonated concrete environment. *Mater. Corros.*, *75*, 1, 39–60, 2024.
67. Tiwari, A.K., Goyal, S., Luxami, V., Influence of corrosion inhibitors on two different concrete systems under combined chloride and carbonated environment. *Structures*, 48, 717–735, 2023, February.

68. Yoo, S.H., Kim, Y.W., Chung, K., Baik, S.Y., Kim, J.S., Synthesis and corrosion inhibition behavior of imidazoline derivatives based on vegetable oil. *Corros. Sci.*, *59*, 42–54, 2012.
69. Asaad, M.A., Ismail, M., Tahir, M.M., Huseien, G.F., Raja, P.B., Asmara, Y.P., Enhanced corrosion resistance of reinforced concrete: Role of emerging eco-friendly Elaeis guineensis/silver nanoparticles inhibitor. *Constr. Build. Mater.*, *188*, 555–568, 2018.
70. Martinez-Molina, W., Torres-Acosta, A., Hernández-Leos, R., Alonso-Guzman, E., Mendoza-Pérez, I., Martinez-Peña, I., The inhibitive properties of Nopal slime on the corrosion of steel in chloride-contaminated mortar. *Anti-Corros. Methods Mater.*, *63*, 1, 65–71, 2016.
71. Wang, W., Song, Z., Guo, M., Jiang, L., Xiao, B., Jiang, Q., Xu, N., Employing ginger extract as an eco-friendly corrosion inhibitor in cementitious materials. *Constr. Build. Mater.*, *228*, 116713, 2019.
72. Pradipta, I., Kong, D., Tan, J.B.L., Natural organic antioxidants from green tea inhibit corrosion of steel reinforcing bars embedded in mortar. *Constr. Build. Mater.*, *227*, 117058, 2019.
73. Asaad, M.A., Hussein, M.N., Alkadhimi, A.M., Saleh, A.T., Commercial and green corrosion inhibitors for reinforced concrete structures: a review. *Int. J. Curr. Sci. Res. Rev.*, 04, 89–99, 2021a.
74. Valdez-Salas, B., Vazquez-Delgado, R., Salvador-Carlos, J., Beltran-Partida, E., Salinas-Martinez, R., Cheng, N., Azadirachta indica leaf extract as green corrosion inhibitor for reinforced concrete structures: corrosion effectiveness against commercial corrosion inhibitors and concrete integrity. *Materials*, 14, 3326, 2021.
75. Ramesh, T., Chauhan, D.S., Quraishi, M.A., Coconut coir dust extract (CCDE) as green corrosion inhibitor for rebar steel in concrete environment. *Int. J. Corros. Scale Inhib.*, 10, 618–633, 2021.
76. Al-Akhras, N. and Mashaqbeh, Y., Potential use of eucalyptus leaves as green corrosion inhibitor of steel reinforcement. *J. Build. Eng.*, 35, 101848, 2021.
77. Kumar, N. and Saharan, S., Use of Lawsonia inermis and sodium nitrite as surface applied corrosion inhibitors. *Int. J. Sustain. Constr. Eng. Technol.*, 13, 1–8, 2022.
78. Xu, Q., Hou, D., Zhang, H., Wang, P., Wu, D., Liu, C., Ding, Z., Zhang, M., Xin, Z., *et al.*, Understanding the effect of vitamin B3, B6 and C as a corrosion inhibitor on the ordinary Portland cement hydration: experiments and DFT study. *Constr. Build. Mater.*, 331, 127294, 2022.
79. Zhang, M., Davidian involucrata leaves extract as green corrosion inhibitor for low carbon steel in concrete pore solution containing chloride ions. *Int. J. Electrochem. Sci.*, 16, 21042, 2021.
80. Liu, C.Y., Trans-1,4-polyisoprene (TPI) extracted from Eucommia bark as natural corrosion inhibitor for carbon steel in the simulated concrete pore solution. *Int. J. Electrochem. Sci.*, 17, 220615, 2022.

81. Naderi, R., Bautista, A., Velasco, F., Soleimani, M., Pourfath, M., Use of licorice plant extract for controlling corrosion of steel rebar in chloride-polluted concrete pore solution. *J. Mol. Liq.*, 346, 117856, 2022.
82. Ben Harb, M., Abubshait, S., Etteyeb, N., Kamoun, M., Dhouib, A., Olive leaf extract as a green corrosion inhibitor of reinforced concrete contaminated with seawater. *Arab. J. Chem.*, 13, 4846–4856, 2020.
83. Subbiah, K., Lee, H.S., Mandal, S., Park, T., Conifer cone (Pinus resinosa) as a green corrosion inhibitor for steel rebar in chloride-contaminated synthetic concrete pore solutions. *ACS Appl. Mater. Interfaces*, 13, 43676–43695, 2021.
84. Gromboni, M.F., Sales, A., Rezende, M.D.M., Moretti, J.P., Corradini, P.G., Mascaro, L.H., Impact of agro-industrial waste on steel corrosion susceptibility in media simulating concrete pore solutions. *J. Clean. Prod.*, 284, 124967, 2021.
85. Hejazian, S. and Ahmadyari-Sharamin, M., Synergistic effect of Arabic and guar gums on corrosion behavior of steel reinforced concrete in a simulator environment. *Case Stud. Constr. Mater.*, 16, e01024, 2022.
86. Jano, A., Lame, A., Kokalari, E., Test of inhibitors for preventing corrosion of steel reinforcement in concrete. *Ovidius Univ. Annal. Chem.*, 32, 110–113, 2021.
87. Wang, Y., Fang, G., Ding, W., Han, N., Xing, F., Dong, B., Self-immunity microcapsules for corrosion protection of steel bar in reinforced concrete. *Sci. Rep.*, *5*, 1, 18484, 2015.
88. Hu, J., Zhu, Y., Hang, J., Zhang, Z., Ma, Y., Huang, H., Wei, J., The effect of organic core–shell corrosion inhibitors on corrosion performance of the reinforcement in simulated concrete pore solution. *Constr. Build. Mater.*, *267*, 121011, 2021.
89. Hu, J., Zhu, Y.Y., Hang, J.Z., Zhang, Z.M., Ma, Y.W., Huang, H.L., Yu, Q.J., Wei, J.X., The effect of organic core-shell corrosion inhibitors on corrosion performance of the reinforcement in simulated concrete pore solution. *Constr. Build. Mater.*, 267, 121011, 2021.
90. Liu, W.L., Li, J.S., Huang, X.Q., Bi, J.Y., Corrosion protection of Q235 steel using epoxy coatings loaded with calcium carbonate microparticles modified by sodium lignosulfonate in simulated concrete pore solutions. *Materials*, 14, 1982, 2021.
91. Wang, Y.Y., Hu, J., Ma, Y.W., Zhang, Z.M., Huang, H.L., Wei, J.X., Yin, S.H., Yu, Q.J., A novel high-efficient MOFs-based corrosion inhibitor for the reinforcing steel in cement extract. *Constr. Build. Mater.*, 317, 125946, 2022c.
92. Song, Z., Liu, Y., Jiang, L., Pei, C., Xu, N., Chu, H., Guo, M., Using ultrasonic wave to trigger microcapsule inhibitor against chloride-induced corrosion of carbon steel in simulated concrete pore solution. *Constr. Build. Mater.*, *311*, 125331, 2021.
93. Standard, A.S.T.M., G109. Standard Test Method for Determining Effects of Chemical Admixtures on Corrosion of Embedded Steel Reinforcement in Concrete Exposed to Chloride Environments. *ASTM Int.*, 2007.

9

Case Studies on Concrete Corrosion

Ali Dashan[1,2], Hedieh Kazemi[1], Bahram Ramezanzadeh[1]* and Ghasem Bahlakeh[2]

[1]*Faculty of Corrosion & Surface Coating, Institute for Color Science and Technology, Tehran, Iran*
[2]*Department of Chemical Engineering, Faculty of Engineering, Golestan University, Aliabad Katoul, Iran*

Abstract

Corrosion in concretes poses a significant challenge for concrete structures, leading to potential environmental and economic damage. To address the inherent weakness in concrete, reinforcing bars (rebar) are used to enhance the tensile strength, thereby increasing the overall durability and resilience of the structure. While the high pH of concrete generally protects rebar from corrosion, various factors, such as carbonation and water infiltration into the porous concrete, can compromise this protection. This can result in fractures within the rebar, weakening the bond between the metal and concrete and diminishing the structural integrity. Several mechanisms have been developed to protect concrete structures and combat these issues. These include cathodic protection methods, such as galvanic sacrificial systems and impressed current systems, as well as the application of anti-corrosion coatings on concrete surfaces. A lot of investigation has also been done on the usage of corrosion inhibitors, which are divided into three groups: anodic, cathodic, and mixed. This chapter will explore the diverse approaches employed to protect concrete from corrosion, reviewing the principles behind each approach and examining relevant case studies. This thorough examination aims to offer a deeper conception of the practical applications and efficacy of various corrosion protection strategies in preserving the integrity and longevity of concrete structures.

***Keywords*:** Concrete, corrosion inhibitor, coating, cathodic protection, rebar

**Corresponding author*: ramezanzadeh-bh@icrc.ac.ir

Ruby Aslam, Qihui Wang, Zhitao Yan and Afroz Aslam (eds.) Corrosion in Concrete Structures: Integrating Advanced Technologies and Sustainable Practices, (257–290) © 2026 Scrivener Publishing LLC

9.1 Introduction

Concrete stands as a cornerstone material in the evolution of various industries, particularly within construction and its associated fields. Concrete plays an essential role in building construction as a composite formed from cement, water, sand, stone, and aggregates. Despite its widespread use, concrete's inherent mechanical properties, precisely its relatively low compressive and tensile strengths, limit its effectiveness when used alone in structures subjected to substantial loads. This limitation poses a risk to the structural integrity of concrete under significant tensile forces, potentially leading to failures if not addressed. Reinforcing bars (rebars) are incorporated into concrete structures to overcome this limitation. Rebars are renowned for their superior tensile strength, which markedly enhances concrete elements' durability and load-bearing volume [1, 2]. The design of rebars featuring continuous ridges, grooves, or indentations is intended to improve their bond with the surrounding concrete, thereby minimizing the risk of slippage and improving overall structural performance. However, a critical vulnerability of rebars is their susceptibility to corrosion. Corrosion of rebars can cause them to expand, generating internal stresses within the concrete that compromise both the rebar and the concrete matrix. This degradation poses significant risks to the safety and longevity of the structure, potentially resulting in severe damage with considerable financial and safety implications. Chloride ions present in the concrete environment or infiltrating through leaks in the concrete structure exacerbate this problem, accelerating corrosion and compromising structural integrity [3]. Under the high pH of the concrete environment, which is mainly caused by the lime compounds in cement, rebars are often provided with a certain level of protection against corrosion. The alkaline conditions promote the development of a passive oxide coating on the surface of the metal, which stays structurally sound as long as the pH of the concrete remains high [4].

Nevertheless, carbonation poses a significant risk to this protective barrier. Carbonation is when carbon dioxide from the atmosphere enters the concrete matrix through its pores and combines with calcium hydroxide to produce calcium carbonate. This chemical process decreases the pH of the concrete from around 13 to a level of around 9, weakening or obliterating the protective oxide coating on the rebar. Therefore, chloride ions can infiltrate and destroy the exposed metal, resulting in expedited corrosion [5]. Factors such as water infiltration further exacerbate this issue, necessitating effective measures to shield concrete structures from these detrimental processes. Several protective strategies are employed to mitigate concrete corrosion [6]. These include the addition of corrosion inhibitors [7], the utilization of cathodic protection (CP) [8], and the use of anti-corrosion coatings [9]. CP can be achieved through two principal methods: applying an electrical current to the rebar or placing the rebar in a passive corrosion zone. Alternatively, sacrificial anodes, made from more reactive metals than the rebar, can corrode preferentially, thereby protecting the rebar. Corrosion inhibitors are integrated into the concrete mixture to counteract corrosive agents such as chloride ions. Moreover, organic coatings can be used to coat either the concrete surface or the rebar, therefore establishing a protective barrier against environmental factors such as carbon dioxide. This barrier effectively repels penetration and subsequent damage. These coatings help reduce the concrete's porosity and protect the rebar from changes in electrochemical potential, thereby delaying the onset of corrosion. The advantages and limits of each corrosion protection approach vary, and the best choice is contingent upon the particular requirements and conditions of the structure. This chapter provides a comprehensive overview of these protective strategies, analyzing empirical research related to each method. It concludes with a discussion of the challenges and perspectives associated with these approaches, focusing on evaluating their effectiveness in safeguarding concrete structures against corrosion.

9.2 Principles of Cathodic Protection: An Examination of Case Studies

9.2.1 Principles of Cathodic Protection

CP typically utilizes two main techniques: applying an electric current and using a sacrificial anode (Figure 9.1). In the current application method, an electric current is directed through the rebar, adjusting the metal's potential to a safe range, thereby preventing corrosion even when pH levels fluctuate. On the other hand, the sacrificial anode technique involves using a metal that is more prone to corrosion than the rebar. This method creates an electrochemical cell by connecting the rebar, such as iron, to a conductive structure and pairing it with a sacrificial anode, often made of zinc. As a result of this process, electrons go from the anode to the cathode, which helps to preserve the iron while causing the zinc to corrode. Individually, each of these approaches possesses distinct benefits and drawbacks. The sacrificial anode method, while simpler and more practical, generates only a small amount of current due to the resistance presented by the concrete

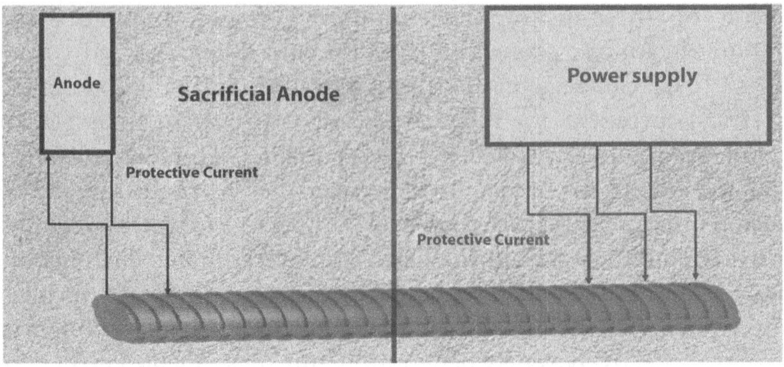

Figure 9.1 CP based on the sacrificial anode and impressed current.

environment. Additionally, sacrificial anodes have a limited lifespan and need to be replaced periodically. There are also limitations in the choice of materials; for example, aluminum and magnesium are generally unsuitable for CP in concrete structures due to the potential issues they can cause [10]. The current application method, although effective, is costly and requires constant monitoring to maintain the appropriate current levels. Moreover, the monitoring systems used in this method are not highly resistant to atmospheric degradation. Despite these challenges, CP systems based on sacrificial anodes offer simpler and more practical solutions than the complex current application method. It is not feasible to definitively state that one approach is generally preferable, as the optimal selection relies on the particular application and the nature of the actual structure.

9.2.2 Case Studies of Cathodic Protection

Increasingly, reinforced concrete structures are employing CP to prevent steel reinforcing bars from corrosion, especially in severe weather conditions. Despite its rising use, the intricate interplay of environmental variables and design parameters specified in national and international standards poses challenges in achieving thorough and effective protection for these structures. In this experimental work, Oleiwi examined the impact of chloride concentrations on the strength of concrete, the rate of corrosion of reinforcing bars, and the effectiveness of CP systems [11]. The study explored various current densities and presented comprehensive findings. The study's objectives were to analyze the relationship between chloride content and concrete strength and evaluate existing CP standards and design recommendations. Notable outcomes are listed below:

1. Identification of a threshold for applying CP to safeguard Portland cement (PC) concrete reinforcements against corrosion. The criterion is meeting either a chloride concentration of 0.31% by weight of cement or a concrete electrical resistivity of 17 kV per square centimeter.
2. The Ag/AgCl/0.5 KCl electrode demonstrated an instant potential reduction of 500 mV, meeting the 100-mV depolarization criterion for reinforced concrete containing up to 3.4% Cl by weight of cement or achieving a concrete resistivity of no less than 6.7 kΩ·cm².
3. A significant relation was discovered between the current demand for concrete pavement and the chloride content and the strength of the concrete. This correlation offers suggestions for modeling these parameters.

In addition, Yao *et al.* did a study on employing a current-based approach to protect bridge cables, which are critical components of bridge infrastructure prone to moisture-induced leakage and corrosion. The research aimed to design and develop a device using impressed current CP (ICCP) for the proactive protection of bridge cables. To evaluate the efficacy of the ICCP system on steel wires in cables exposed to acid rain, a speed-up corrosion test was done. The study examined the corrosion resistance of ICCP in bridge cables and found that higher protection effectiveness is associated with a larger negative ICCP voltage. However, an exceedingly negative voltage might lead to H_2 embrittlement in metal wires, therefore reducing their quality. An optimal voltage area of -1.13 to -1.15 V was identified for effective protection, where ICCP and sacrificial anode CP (SACP) work synergistically. While the effectiveness of SACP diminishes as corrosion advances, the ICCP system's performance improves over time. This research highlights the progression of ICCP techniques from a combination with sacrificial anodes to a primary reliance on impressed current, with the choice of protection method varying based on the specific application and type of concrete structure [12]. In order to offer CP for reinforced concrete structures, especially in marine conditions, Lee *et al.* devised a Zn mesh SACP system. Their research identified a notable gap in understanding the application of zinc mesh sacrificial anodes in such settings. Specifically, detailed information on how effectively these anodes transfer protective current to the reinforced concrete is lacking, especially under varying temperature conditions. Lee *et al.* explored the distribution of electric potential and current within a horizontal beam model, employing a single long bar and multiple reinforcing bars to replicate conditions found in concrete bridges or port structures. The study utilized vertical pile specimens equipped with zinc mesh sacrificial anodes. The target of the study was to assess the effect of CP by investigating the CP potential and current and performing a 100-mV depolarization test. The findings indicated that the protective current from the anodes could extend up to 10 cm from the waterline, though its effectiveness was influenced by resistance and temperature variations [13]. Through comparative studies, Wang *et al.* evaluated the effectiveness of sacrificial anodes fabricated from Zn alloy, Mg alloy, and Al alloy on Cl^--contaminated reinforced concrete. They constructed a large concrete slab (1.0 × 1.0 × 0.06 m) embedded with steel mesh to simulate practical conditions. Calculations of half-cell potentials, anode excitation currents, electrochemical impedance spectroscopy (EIS), and chloride ion levels were done at several time intervals in the study. The results showed that aluminum alloy anodes were ineffective due to prolonged activity loss, whereas both zinc and magnesium alloy anodes demonstrated highly efficient electrochemical activity and robust stability. Among these, magnesium alloy anodes provided superior protection to zinc alloy anodes. The effectiveness of these anodes was validated through changes in the characteristic frequency associated with the highest phase angle in the EIS Bode plot, with Cl^- ion immigration noted as a factor influencing protection efficacy. A comparative analysis was undertaken by Wilson *et al.* to evaluate the merits and drawbacks of sacrificial anode protection versus current-based CP technology. Key factors influencing the choice between current-based CP and SACP include the structural condition, budget constraints, and the anticipated lifespan of the repair. SACP is generally suited for minor, localized repairs with a shorter expected lifespan (around 10 years) and limited budgets. Conversely, ICCP is better suited for dealing with substantial corrosion problems in large structures with extended lifespans (over 25 years), especially when managing access is intricate and expensive [14]. In a study by Fatemi *et al.*, two systems with varying acid concentrations were developed to assess their properties in preventing concrete corrosion through current application. Acidification near the anodic regions due to anodic reactions can deteriorate concrete, reducing its effectiveness (Figure 9.2). The researchers used HP_1 and HP_2, hybrid geopolymers, for enhanced acid resistance. The composition of these mortars consists of less than 10% ordinary PC, together with other components such as slag, fly ash, and silica fume. Both mixtures include various fibers (like PVA and polypropylene) and additives, but HP_2 is distinct because it is activated with sodium silicate and sodium hydroxide. The laboratory results demonstrated that these hybrid geopolymer mortars, particularly HP_2, exhibited high resistance to hydrochloric acid. HP_2,

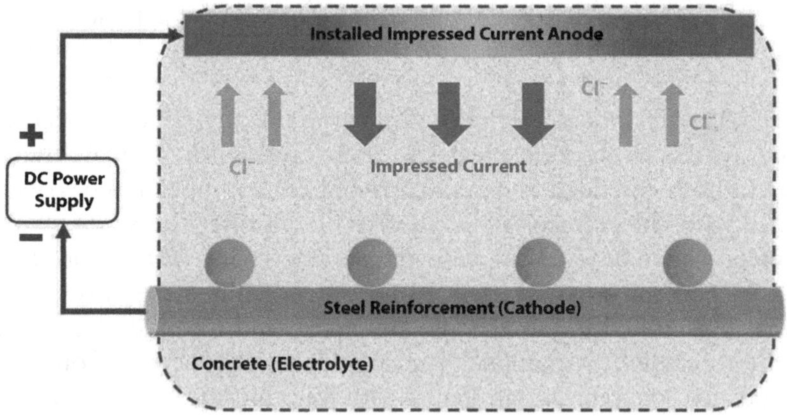

Figure 9.2 Schematic representing a conventional ICCP system [15].

with its alkaline activators, had better acid resistance, showing less mass loss, shallower acid penetration, and less compressive strength than HP_1. The testing results indicated that the geopolymer mortars performed better than those exposed to less harsh conditions when exposed to a 32% hydrochloric acid environment. This can be associated with developing a gel-like substance that functions as a barrier during deterioration. Furthermore, HP_2, when activated with $(NaO)_x.(SiO_2)_y$ and NaOH, exhibited enhanced adhesion to substrates, as evidenced by the absence of delamination among the repair mortar and deteriorated OPC concrete during the joint integrity tests. This suggests a strong interface and significantly improved durability compared to traditional OPC mortars, making them proper for long-term use in acidic environments. An assessment was conducted to define the impact resistance of the mortars used in reinforced concrete bridge foundations, which are exposed to tidal forces and splashing. Both geopolymer mortars displayed comparable impact failure patterns, with HP_2 displaying greater maximum peak stress. Upon achieving favored laboratory reliability, the HP_1 acid-resistant linked geopolymer was applied as a backfill mortar atop MMO Ti strip mesh anodes in a real Pons construction project. Testing on the samples taken from the bridge corroborated the lab results, confirming that the supplementary cementitious materials (SCMs) in the geopolymer helped control acid diffusion within the mortar. An observational assessment verified that all of the visible components of the ICCP system were in an excellent state, without any indications of acidity. Rebar potentials were confirmed to meet the protective parameters specified in applicable standards by depolarization potential tests [15].

In a study conducted by Astuti *et al.*, environmental parameters for example temperature and moisture were investigated for their impact on the performance and longevity of SACP systems, where current control is limited. These factors are critical in designing effective CP using sacrificial anodes. In this study, zinc anodes were investigated to regulate the corrosion of both corroded and non-corroded rebars implanted in concrete polluted with chloride and concrete free of chloride. The study involved reinforced concrete (RC) beam structures measuring 580 mm long with a 100 × 150 mm cross-section. Two steel rods, the same surface quality, were positioned in parallel, 40 mm apart, and inserted 30 mm deep into the concrete. A rectangular zinc anode was conducted on one of the steel rods for concrete contaminated with free chloride. The samples were exposed to three different conditions during testing: air curing at 20°C, wet and dry cycles, and dry air in the laboratory. A linear polarization resistance (LPR) corrosion meter was used to measure corrosion activity. The findings indicated that the corrosion rate of corroded and stainless steel bars declined while exposed to dry conditions but escalated considerably during wet and dry cycles, most likely attributable to the elevated humidity value in the concrete. In chloride-free environments, corrosion on the steel rods was significantly reduced. Utilizing stainless steel bars in patch repair environments has shown superior efficacy in mitigating corrosion. Nevertheless, in specific circumstances, the steel rod treated with the sacrificial zinc anode exhibited a higher corrosion rate than the unprotected rod in regions devoid of active chloride exposure. Temperature displayed a significant role in influencing corrosion and the efficiency of CP systems. All processes associated with corrosion are affected, including anodic and cathodic reactions, migration of corrosive chemicals to the steel surface, and immigration of ions *via* the concrete matrix. For sustained and effective protection, it is crucial to carefully consider both the type and position of sacrificial anodes when designing a CP device. The output of sacrificial anodes is significantly influenced by environmental factors, with anodes placed in damp or humid conditions typically generating higher currents, which are then directed to the reinforcing steel for protection [16]. In an investigation done by Astuti *et al.*, the behavior of sacrificial anodes in concrete exposed to elevated temperatures (approximately 40°C) was analyzed. The investigation involved embedding anodes in Cl-free and Cl-contaminated concrete, with reinforcing steel connected and disconnected from the anodes. The experiment was conducted in a controlled setting with a temperature of 40°C and a relative moisture of 98%.

A performance evaluation of the anodes was conducted using electrochemical testing. The depolarization tests demonstrated that the sacrificial anodes successfully induced polarization in the steel rebar to the protective level, resulting in a protective current that complied with the design criteria for CP. Corrosion rate measurements conducted *in situ* further demonstrated that SACP systems promoted rebar passivation, thereby improving the steel's resistance to corrosion. Even under high-temperature conditions, the results confirmed the successful transfer of protective current from the anodes to the rebar in the concrete [17]. The study of Cheng *et al.* emphasized the importance of the space between the anode and cathode in optimizing the performance of SACP systems. To improve the efficiency of SACP in preventing steel degradation, a model was developed to account for chloride ingress in concrete. The model simulated and assessed how factors such as anode positioning, spacing, and size influence the reduction of corrosion currents. The results showed that smaller anodes placed closer together achieved a greater reduction in corrosion current compared to larger, more distantly spaced anodes. Throughout the SACP process, about 25%–30% of the zinc current was used to neutralize the reaction at the cathode, while 70%–75% of the current was directed toward protecting the steel reinforcement. However, more than half of this protective current was required to offset the increased cathodic current in the steel due to cathodic polarization. As a result, less than 30% of the zinc current was directly utilized to prevent steel corrosion. While increasing the zinc current in the SACP system further reduced the steel corrosion current, it also intensified the cathodic polarization, accelerating zinc anode consumption. Upon conducting a comprehensive analysis, it was determined that the most effective approach to minimizing zinc anode consumption and optimizing corrosion current reduction was to apply a zinc current that was 5.0–7.5 times the corrosion current of the steel [18].

Research on mitigating corrosion in concrete structures extends beyond practical methodologies, with computational approaches critically optimizing CP systems. These predictive models often achieve remarkable accuracy and are instrumental in bridging theoretical insights with practical applications. Gong *et al.* suggested a numerical pattern to study the interplay of corrosion kinetics and ion transfer, aiming to analyze the distribution of current and Cl^- ions within CP systems. The study evaluated several key parameters, including current density, structure strength, Cl^- ion concentration, the order of rebars and anodes, and the thickness of protective layers. The findings revealed that rebars positioned closer to the anode received a significant protection current and exhibited the lowest chloride ion concentrations, indicating superior corrosion prevention in these areas. However, the irregularity of current distribution was found to diminish with increases in applied current, concrete strength, or the number of rebars as well as reductions in chloride concentration, anode count, or protective layer thickness.

Moreover, the study highlighted that the efficiency of CP in mitigating Cl^- ions near reinforcement bars depends on the distribution of protective current. Higher resistance, under constant current, amplified the potential diversity among the anode and rebars, thereby enhancing the repulsion of chloride ions. While strategies such as using numerous anodes and increasing the thickness of the passive layer improved current uniformity, these adjustments were found to impede the system's ability to extract Cl^- ions around the reinforcement.

This analysis emphasizes the importance of a holistic design approach for ICCP systems (Figure 9.3). Effective designs must balance the current distribution with the removal of chloride ions to maximize protection efficacy and ensure the durability of reinforced concrete structures [19].

Protecting concrete structures in marine environments presents a significant challenge because of the high content of Cl^- ions and the dynamic forces exerted by waves. These conditions can accelerate the degradation of the concrete layer, exposing the reinforcing steel (rebar) to corrosive electrolytes, thereby initiating rapid corrosion and compromising the structural integrity. Among the protection strategies employed, organic coatings are widely used due to their cost-effectiveness and ease of usage. However, the harsh marine environment can undermine their effectiveness. The

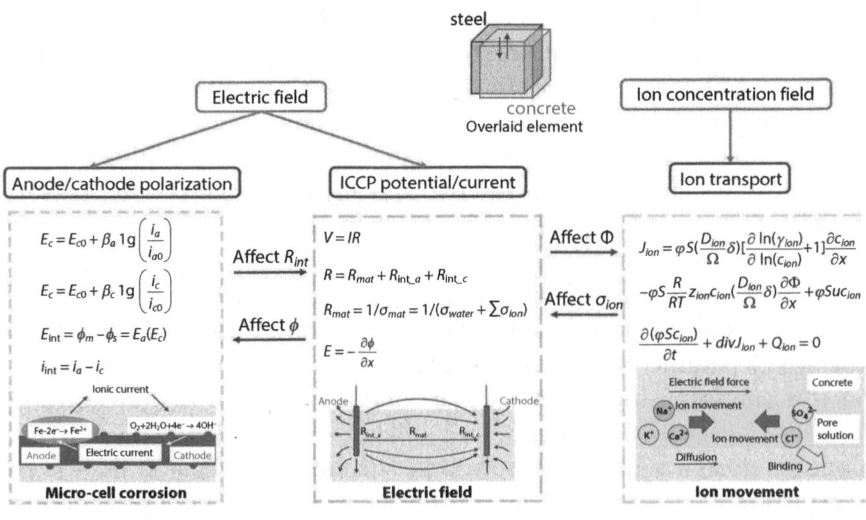

Figure 9.3 Representation of the relation between the electrical field and ion content field within the durability concrete model simulation framework [19].

mechanical impact of waves may cause coatings to deteriorate, and issues such as uneven application due to gravity can lead to reduced mechanical performance. Furthermore, while increasing coating thickness may seem advantageous, it often fails to provide the desired resistance to wave-induced scouring.

To address these limitations, a shift from passive to active protection strategies has emerged as a promising avenue. One innovative approach involves integrating triboelectric nanogenerators (TENGs) with surface coatings. TENGs, capable of harvesting mechanical energy and converting it into electrical energy, offer a dual-function solution. On the one hand, the elastic structure of TENGs enhances the anti-scouring properties of concrete by dissipating wave impact energy. On the other hand, the electrical energy generated by TENGs can be utilized in ICCP systems, providing an active method to safeguard reinforced concrete structures. Vertical contact–separation mode TENGs are particularly promising for this application due to their elastic design and efficient energy conversion. However, integrating these devices into marine coatings poses two critical challenges. First, the rigidity of traditional TENG devices limits their adaptability to curved structures, such as bridge piers, where perfect alignment of electrodes is required. Second, the moisture-rich and ion-laden marine environment can compromise the durability and output performance of TENGs, raising concerns about their long-term functionality. Addressing these challenges will require innovative material and structural designs for TENG devices, ensuring flexibility, durability, and resistance to marine conditions. Advances in these areas could pave the way for developing robust, multifunctional coatings that actively protect concrete structures in the most demanding marine environments. Gao *et al.* significantly enhanced marine reinforced concrete protection by designing a self-charging protective layer incorporating an arc-shaped TENG sealed within a flexible protective coating. This innovative system leverages wave energy to generate electrical power for CP while simultaneously serving as a physical barrier against corrosive elements. Performance evaluations comparing sealed and unsealed arc-shaped TENGs demonstrated that flexible sealing had less effect on the device's efficiency. At a wave frequency of 1 Hz, the sealed TENG displayed minor reductions in short-circuit current (Isc), open-circuit voltage (Voc), and charge transfer (Qc) by 8.4%, 9.5%, and 13.8%, respectively, figured out to the unsealed device. These reductions were deemed negligible, confirming that the sealing process preserves the TENG's functionality while offering additional protection against environmental damage.

In the simulation of sea environments, the self-charging protective structure showcased its dual-functionality by effectively blocking ion migration and converting mechanical wave energy into electrical energy (Figure 9.4). This electrical output was subsequently harnessed for automatic CP of the embedded reinforcing bars. The open-circuit potential (OCP) of A3 substrate, a commonly used rebar material, significantly improved under the influence of self-powered CP. The OCP shifted from -584 mV (vs. SCE) in the absence of CP to an impressive -1,121 mV (vs. SCE) with the self-charging system, underscoring the substantial anti-corrosion efficacy of this innovative solution.

This integrated protective system enhances the durability of marine reinforced concrete by mitigating wave-induced scour and corrosion and provides a sustainable approach to structural resilience. By addressing protection at both the concrete and rebar levels, the self-charging protective structure offers a comprehensive and forward-looking strategy for safeguarding marine infrastructure in highly corrosive environments [20].

In general, here we conducted some investigations on both protection methods, and finally, we reported a comparative study. As mentioned, to choose the correct method, it is imperative to examine various factors such as cost, environmental conditions, concrete conditions, etc.

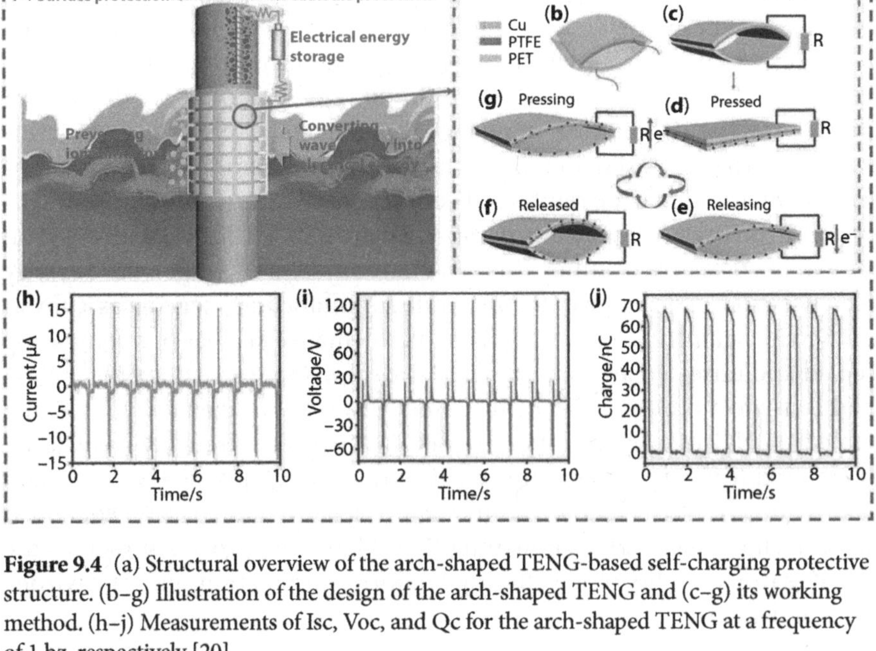

Figure 9.4 (a) Structural overview of the arch-shaped TENG-based self-charging protective structure. (b–g) Illustration of the design of the arch-shaped TENG and (c–g) its working method. (h–j) Measurements of Isc, Voc, and Qc for the arch-shaped TENG at a frequency of 1 hz, respectively [20].

9.3 Corrosion Protection Based on Anti-Corrosion Coating: An Examination of Case Studies

9.3.1 Corrosion Protection Through Anti-Corrosion Coatings

When metals come into contact with environmental elements, their surface potential can shift, leading to the onset of corrosion. To combat this, anti-corrosion coatings are crucial. These coatings function as a physical barrier on the metal surfaces, greatly decreasing their contact with ambient elements and hindering the electron transfer mechanisms that cause anode corrosion reactions (Figure 9.5). By disrupting electron transfer, these coatings help maintain the stability of the metal's surface potential, thereby preventing the corrosion process from initiating. In concrete structures, one effective method for protecting rebar from corrosion is the application of protective coverings. These coatings effectively minimize the impact of corrosive agents and fluctuations in concrete pH on the rebar, thus preserving its integrity. By controlling the passivation of the metal, these coverings provide a significant defense against corrosion. Similarly, concrete coating procedures are employed to safeguard concrete surfaces. Applying a protective layer to concrete prevents the ingress of carbon dioxide and moisture, which helps maintain a stable pH level. This is crucial because chlorine or other corrosive substances can otherwise compromise the oxide layer on the embedded rebar. This study explores various anti-corrosion strategies, focusing on applying coatings for both concrete and rebar to enhance their resistance to environmental degradation [9].

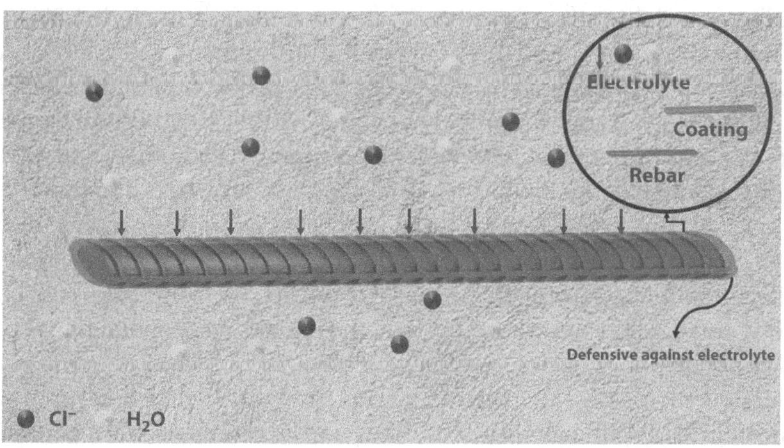

Figure 9.5 Anti-corrosion coating protection mechanism.

9.3.2 Case Studies of Anti-Corrosion Coatings

In their research on coatings that protect rebars from corrosion, Dong *et al.* compared how well different types of rebars resist corrosion in marine-based concrete buildings. The study focused on black rebar, epoxy-coated reinforcement, galvanized rebar, and zinc/epoxy-coated reinforcement. The findings revealed significant corrosion resistance differences among the different rebar types. The black rebar exhibited substantial corrosion after six months of immersion in a marine environment. Specifically, the black rebar's corrosion potential (E_{corr}) was enhanced from 0.2 to 0.8 V_{SCE}, and the corrosion current density (i_{corr}) rose from 1.0 to 10 $\mu A/cm^2$. This means that black rebars are not very resistant to rusting and start rusting quickly.

In contrast, the epoxy-coated rebar demonstrated excellent corrosion resistance. However, the protective performance of the epoxy coating was notably compromised when subjected to mechanical damage. Galvanized rebar maintained its corrosion resistance effectively, even when the coating experienced damage. Notably, the galvanized coatings were highly active in alkaline environments. The study found that the epoxy/zinc double-coating provided superior protective performance in undamaged and mechanically damaged conditions. This combination coating effectively mitigated corrosion, demonstrating resilience and robustness under various environmental conditions [21]. Sajid *et al.* investigated the development of novel anti-corrosion coatings derived from soybean-based chemicals, tailored to protect reinforcement materials like rebar from corrosion. The research focused on formulating soy-based coatings by blending denatured soy protein isolate (SPI) with corundated sorbitol softener in different weight ratios. Specifically, the formulations included 5%–15% SPI in deionized water and 10%–30% SPI in the corundated sorbitol softener. This study evaluated soy-based coatings' physicochemical efficiency and efficacy in safeguarding rebars covered with SPI and cement mortar. Characterization tests, such as viscosity and wear resistance evaluations, demonstrated that the soybean-based coatings provided adequate wear resistance and contained the essential functional groups necessary for corrosion protection. Potentiodynamic polarization experiments revealed a significant reduction in corrosion current density, up to 96% decrease compared to uncoated rebar, when soy protein coatings were applied in a simulated concrete pore solution (CPS) of 3.5 wt% NaCl. In addition, macrocell accelerated corrosion experiments demonstrated that rebars covered with soy protein had reduced corrosion rates and maintained more consistent corrosion potentials than steel bars without coatings. The study concluded that the formulation containing 15% SPI and 30% sorbitol provided the most effective corrosion inhibition and exhibited superior wear resistance. The soy protein coating material exhibits promise for the on-site repair of impaired rebar coatings in extremely corrosive chloride settings.

Additionally, it has the potential to be employed as an independent coating in less corrosive situations [22]. Sharma *et al.* carried out groundbreaking research on rebar coatings aimed at preventing corrosion in concrete structures, focusing on an innovative epoxy resin coating enhanced with graphene oxide (GO) nanoparticles, reduced GO (rGO), carbon nanotubes (CNT), and their hybrid combinations. The study's results are categorized into several significant areas:

- Enhanced corrosion resistance: The corrosion resistance of epoxy coatings was much improved and longer-lasting when modified with a combination of GO and CNTs, compared to coatings composed only of epoxy or those modified only with rGO and CNTs. This enhanced protection was particularly notable in environments with high chloride content, where corrosion presents a more significant challenge.
- Performance of graphene oxide (GO): GO was more effective than rGO in corrosion inhibition, mainly when used in low concentrations (0.4% of the epoxy resin). GO creates a dense, impermeable network that effectively blocks corrosive ions, thereby providing strong protection against corrosion.
- Improved mechanical properties: The hybrid coatings not only enhanced the mechanical strength of the rebar but also offered better resistance to mass loss and a reduction in tensile strength caused by corrosion. This suggests that these coatings can help preserve the structural integrity of the rebar over an extended period.
- Validation through testing: The outstanding performance of the GO/CNT-coated rebars was confirmed through both non-destructive and destructive testing methods. Non-destructive techniques included corrosion flow analysis and ultrasonic-guided wave monitoring, while destructive tests evaluated mass loss and tensile strength. Both testing approaches affirmed the hybrid coatings' superior corrosion resistance and durability.
- Increased structural durability: Furthermore, the altered coatings enhanced corrosion resistance and strengthened the adhesive connection between the rebar and the adjacent

concrete. This more robust bond can improve load-bearing efficiency and prolong the lifespan of rebar in concrete structures, decreasing long-term maintenance expenses.

In summary, the research of Sharma *et al.* underscores the effectiveness of epoxy coatings infused with graphene oxide and carbon nanotubes in substantially enhancing the longevity and resistance to rebar corrosion in concrete constructions, particularly in aggressive environmental conditions. As previously mentioned, concrete coatings effectively mitigate rebar corrosion by preventing the ingress of Cl^- and water into the concrete structure [23].

In order to further reduce the permeability of water and chloride in concrete, Shariatmadar *et al.* developed an innovative method that involves synthesizing a thin hydrophobic layer using polyisocyanate-functionalized GO (PIF-GO) (Figure 9.6). To create this hydrophobic barrier, GO 2D-sheets were chemically modified with a silicon-modified polyacrylate, which was covalently bonded through polyisocyanate (PI) bridges. The characterization confirmed the successful attachment of PI and silicone-modified polyacrylate onto the GO nanosheets. The hydrophobicity of the GO was notably enhanced by this modification, as evidenced by the fact that the water contact angle of GO was enhanced from 43° to 102° post-modification. This improvement in the coating's water-repellent

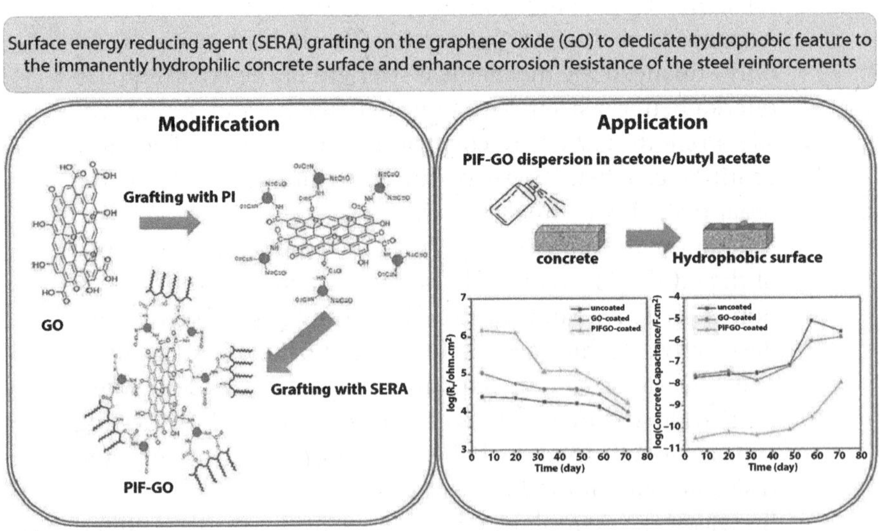

Figure 9.6 Mechanism of GO modification and applied hydrophobic coating on concrete [24].

properties is considerable. The water resistance of concrete coated with the PIF-GO layer was evaluated through capillary water absorption tests. The outcomes displayed that the capillary water penetration rate for PIF-GO coated concrete was significantly reduced (60.4 $g{\cdot}cm^{-2}{\cdot}h^{-1/2}$) compared to the unmodified GO-coated concrete (684.8 $g{\cdot}cm^{-2}{\cdot}h^{-1/2}$), highlighting the superior waterproofing capabilities of the PIF-GO coating. EIS was employed to investigate the corrosion characteristics of reinforcement that was inserted into concrete following a 71-day immersion in a saline solution. The EIS results demonstrated a marked decrease in the penetration of Cl^- and H_2O into the concrete, indicating that the PIF-GO coating effectively protects the rebar from corrosion, even in harsh environments. The study illustrates that applying a thin PIF-GO layer on concrete surfaces significantly improves resistance to water and chloride penetration, thereby offering enhanced protection against rebar corrosion [24].

Zinc-rich coatings (ZRE) are commonly employed in concrete structures to combat corrosion. These deposits provide CP as a sacrificial anode, where a substantial quantity of zinc safeguards the steel underneath. In addition to their cathodic role, these coatings contribute to anodic protection by passivating the surface and facilitating electron transfer between the coating and its environment. According to a study by Ge *et al.*, a ZRE was enhanced by incorporating carbon nanofibers to develop ZRE coatings. The introduction of carbon fibers significantly increased the electrical conductivity (E.C.) of the coatings, which enhanced CP's efficiency.

Furthermore, the inherent mechanical strength of carbon fibers contributed to improved mechanical properties of the final coating. To enhance the E.C. and mechanical performance of the ZRE coating, the work investigated using carbon fibers of 2, 5, and 10 mm lengths to establish conductive routes for zinc particles. The results showed that ZRE-10, which had 10-mm fibers, offered efficient shielding for a copper substrate for 50 days in a sea simulation, surpassing the efficiency of conventional ZRE coatings. Additionally, ZRE-10 exhibited a 14% water absorption rate, compared to 20% for the standard ZRE, and its adhesion strength increased by 65%. These findings highlight the significant enhancement in anti-corrosion performance and mechanical durability achieved by incorporating longer carbon fibers into the ZRE [25]. A study carried out by Hayatdavoudi *et al.* investigated the influence of graphene 2D-sheets on the corrosion resistance and protective properties of ZRE. Their results demonstrated that adding 0.4% by weight of graphene nanosheets greatly enhanced the protective effectiveness of the ZRE. Nevertheless, lesser quantities of graphene had an adverse impact, diminishing the efficacy of the covering. The improved corrosion resistance observed with an optimal graphene concentration was attributed to the enhanced homogeneous activation of Zn particles and the remarkable barrier characteristics of graphene 2D-sheets. Graphene's high impermeability to corrosive agents helped to create a more stable environment, reducing the exposure of zinc particles to aggressive species. The improved barrier effect of the ZRE coating, when an adequate graphene loading is present, directly led to its higher durability and corrosion protection. Therefore, this work emphasizes the crucial significance of graphene nanosheets in enhancing the consistent activation of zinc and the general barrier characteristics of ZRE coatings. It emphasizes the need to tailor the graphene concentration for the best protective performance [26]. Nguyen *et al.* explored the effectiveness of nanoparticle-enhanced epoxy coatings in mitigating steel corrosion within a concrete simulation environment. The epoxy, fortified with nanoparticles, was used in their study to coat steel reinforcement that was inserted in cement mortar polluted with chloride. Over 105 days of immersion in a 0.1-M sodium hydroxide solution, the study demonstrated that nanoparticles significantly bolstered the resistance against corrosion of the coated steel. Of the tested nanoparticles, nano-Fe_2O_3 particles delivered the most notable results. Through polarization testing, it was shown that the corrosion current of the steel rebar decreased by a factor of 2 to 6 after 105 days, attributed to the incorporation of nanoparticles. EIS also indicated that the nanoparticles enhanced the coating's resistance, improving its overall barrier properties. This combined mechanism of action contributed to the superior corrosion protection provided by the epoxy coating. The researchers directly incorporated nanoparticles into fresh cement mortar, using 1% by weight of cement and their utilization in coatings. Electromigration tests revealed that the nanoparticles markedly increased the mortar's resistance to chloride ion penetration. Specifically, the apparent diffusion coefficient for chloride ions (*D*Cl−) dropped by 83.45% for nano-SiO_2 and by 81.4% for nano-Fe_2O_3, compared to control samples, with values declining from 7.31×10^{-11} m^2/s to 1.21×10^{-11} and 1.36×10^{-11} m^2/s, respectively. The morphology of cement hydration products was altered, and the mortar matrix became denser due to the addition of nanoparticles, as illustrated by FE-SEM. Moreover, compressive strength tests on 28-day-old nano-modified mortar samples indicated a strength increase of up to 10.8%, enhancing both durability and mechanical performance [27].

Hydrophobic concrete is the most effective method of preventing rebar corrosion in concrete. Gu *et al.* developed a multi-modal superhydrophobic system by incorporating a coating composed of multi-walled CNT (MWCNTs) and PDMS known as polydimethylsiloxane (Figure 9.7). This system demonstrated exceptional anti-corrosion, anti-fouling, anti-icing,

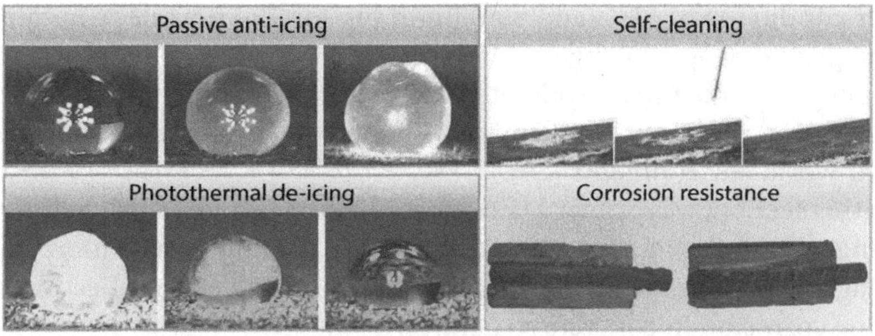

Figure 9.7 Schematic of multi-functional coating based on anti-icing/self-cleaning/photothermal de-icing/anti-corrosion [28].

photothermal anti-icing, and self-cleaning applications. Under infrared light irradiation, ice on the coating quickly melted into H_2O, which was quickly removed from the surface.

In addition to its excellent chemical and mechanical durability, the coating exhibited remarkable corrosion resistance. This innovation advances superhydrophobic concrete technology and facilitates its application in real-world structures. When applied to concrete, the coating produced a contact angle of approximately 162° with a 5-μL H_2O droplet. This high contact angle enabled the water droplets to effortlessly remove dust from the surface, preserving the coating's self-cleaning and anti-fouling capabilities [28].

The outer surface of building walls in outdoor environments is consistently exposed to challenges such as ultraviolet (UV) radiation, temperature fluctuations, abrasion, erosion, dust accumulation, and microbial activity. These elements contribute to a progressive decline in the durability and structural integrity of external walls over time. To mitigate these issues, superhydrophobic coatings offer a promising solution for extending the lifespan and enhancing the durability of concrete walls. In this context, Boutamart *et al.* designed a fluorine-free superhydrophobic coating by applying the sol-gel approach, which was successfully used in concrete through an immersion technique. The coating formulation incorporated PDMS, a low-surface-energy material, alongside polymeric silica (PS) to refine the surface characteristics and microstructure of the concrete. The treated concrete surface exhibited a water contact angle (WCA) exceeding 150°, signifying its superhydrophobic nature. Furthermore, the coating's self-cleaning ability, mechanical durability, UV radiation stability, and tolerance to temperature and humidity were evaluated. The findings revealed that the coated surface resisted dust adherence and endured over 300 cycles of abrasion testing. The coating also demonstrated stability under conditions of 45°C and 80% humidity for 720 h while maintaining its integrity during prolonged UV exposure for 1,440 h. Outdoor natural aging experiments further validated the coating's exceptional resistance to atmospheric conditions, confirming its potential for long-term application in real-world environments [29].

In this section, we reviewed various studies that explored using different coatings. Overall, whether applied as a rebar coating or a reinforcement coating, these approaches can be considered effective strategies for preventing corrosion in concrete structures.

9.4 Utilizing Corrosion Inhibitors: An Examination of Case Studies

9.4.1 The Utilization of Corrosion Inhibitors in Material Preservation

Corrosion inhibitors are vital in preventing corrosion in different conditions, including neutral, acidic, and alkaline environments like concrete pavements. These inhibitors have become essential components in concrete mixtures, but their application must adhere to specific guidelines and standards. The improper use of corrosion inhibitors can lead to issues such as reduced concrete efficiency, while inadequate amounts may compromise their effectiveness. Research on corrosion inhibitors generally falls into two main areas: the study of their use within concrete structures and their performance in simulated concrete environments.

Inhibitors of corrosion can be classified as organic versus inorganic, environmentally friendly versus non-environmentally friendly, and anodic versus cathodic inhibitors. Among these, a particularly insightful classification is based on their mechanism of action, which distinguishes between cathodic and anodic inhibitors. Anodic inhibitors operate by oxidizing the contact surface of the metal and creating a complex with it, making the metal surface non-reactive. In contrast, cathodic inhibitors work by mitigating environmental factors that promote corrosion, such as by controlling cathodic reactions and thereby hindering the progress of corrosion (Figure 9.8). Disrupting either the anodic or cathodic reactions can influence the rate and control of corrosion. Additionally, mixed inhibitors employ combined mechanisms to control both anodic and cathodic reactions. Green inhibitors can be either anodic or cathodic. Numerous studies have investigated various types of corrosion inhibitors within concrete environments, some of which are discussed in the following section [30].

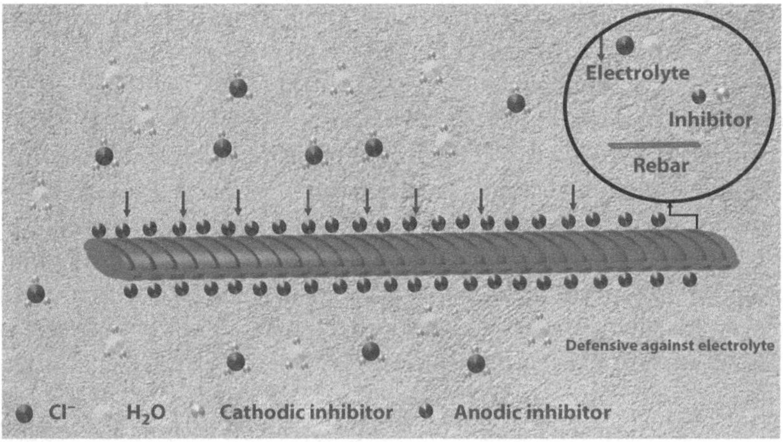

Figure 9.8 Mechanism of corrosion inhibitor protection.

9.4.2 Case Studies of Corrosion Inhibitors

Verbruggen *et al.* undertook a study to assess the efficacy of 1H-benzotriazole as one of the corrosion inhibitors for rebar in a concrete simulation setting. The simulation environment employed a combination hydroxide of sodium, calcium hydroxide, sodium sulfate, and sodium chloride to replicate the chloride attack. Carbonation was simulated using calcium carbonate and sodium sulfate, with the pH adjusted to 13 for chloride conditions and 9 for carbonation. The performance of four different inhibitors was assessed. According to Verbruggen *et al.*, all inhibitors except 1H-benzotriazole demonstrated satisfactory corrosion resistance. Sodium molybdate, even at low concentrations, showed powerful inhibitory performance. Its mechanism involves the adsorption of molybdate onto an existing porous oxide layer on the steel surface, stabilizing it and enhancing its corrosion resistance. Although molybdate has limited adsorption in high-pH environments, making detection challenging, it effectively mitigates corrosion by forming $FeMoO_4$ *in situ*, even at the early stages of pitting. This early-stage repassivation likely explains the absence of pitting observed in subsequent detailed imaging analyses. Sodium molybdate is also noted for its non-toxic nature and suitability in organic coatings.

In conclusion, sodium molybdate proved to be the most effective inhibitor among those tested [31]. In a study focused on steel corrosion prevention in a concrete simulation environment, Jiang and his team evaluated the effectiveness of DNA green corrosion inhibitors. Known for its environmentally friendly properties, the DNA inhibitor was tested for its performance in controlling steel corrosion. Initially, the DNA inhibitor functioned as a cathodic corrosion inhibitor, effectively mitigating corrosion by addressing the cathodic sites. However, as the corrosion progressed and salt concentration increased, the DNA inhibitor adsorbed onto the steel surface and began to act as an anodic inhibitor. This dual functionality highlights the DNA inhibitor's capability as a mixed corrosion inhibitor, demonstrating its effectiveness in preventing steel corrosion in concrete environments [32]. In a comparable investigation, Chen *et al.* examined the efficacy of DNA-based corrosion inhibitors obtained from salmon fish, particularly when exposed to sulfate ions. Their findings indicated that sulfate ions negatively impacted the inhibitor's effectiveness. High NaCl content further diminished the efficacy of the DNA inhibitor in conjunction with sulfate ions. Despite these challenges, the DNA inhibitor outperformed commercial sulfate-based corrosion inhibitors, suggesting that environmentally friendly green inhibitors warrant further investigation and potential industrial application [33]. In a study on green corrosion inhibitors, Rahmani *et al.* conducted an extensive survey, including both experimental and molecular analyses to assess the efficacy of mango extract in controlling corrosion. Their findings revealed that mango extract operates through a combined inhibition mechanism, affecting both anodic and cathodic reactions and leading to significant corrosion suppression. Specifically, the polarization efficiency improved to 91% with the addition of 2% by weight of mango extract. The EIS results further demonstrated that steel corrosion was reduced by approximately 98% after 48 hours of exposure to a corrosion protection system containing 2% mango extract. Surface analysis using FE-SEM confirmed that the application of 2% mango extract effectively mitigated corrosion on the metal surface. Analysis using spectroscopic techniques such as Raman spectroscopy, Fourier transform infrared (FT-IR), and X-ray diffraction (XRD) revealed that chemicals derived from the mango extract were effectually adsorbed onto the steel surface, indicating successful chelation with the metal substrate. It was shown by Density Functional Theory (DFT) simulations that the adsorption of mango extract molecules and calcium cations on the steel surface mostly takes place through electron-rich locations, which promote donor–acceptor interactions. Substantial negative adsorption energies for the organic–inorganic complex inhibitors were shown in molecular dynamics (MD) and Monte Carlo (MC) simulations impact in both aqueous and gas phases. These outcomes were consistent with the observed corrosion inhibition effect of the mango extract. These comprehensive results underscore the effectiveness of the extract of mango as a green inhibitor and demonstrate a strong correlation between theoretical predictions and experimental observations of its anti-corrosion properties [34].

Enhancing the corrosion resistance of rebar is crucial for prolonging the lifespan of rebar concrete structures, especially in aggressive media. In recent advancements, a novel combination of sodium phytalate and Na_2MoO_4 has emerged as a promising corrosion inhibitor for Q235 carbon steel in concrete. This innovative approach leverages the synergistic effects of the two inhibitors, applied in an optimal inhibitor content ratio, to mitigate Cl^--induced corrosion effectively. Electrochemical experiments have revealed that this mixture of inhibitors significantly improved the passivation process by enhancing the thickness of the passivation layer. While the direct passivation capability of the steel was not definitively demonstrated, the results indicated a notable enhancement in resistance to corrosion.

Additionally, the organization of a robust, three-layered passivation film played a significant role in increasing the steel's resistance to pitting corrosion. After 30 days of exposure to chloride ions, the inhibitor displayed remarkable efficiency in protecting the steel. These findings suggest that the combined sodium molybdate–sodium phytate inhibitor offers a reliable solution for mitigating corrosion in RC structures. The proposed synergistic inhibition mechanism further underscores the potential of this approach to address long-standing challenges in corrosion control. By providing enhanced protection and durability, this strategy could pave the way for more sustainable and resilient concrete infrastructure [35]. This study, carried out by Gharieb *et al.*, focuses on the impact of organic polyamine-based corrosion inhibitors on the efficiency of cement paste concrete and the corrosion resistance of steel reinforcement. To attain this, various blends were collected by incorporating the inhibitors at concentrations of 0, 0.5, 0.75, 1, 1.25, 1.5, 1.75, 2, and 3 wt% relative to the weight of OCP. Key parameters such as rheological behavior, setting times, and the physical and mechanical characteristics of these mixtures were systematically assessed. Additionally, the hydration products formed during the process were analyzed using XRD and thermogravimetric analysis (TGA/DTG).

The study also delved into the localized corrosion resistance of steel rebar inserted in concrete, employing electrochemical techniques alongside surface characterization methods like X-ray photoelectron spectroscopy (XPS). These analyses aimed to determine the formation of a protective film and evaluate the efficiency of the inhibitor. The results revealed that incorporating corrosion-inhibiting admixture polyamine (SEDREEN-SS) organic polyamine inhibitors at dosages up to 2 wt% significantly enhanced the mechanical performance of cement pastes and hardened concrete. Furthermore, the optimal inhibitory performance was observed at a concentration of 1.25 wt%, where the inhibitor effectively minimized corrosion.

XPS analysis confirmed the ability of the SS inhibitor to diffuse through concrete pores and form a robust protective layer on the steel surface. These findings, as demonstrated by Gharieb *et al.*, highly spot the potential of SEDREEN-SS inhibitors as a favorable solution for mitigating the corrosion of rebar in concrete structures, particularly under aggressive environmental conditions [36].

In addition to published research articles, various methods to prevent concrete corrosion that are suitable for industrial application can also be found in patents. For example, Kyong-Chul Kim and colleagues proposed an innovative marine concrete composition utilizing dechlorinating microorganisms. This system effectively removes chlorides generated by seawater through an electrochemical mechanism. It employs electricity-producing microorganisms that emit electrons, which are transferred *via* the steel fibers embedded in ultra-high-performance concrete (UHPC) or high-performance fiber-reinforced cement composites (HPFRCC). The concrete incorporates a microbial encapsulation carrier capable of both dechlorination and self-healing, allowing it to repair cracks in the material. This approach not only addresses fundamental challenges related to reduced durability caused by salt damage in UHPC or HPFRCC but also enhances the material's suitability for applications in marine environments. By incorporating the dechlorinating microbial carrier, the system ensures long-term durability and structural integrity in such aggressive conditions. Additionally, the study describes a method for fabricating marine concrete structures using this novel composition.

This patented solution, as reported by Kyong-Chul Kim and colleagues, represents a significant advancement in concrete technology by integrating dechlorinating and self-healing functionalities, offering a practical and sustainable approach to extending the service life of marine infrastructure [37].

According to an investigation conducted by Fazayel *et al.*, the efficacy of various polycarboxylate derivatives, including comonomers and different functional groups, in mitigating steel corrosion was assessed. The study utilized a highly alkaline CPS with chloride to simulate pitting corrosion. The research focused on determining the method underlying the inhibitory effects of these compounds. The study found that the inhibition performance and effectiveness in controlling pitting corrosion varied among the methacrylate-copolymer derivatives. The polymers were ranked in the following order of efficacy: polymethacrylate-co-acrylamide > polymethacrylate-co-2-acrylamido-2-methylpropanesulfonic acid > polymethacrylate-co-hydroxyethyl methacrylate. Notably, polymethacrylate-co-acrylamide emerged as the most effective inhibitor, demonstrating a consistent trend in corrosion potential across different NaCl concentrations in a saturated calcium hydroxide solution. Polymethacrylate-co-acrylamide achieved an impressive inhibition efficiency of 92.35% in a saturated calcium hydroxide solution with 1.8% chloride by weight, significantly reducing the corrosion rate. This inhibitor effectively mitigated the adverse effects of Cl ion rush on the steel surface and passive film, even at a higher chloride concentration of 3.5 wt% NaCl. The placement of a passive barrier layer on the metal surface is responsible for the observed inhibitory effect, which aids in the prevention of corrosion [38]. In an evaluation by Goyal *et al.*, the impact of six diverse corrosion inhibitors on steel corrosion within a high-chloride concrete environment was examined. The study evaluated six different types of inhibitors: MCI_1 and MCI_2, which were silane-based and contained organic functional groups; MCI_3 and MCI_4, which were formulated with amino alcohols; and MCI_5 and MCI_6, which were composed of surfactants and amine salts. The results demonstrated that all inhibitors significantly reduced the corrosion rate compared to the untreated control samples. Of the inhibitors tested, the silane-based compounds, MCI_1 and MCI_2, showed the greatest reduction in both corrosion rate and potential values. Additionally, these inhibitors demonstrated the lowest water absorption and the maximum electrical resistance, which underscored their exceptional barrier capabilities. The study's final finding was that silane-based inhibitors containing organic functional groups exhibited superior corrosion resistance and barrier protection compared to inhibitors based on amino alcohol and surfactants. The greater efficacy of the silane-based inhibitors was ascribed to their hydrophobic characteristics, which augmented their capacity to block corrosion [39]. The application of nanomaterials in corrosion prevention has gained significant traction, with materials like metal-organic frameworks (MOFs) showing considerable promise. ZIF-8, a particularly intriguing MOF, has been explored for its potential in corrosion mitigation, both as a component in polymer coatings and as a direct corrosion inhibitor in various environments. According to a study undertaken by Wang *et al.*, the efficacy of ZIF-8 as a corrosion inhibitor was investigated using a concrete extract simulating a corrosive environment. The study successfully synthesized ZIF-8 with an approximate diameter of 75 nm. The results demonstrated that ZIF-8, when introduced into a chloride-containing cement extract, effectively adsorbed onto the surface of steel reinforcements. The main effect of this adsorption was to suppress the anodic process by reducing the disintegration of the steel, resulting in significant enhancements in charge transfer resistance and polarization resistance. Initially, ZIF-8 exhibited an impressive inhibition efficiency of over 95% within the first 24 h. Nevertheless, the dissociation of ZIF-8 and the breaking of metal–nitrogen coordination interactions among the released 2-MeIm and the rebar were induced by a reduction in pH resulting from localized corrosion. Despite this, the inhibitor significantly delayed the onset of corrosion, and its effectiveness remained high, with an inhibition efficiency of 87.8% even after 168 h at a ZIF-8 content of 0.04 wt% (Figure 9.9) [40].

Among the diverse nanomaterials employed as corrosion inhibitors, layered double hydroxides (LDHs) stand out due to their unique two-dimensional structure and inherent protective capabilities. LDHs consist of two types of cations interleaved with anions, which allows them to function actively in corrosion protection through ion exchange processes. When exposed to a corrosive environment, LDHs can swap their interlayer anions with environmental ions such as hydroxides or chlorides. If an anti-corrosion inhibitor is incorporated between their layers, it can be released into the environment to counteract corrosive agents, while the LDHs themselves absorb these agents. In an innovative study, Li *et al.* explored the use of LDHs loaded with three different phosphate-based anionic inhibitors to assess their effectiveness in mitigating steel corrosion in a high-chloride concentration cement environment. The study demonstrated that the capacity of LDHs to accommodate phosphate anions in their interlayer structure was positively correlated with the anionic charge of the phosphate. Specifically, higher anionic charges led to a greater amount of phosphate being inserted between the LDH layers, which in turn enhanced the material's ability to exchange anions and reduced its tendency to absorb corrosive Cl^-. The outcomes showed that the inhibition efficiencies of various phosphate anions were as follows: R-$H_2PO_4^-$ achieved

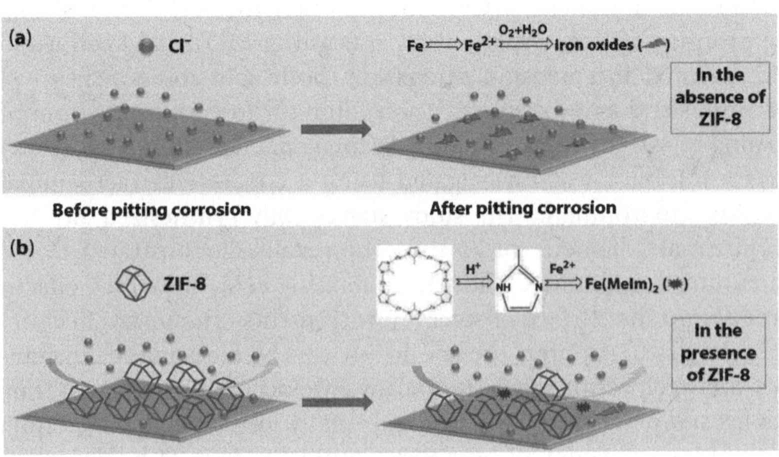

Figure 9.9 Corrosion protection mechanism of ZIF-8 in cement extract [40].

an inhibition rate of 78%, R-PO_3F^{2-} 36%, and PO_3^{3-} 5% (Figure 9.10). This variation underscores the importance of anionic charge in the effectiveness of LDHs as corrosion inhibitors [41].

In a study exploring corrosion inhibitors in a concrete-simulating environment in order to decrease rust of steel in CPS, Mohammadkhah *et al.* suggested the use of monosodium glutamate (GT) as a promising choice for long-term corrosion inhibition. They investigated the electrochemical and morphological behavior of GT at concentrations ranging from 1% to

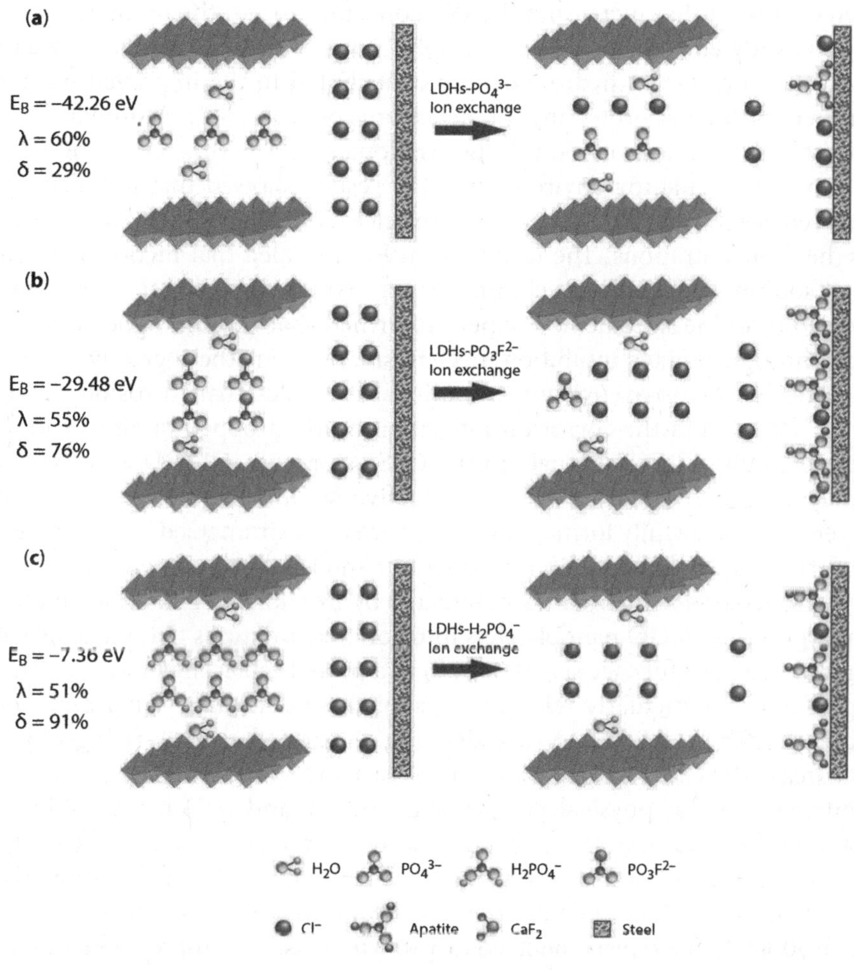

Figure 9.10 The illustration represents the influence of diverse anionic charges on the anti-corrosion mechanisms of LDHs containing various anions: (a) LDHs-PO_4^{3-}, (b) LDHs-PO_3F^{2-}, and (c) LDHs-$H_2PO_4^-$ [41].

5% by weight in simulated concrete pore environments. The EIS results indicated that 4 wt% GT effectively reduced the corrosion rate of steel by 86%, demonstrating an efficient inhibition mechanism. Polarization measurements further revealed a significant drop in corrosion I_{corr} to 0.169 μA/cm^2 since adding 4 wt% GT to the corrosive medium. The investigation also used FE-SEM to verify the development of a graphene-titanium layer on the surface of the steel. Analysis using spectroscopic methods such as Raman and grazing incidence X-ray diffraction (GIXRD) verified the effectual adsorption of GT structure onto the metal substrate, therefore improving corrosion resistance. Moreover, contact angle measurements demonstrated that increasing the GT concentration to the optimal 4 wt% significantly enhanced surface hydrophobicity, raising the contact angle to 62°. This increase in hydrophobicity contributed to the improved barrier properties, further protecting the steel from corrosion [42]. A similar study investigated the use of sodium lignosulfonate as a corrosion inhibitor in a concrete-simulating environment. The results showed that it effectively delayed steel corrosion, with performance consistently improving with higher concentrations. The results from EIS revealed that incorporating 1 wt% sodium lignosulfonate eliminated approximately 95% of the corrosion. Polarization measurements further confirmed that sodium lignosulfonate acts through a mixed inhibition mechanism, reducing the aggressiveness of steel in the corrosive environment. FE-SEM revealed that corrosion results were absent, and the corrosion rate significantly dropped upon the addition of sodium lignosulfonate to the CPS. Moreover, GIXRD and Raman analysis results indicated that a protective sodium lignosulfonate-based layer was successfully formed on the surface of the immersed steel samples, contributing to the reduction in corrosion and the enhancement of protective properties [43]. The work conducted by Rezakhani *et al.* evaluated the incorporation of GO nanoplatelets into concrete mixtures that make use of granulated blast furnace slag (GGBFS) as a partial replacement for PC. This approach is particularly relevant for applications in saline environments, such as offshore structures, road de-icing scenarios, or saline-rich geological areas. The effort assessed several performance parameters of the altered concrete, such as physical, mechanical, and Cl^- and H_2O permeability at varying temperatures and water pressures. The incorporation of GGBFS is recognized to improve the mechanical characteristics of concrete. The results indicated that including over 0.1% by weight of GO into concrete with 50% GGBFS replacement led to a 40% increase in compressive strength and a 60% enhancement in flexural strength after the curing process.

Additionally, the chloride permeability, indicated by E.C., decreased significantly from approximately 4,000 C in plain concrete to about 1,100 C in the modified concrete, representing a 70% reduction. This improvement in performance is attributed to speeding up C-S-H gel formation and calcium hydroxide crystallization due to the addition of GO, alongside changes in pore construction from the finely ground slag. The mean pore diameter observed in PC concrete was reduced from 11.4 to 6.2 nm, indicating a reduction in porosity. Enhanced strength, reduced water absorption, and greater resistance to chloride penetration are attributed to the decrease in pore size. Additional investigation using conductivity calorimetry, thermal analysis (TGA and DTG), mercury intrusion porosimetry (MIP), linear polarization corrosion tests, and EIS has verified that including 0.1 wt% GO greatly improves the physical characteristics, durability, and resistance to corrosion of gray-colored GGBFS [44]. This study highlights that nanoparticles, including MOFs and LDHs, exhibit inherent corrosion inhibition properties.

In contrast, graphene nanoparticles, although not active corrosion inhibitors themselves, increase the corrosion resistance of cement by changing the basic properties of cement. This enhancement occurs alongside improvements in the mechanical strength and durability of cement. These findings pave the way for novel applications of nanomaterials, not only as corrosion inhibitors but also as cement fillers and additives, opening up new avenues in the field of material science.

9.5 Comparative Analysis, Challenges, and Outlooks

Numerous studies have examined a range of protective strategies to combat corrosion, including CP, anti-corrosion coatings, and corrosion inhibitors. The efficiency of any approach can be influenced by its unique benefits and obstacles, which are contingent upon the particular application.

- CP: CP is a widely used technique to mitigate corrosion, and it primarily involves two approaches: ICCP systems and SACP systems.
  - ICCP: The method entails the usage of an external electric current to oppose the electrochemical processes responsible for corrosion. Despite their ability to provide accurate regulation of current flow and their effectiveness in large-scale or essential structures, increased expenses for installation and maintenance accompany current systems. The system's complexity requires careful monitoring and adjustment to ensure optimal performance.

   - Sacrificial anode systems: SACP is made from more reactive metals that corrode preferentially, thereby protecting the underlying structure. These systems are simpler and more cost-effective compared to impressed current systems. However, they are less controllable and require periodic replacement of the anodes. They are also less effective when a higher current is needed, as they provide less current to the metal surface.

     Considerations such as cost, complexity, and the particular needs of the application determine the selection among various approaches. Each method has its specific strengths, and recognizing these nuances is essential for effective corrosion management.
- Protective coatings: An organic coating functions as a defensive barrier that renders the metal surface inactive and hinders the infiltration of corrosive substances. These coatings can be further enhanced with the integration of various fillers to increase their mechanical and anti-corrosion ability. ZRE: ZRE provides dual protection through a sacrificial mechanism and a barrier effect. Zinc corrodes preferentially, thus protecting the underlying metal. Moreover, the inclusion of zinc substantially improves the overall CP of the coated structure. By including conductive elements like graphene and CNTs, the efficiency of these coatings can be greatly enhanced. The use of these additives serves to augment the E.C. of the coatings, therefore strengthening the CP and improving the overall efficacy of the coating.
- Corrosion inhibitors: Corrosion inhibitors are chemical substances that disrupt the corrosion process and are classified according to their determining mechanism of action:
   - Anodic inhibitors: Their mechanism of action is the deactivation of the metal surface and the inhibition of oxidation processes. They serve to provide a defensive coating on the metal surface that inhibits more oxidation.
   - Cathodic inhibitors: Cathodic inhibitors work by suppressing the cathodic reactions that occur during corrosion. They help in reducing the rate of electron transfer and thus limit the overall corrosion process.
   - Mixed inhibitors: Mixed inhibitors affect both anodic and cathodic reactions, offering comprehensive protection against corrosion.

Green inhibitors, which are derived from environmentally friendly sources and are cost-effective, have gained significant attention. They are particularly valued in concrete environments for their environmental sustainability and practical benefits. However, the effectiveness of corrosion inhibitors in concrete is often constrained by their concentration levels, with many inhibitors being ineffective at lower concentrations.

- Nanotechnology in corrosion prevention: Recent advancements in nanotechnology have opened new avenues for corrosion prevention. Nanoparticles, such as MOFs, show promise as effective corrosion inhibitors due to their unique properties and high surface area. These materials can interact with corrosive agents at the molecular level, providing enhanced protection. LDHs: LDHs are two-dimensional nanomaterials with ion-exchange capabilities. They can absorb harmful chloride ions from the environment while releasing corrosion inhibitors into the concrete matrix. This dual action makes LDHs effective in both preventing and mitigating corrosion. Despite their potential, research on the use of nanomaterials in highly alkaline concrete environments is less advanced compared to the studies in neutral environments. Self-healing systems and advanced nanomaterials, such as 2D materials, are still underexplored in the context of alkaline environments. These fields indicate potential topics for future study, as improvements could significantly improve the resilience and lifespan of concrete structures under challenging environments.

AI disclosure statement: The author acknowledges the use of AI-based tools for minor language editing or grammatical correction. However, all intellectual contributions, analyses, and conclusions presented in this chapter are entirely the author's own.

References

1. Makul, N., Modern sustainable cement and concrete composites: Review of current status, challenges and guidelines. *Sustain. Mater. Technol.*, 25, e00155, 2020.
2. James, A., Bazarchi, E., Chiniforush, A.A., Aghdam, P.P., Hosseini, M.R., Akbarnezhad, A., *et al.*, Rebar corrosion detection, protection, and rehabilitation

of reinforced concrete structures in coastal environments: A review. *Constr. Build. Mater.*, 224, 1026–39, 2019.

3. Xu, H. and Chen, J.-K., Coupling effect of corrosion damage on chloride ions diffusion in cement based materials. *Constr. Build. Mater.*, 243, 118225, 2020.
4. Bagheri, A. and Rastegar, M.M., Investigation of passive layer formation on steel rebars in foamed concrete. *Mater. Corros.*, 70, 1252–61, 2019.
5. Silva, R., Neves, R., De Brito, J., Dhir, R., Carbonation behaviour of recycled aggregate concrete. *Cem. Concr. Compos.*, 62, 22–32, 2015.
6. Ahmad, S., Reinforcement corrosion in concrete structures, its monitoring and service life prediction—-a review. *Cem. Concr. Compos.*, 25, 459–71, 2003.
7. Lee, H.-S., Saraswathy, V., Kwon, S.-J., Karthick, S., Corrosion inhibitors for reinforced concrete: A review. *Corros. Inhibitors Princ. Recent Appl.*, 95–120, 2018.
8. Liu, Y. and Shi, X., CP technologies for reinforced concrete: introduction and recent developments. *Rev. Chem. Eng.*, 25, 339–88, 2009.
9. Yan, L., Deng, W., Wang, N., Xue, X., Hua, J., Chen, Z., Anti-corrosion reinforcements using coating technologies—A review. *Polymers*, 14, 4782, 2022.
10. Angst, U.M., A critical review of the science and engineering of CP of steel in soil and concrete. *Corrosion*, 75, 1420–33, 2019.
11. Oleiwi, H.M., Wang, Y., Curioni, M., Chen, X., Yao, G., Augusthus-Nelson, L., *et al.*, An experimental study of CP for chloride contaminated reinforced concrete. *Mater. Struct.*, 51, 1–11, 2018.
12. Yao, G., He, X., Liu, J., Guo, Z., Chen, P., Test study of the bridge cable corrosion protection mechanism based on impressed current CP. *Lubricants*, 11, 30, 2023.
13. Lee, D. and Jeong, J.-A., Investigation of the Effective Range of CP for Concrete Pile Specimens Utilizing Zinc Mesh Anode. *Corros. Sci. Technol.*, 23, 195–202, 2024.
14. Wang, F., Xu, J., Xu, Y., Jiang, L., Ma, G., A comparative investigation on CPs of three sacrificial anodes on chloride-contaminated reinforced concrete. *Constr. Build. Mater.*, 246, 118476, 2020.
15. Fatemi, H., Hadigheh, S.A., Tao, Y., Adam, G., Development of a novel and specialised cementitious matrix overlay for anode embedment in impressed current CP (ICCP) systems for reinforced concrete bridges. *Case Stud. Constr. Mater.*, 20, e02908, 2024.
16. Astuti, P., Rafdinal, R.S., Yamamoto, D., Hamada, H., Corrosion rate of deteriorated steel bar protected by sacrificial anode CP. *AIP Conference Proceedings*, AIP Publishing, 2023.
17. Astuti, P., Rafdinal, R.S., Yamamoto, D., Hamada, H., Application of sacrificial anode CP in reinforced concrete exposed to high-temperature environment. *AIP Conference Proceedings*, AIP Publishing, 2024.
18. Cheng, X., Xia, J., R-j, W., Jin, W.-., Pan, C.-., Optimisation of sacrificial anode CP system in chloride-contaminated reinforced concrete structure. *J. Build. Eng.*, 45, 103515, 2022.

19. Gong, F., Peng, Y., Wang, Z., Fujiyama, C., Jin, W., Zhao, Y., *et al.*, Computational approach to the current and chloride ions distribution of reinforced concrete CP system. *Constr. Build. Mater.*, 438, 137078, 2024.
20. Gao, Y., Dong, B., Wang, Y., Hong, S., Fang, G., Zhang, Y., Self-charging protective layer for marine reinforced concrete based on arch-shaped triboelectric nanogenerator. *Mater. Des.*, 237, 112536, 2024.
21. Dong, S., Zhao, B., Lin, C., Du, R., Hu, R., Zhang, G.X., Corrosion behavior of epoxy/zinc duplex coated rebar embedded in concrete in ocean environment. *Constr. Build. Mater.*, 28, 72–8, 2012.
22. Sajid, H.U., Kiran, R., Bajwa, D.S., Soy-protein and corn-derived polyol based coatings for corrosion mitigation in reinforced concrete. *Constr. Build. Mater.*, 319, 126056, 2022.
23. Sharma, N., Sharma, S., Sharma, S.K., Mahajan, R.L., Mehta, R., Evaluation of corrosion inhibition capability of graphene modified epoxy coatings on reinforcing bars in concrete. *Constr. Build. Mater.*, 322, 126495, 2022.
24. Shariatmadar, M., Mahdavian, M., Ramezanzadeh, B., Designing a novel waterproof thin-layer based on silicon-modified polyacrylate grafted 2D-graphene nanosheets for chloride-induced corrosion protection of rebar in concrete structures. *J. Taiwan Inst. Chem. Eng.*, 149, 104987, 2023.
25. Ge, T., Zhao, W., Wu, X., Lan, X., Zhang, Y., Qiang, Y., *et al.*, Incorporation of electroconductive carbon fibers to achieve enhanced anti-corrosion performance of zinc rich coatings. *J. Colloid Interface Sci.*, 567, 113–25, 2020.
26. Hayatdavoudi, H. and Rahsepar, M., A mechanistic study of the enhanced CP performance of graphene-reinforced zinc rich nanocomposite coating for corrosion protection of carbon steel substrate. *J. Alloys Compd.*, 727, 1148–56, 2017.
27. Nguyen, T.A., Pham, T.L., Dinh, T.M.T., Thai, H., Shi, X., Application of nano-SiO2 and nano-Fe2O3 for protection of steel rebar in chloride contaminated concrete: epoxy nanocomposite coatings and nano-modified mortars. *J. Nanosci. Nanotechnol.*, 17, 427–36, 2017.
28. Gu, Z., Zhao, M., Liu, Q., Mao, C., Zhang, L., Sun, X., *et al.*, Photothermal superhydrophobic coating for concrete: Highly effective anti/deicing performance and corrosion resistance. *Mater. Today Commun.*, 41, 110623, 2024.
29. Boutamart, M., Rafqah, S., Hadri, A., Nouneh, K., Zaidi, S., Bichara, H., *et al.*, Design of fluorine-free superhydrophobic coating for fibred architectonic concrete. *Constr. Build. Mater.*, 425, 136086, 2024.
30. Kadhim, A., Al-Amiery, A., Alazawi, R., Al-Ghezi, M., Abass, R., Corrosion inhibitors. A review. *Int. J. Corros. Scale Inhibition*, 10, 54–67, 2021.
31. Verbruggen, H., Terryn, H., De Graeve, I., Inhibitor evaluation in different simulated concrete pore solution for the protection of steel rebars. *Constr. Build. Mater.*, 124, 887–96, 2016.
32. Jiang, S., Jiang, L., Wang, Z., Jin, M., Bai, S., Song, S., *et al.*, Deoxyribonucleic acid as an inhibitor for chloride-induced corrosion of reinforcing steel in simulated concrete pore solutions. *Constr. Build. Mater.*, 150, 238–47, 2017.

33. Chen, C., Jiang, L., Guo, M.-Z., Xu, P., Chen, L., Zha, J., Effect of sulfate ions on corrosion of reinforced steel treated by DNA corrosion inhibitor in simulated concrete pore solution. *Constr. Build. Mater.*, 228, 116752, 2019.
34. Rahmani, M.H., Dehghani, A., Salamati, M., Bahlakeh, G., Ramezanzadeh, B., Mango extract behavior as a potent corrosion inhibitor against simulated chloride-contaminated concrete pore solution; coupled experimental and computer modeling studies. *J. Ind. Eng. Chem.*, 130, 368–81, 2024.
35. Liu, Y., Guan, X., Shi, J., Synergistic inhibition of molybdate and phytate on chloride-induced corrosion of carbon steel in simulated concrete pore solutions. *Cem. Concr. Compos.*, 145, 105366, 2024.
36. Gharieb, M., Aboutaleb, S.A., Dokaila, A.M., El-Attar, M., Impact of combined organic chemical inhibitor on mechanical properties of cement pastes and steel reinforcement corrosion in concrete upon exposure to aggressive environment. *Constr. Build. Mater.*, 420, 135567, 2024.
37. Kim, K.-C., Lim, K.-M., Kyung-Taek, K., Gum-Sung, R., Park, S.Y., Jae-Yoon, K., *et al.*, Marine concrete composition using dechlorination microorganism, and construction method of marine concrete structure for the same. *Google Pat.*, 2024.
38. Fazayel, A., Khorasani, M., Sarabi, A., The effect of functionalized polycarboxylate structures as corrosion inhibitors in a simulated concrete pore solution. *Appl. Surf. Sci.*, 441, 895–913, 2018.
39. Goyal, A., Ganjian, E., Pouya, H.S., Tyrer, M., Inhibitor efficiency of migratory corrosion inhibitors to reduce corrosion in reinforced concrete exposed to high chloride environment. *Constr. Build. Mater.*, 303, 124461, 2021.
40. Wang, Y., Hu, J., Ma, Y., Zhang, Z., Huang, H., Wei, J., *et al.*, A novel high-efficient MOFs-based corrosion inhibitor for the reinforcing steel in cement extract. *Constr. Build. Mater.*, 317, 125946, 2022.
41. Li, J., Luo, M., Chen, Z., Zhuang, E., Yu, B., Chen, Y., *et al.*, Anti-corrosion mechanism of MgAl-LDHs inhibitors with varying anionic charge on reinforcing steel in simulated concrete pore solutions. *Constr. Build. Mater.*, 363, 129882, 2023.
42. Mohammadkhah, S., Dehghani, A., Ramezanzadeh, B., Adsorption and mitigation impact of the monosodium glutamate (C5H8NO4Na) bio-molecules on the steel rebar corrosion in the chloride-contaminated simulated concrete pore solution. *Sci. Rep.*, 13, 11130, 2023.
43. Mohammadkhah, S., Dehghani, A., Ramezanzadeh, B., Sodium lignosulfonate biomolecule application as an effective steel-rebar corrosion inhibitor in a chloride induced–simulated concrete pore solution. *Biomass Convers. Biorefin.*, 1–13, 2023.
44. Rezakhani, D., Jafari, A., Hajabassi, M., Durability, mechanical properties and rebar corrosion of slag-based cement concrete modified with graphene oxide, in: *Structures*, pp. 678–97, Elsevier, 2023.

10

Innovations in Concrete Corrosion Control

Ruby Aslam[1,2], Zhitao Yan[1,3*], Qihui Wang[1,4], Yi Sun[1], Afroz Aslam[5] and Jeenat Aslam[6]

[1]*School of Civil and Hydraulic Engineering, Chongqing University of Science and Technology, Chongqing, China*
[2]*Engineering Research Centre for Waste Oil Recovery Technology and Equipment, Ministry of Education, Chongqing Technology and Business University, Chongqing, China*
[3]*Chongqing Industry & Trade Polytechnic, Chongqing, China*
[4]*School of Mechanical and Intelligent Manufacturing, Chongqing University of Science and Technology, Chongqing, China*
[5]*Department of Chemistry, Constituent Government College, Thakurdwara, Moradabad, MJP, Rohilkhand University, Bareilly, U.P., India*
[6]*Department of Chemistry, College of Science, Taibah University, Yanbu, Al-Madina, Saudi Arabia*

Abstract

Concrete deterioration is a significant problem affecting infrastructure's durability and safety worldwide. Innovative approaches to concrete corrosion prevention are addressed in this chapter, including the development of advanced materials through materials science and engineering. It provides an overview of corrosion-resistant reinforcements, including innovative stainless steel reinforcements, fiber-reinforced polymers, and coated rebars that provide additional protection. A significant concentration is on self-healing concrete, which involves filling concrete composition with mechanisms for repair (autogenic repair, hollow fibers, microencapsulation, alkaline-activated geopolymer binders without cement content, biomineralization process, etc.). The chapter further discusses the contributions of advanced coating and sealants in combating corrosion. It highlights the efficacy of chemical admixtures such as innovative corrosion inhibitors and supplementary cementitious

**Corresponding author*: yanzhitao@cqu.edu.cn

Ruby Aslam, Qihui Wang, Zhitao Yan and Afroz Aslam (eds.) Corrosion in Concrete Structures: Integrating Advanced Technologies and Sustainable Practices, (291–316) © 2026 Scrivener Publishing LLC

materials (SCMs) in improving concrete durability within the smart infrastructure context. Chloride binding innovations are also explored. Moreover, advanced methods such as electrochemical corrosion monitoring technologies and integration with artificial intelligence (AI) and machine learning (ML) techniques in designing corrosion processes, such as resistant concrete, are also outlined. Afterward, the chapter ends with future perspectives and the hurdles yet to be overcome in the progression of corrosion prevention approaches to concrete structures.

Keywords: Concrete corrosion, coated rebars, self-healing concrete, autogenic repair, hollow fibers, microencapsulation

10.1 Introduction

Reinforced concrete is one of the most widely used, versatile, and durable materials for constructing many kinds of structures [1]. In normal conditions, the reinforcing steel embedded in concrete is passive due to an inert oxide layer formed on its surface. The formation and breakdown of this protective film are controlled primarily by the pH and chloride content of the surrounding concrete [2]. However, when the passive condition is not preserved at the rebar–concrete interface, corrosion may occur as uniform or localized (e.g., pitting). Corrosion of reinforcing steel is increasingly problematic worldwide due to the widespread use of reinforced concrete infrastructure. A 2002 study by the US Federal Highway Administration (FHWA) and the National Association of Corrosion Engineers (NACE) estimated that corrosion has an annual direct cost to the US economy of \$276 billion (3.1% of the nation's GDP) [3]. When indirect costs such as productivity loss are taken into account, the total rises to approximately \$552 billion, or more than 6% of GDP [4]. Interestingly, up to 30% of this comes from the corrosion of concrete structures. Due to increased maintenance costs, engineers and building owners focus more than ever on protecting existing concrete infrastructure from corrosion. That, in turn, has created a rising demand for corrosion mitigation systems designed to defer the need to make expensive repairs. The choice of an appropriate corrosion mitigation strategy will depend on a variety of factors, such as the magnitude of contamination (i.e., chloride ingress, carbonation), the degree of concrete damage, the type and extent of corrosion, the environmental exposure conditions, the anticipated service life of the structure, and the cost of the mitigation system. Codes, specifications, and standards are being developed, but technology is advancing rapidly, and these codes, specifications, and standards cannot always keep up with the technology. Cutting-edge materials, processes, and corrosion control technologies have allowed designers and architects to meet performance-based specifications at lower life cycle costs. This chapter concentrates on the latest developments and the applications of emerging technologies for repairing and rehabilitating reinforced concrete structures.

10.2 Innovative Materials for Corrosion Control

10.2.1 Corrosion-Resistant Reinforcements

Using stainless steel, fiber-reinforced polymers (FRP), and coated rebars in concrete construction offers significant durability and corrosion resistance advantages. Each material presents unique benefits that enhance reinforced concrete's longevity and structural integrity, making it increasingly popular in modern construction practices. Advancements in these materials have revolutionized construction technology by improving durability, corrosion resistance, and sustainability. These innovations play a crucial role in modern engineering by addressing key challenges such as structural longevity, weight reduction, and environmental impact.

10.2.2 Stainless Steel Innovations

Recent breakthroughs in stainless steel manufacturing have led to enhanced production processes and expanded applications [5]. These advancements have improved the structural performance of stainless steel, making it a preferred choice in critical infrastructure projects due to its exceptional corrosion resistance and reduced maintenance requirements.

10.2.3 Fiber-Reinforced Polymers (FRP)

FRP composites, particularly glass fiber-reinforced polymer (GFRP) and carbon fiber-reinforced polymer (CFRP), offer excellent resistance to harsh environments, including marine and high-chloride conditions [6]. Integrating FRP with stainless steel tubes has resulted in hybrid composite structures that optimize material efficiency by reducing weight and cost while maintaining high mechanical performance [7].

10.2.4 Coated Rebars

Coated reinforcement bars, such as epoxy-coated and galvanized rebars, are widely used to enhance corrosion resistance in concrete structures.

However, their effectiveness varies depending on coating quality, manufacturing techniques, and environmental exposure [8]. Ongoing research focuses on improving coating durability and optimizing application methods to ensure long-term performance in aggressive conditions. While these innovations have significantly improved construction materials, challenges remain in ensuring consistent performance across diverse environmental conditions. Further research is essential to refine these materials and expand their applications in sustainable and resilient infrastructure development.

10.2.5 Self-Healing Concrete

The significant problems of concrete are cracking and local damage due to its microstructure properties and the applied external loads [9]. Due to the material's self-weight and environmental conditions, cracking is often inevitable, where microcracks form mainly as the mixture undergoes a contraction process because of the rapid loss of water. The treatment of concrete cracking has historically been a focus of interest in both engineering and academic domains. The self-healing feature developed by concrete lets the material withstand environmental effects and extend its service life, thus improving the overall safety, longevity, and durability of the whole structure. Various influential factors in the surrounding environment (e.g., temperature or humidity changes) and mechanical loads can induce microcracks or other microstructural damage. This greatly hampers the load-bearing capacity, the durability, and the impermeability of the concrete. Left unaddressed or poorly addressed, these cracks can dramatically impact the structure's performance, shorten its useful life, and even result in catastrophic failures, endangering lives and road infrastructure safety. However, the healing of the microscopic internal cracks cannot be achieved with traditional repair approaches [10]. With the continuous improvement of road infrastructure, the damage of microcracks in road infrastructure is a significant problem. The research and development of self-healing concrete materials and related technologies has become an important research topic in the field of construction, which can effectively improve the stability of concrete performance and extend the service life of concrete and road infrastructure projects. The continued advancement of dependable *in-situ* concrete self-healing mechanisms and the insights learned from such cracks and repairs will only enhance concrete durability and prolong concrete lifetime. For example, concrete used in road infrastructure has a high compressive strength and high durability and reliability and requires minimal maintenance throughout its useful life. Cracks in concrete can be treated in one of two ways: active or passive [10]. Therefore, passive treatments can only repair superficial cracks, while active methods can have external and internal effects [10]. The passive treatments usually consist of coatings of chemicals and polymers that are deposited directly on the surface of the concrete, preventing the intrusion of aggressive substances and making the concrete more durable. Alternatively, sealants can be injected or sprayed into the observed cracks. These sealants typically comprise epoxy resins, chlorinated rubbers, waxes, polyurethane, acrylics, and siloxane materials. However, while beneficial, these treatments have weaknesses, such as low resistance to atmospheric conditions, moisture sensitivity, low heat resistance, and differences in expansion coefficients between the concrete and sealants. Active treatments, also known as self-repair, self-sealing, or self-healing methods, can be activated automatically based on external conditions, regardless of the crack position. Their treatments are formulated to trigger whenever a crack occurs, enabling the material to seal itself autonomously. A generic dividing line can be drawn which separates materials that act exclusively as sealants, merely covering the crevice to prevent it from environmental exposure (usually passive treatments), from agents that possess repair capabilities to react with the cementitious matrix and regenerate its properties, either partially or totally (active treatments) [11]. Below are several ways by which the self-healing mechanism is activated.

10.2.5.1 Autogenic Repair

Autogenic repair is a natural mechanism of repair that occurs in standard concrete *via* the hydration of unhydrated cement particles or the carbonation of dissolved calcium hydroxide [12]. The mechanisms of autogenic self-repair include the following: (a) the deposition of calcium carbonate or calcium hydroxide crystals, (b) the blockage of cracks due to the accumulation of impurities from water and concrete particles generated during cracking, (c) the hydration of unreacted cement or other cementitious materials remaining in the concrete, and (d) the propagation of cement hydration around the crack edges [13]. Of these mechanisms, calcite crystallization is the primary contributor to autogenic self-healing.

10.2.5.2 Hollow Fibers

Hollow fibers include adding fibers that hold repair agents to the concrete mix. This causes the fibers to break and the concrete to crack, at which point the contents of the fibers are released to begin the repair process.

Specifically, hollow fibers are used to store a recovery agent inside that is evenly distributed throughout a concrete matrix. They break when cracks form, and damage occurs under specific conditions, releasing the agent contained within. The agent seeps in the cracks, repairing them in the process. The hollow polypropylene fibers filled with methyl methacrylate as a repair agent have been studied by Dry [14, 15]. When the fibers rupture, the agent releases out, and as the damage progresses, it fills the fissures. Ethyl cyanoacrylate has also been used as the repair agent in other studies with similar positive results [16]. These studies have been carried out in two aspects: on one side, the ability of the repair agent to penetrate cracks because of capillary action and gravity due to the bent shape with respect to the local gravity field; on the other side, the analysis of mechanical parameters like stiffness, peak force, and ductility after repairing them.

10.2.5.3 Microencapsulation

In microencapsulation, self-repairing or sealing agents are encapsulated in small spheres released once cracks form in the concrete. This approach consists of encapsulating the healing catalyst particles in microcapsules, which were uniformly distributed throughout the matrix. The microcapsule shell breaks when cracks form, releasing the repair agent to mend the crack. While hollow fibers introduce the agent as an independent part within a composite, microencapsulation isolates repair agents within microcapsules. Spherical, lamellar, and tubular carrier polymer microcapsules, mesoporous silica, layered double hydroxides and nanotubes release agents in response to pH, heat, or ions. Although microencapsulation can partially reinstate corrosion resistance, its effect on the mechanical properties of the concrete is still undetermined. The systematic design of multifunctional core materials and enhanced compatibility between the core and shell materials are promising approaches, but the preparation is complex and expensive. Future research shall concentrate on easy preparation procedures, low cost, and efficiency in inhibiting corrosion.

10.2.5.4 Alkaline Activated Geopolymer Binder without Cement Content

Sodium silicate (Na_2SiO_3), and sodium hydroxide (NaOH) form the alkaline activator for geopolymer materials. An alumino-silicate gel stabilizes the matrix and aggregates, giving the material strength when silicon and aluminum react with the alkaline solution [17]. Geopolymer binders, without any cement, are in a paste and then hardened by heat treatment (50°C –

80°C for 4–48 h) [18]. After 7 days, its apparent dry density, bending tensile strength, and compressive strength are 1,200–1,400 kg/m^3, 1.3–6.5 N/mm^2, and 5.1–35.0 N/mm^2, respectively [18, 19]. The raw materials used are class F power plant ash and alkaline activator. The ash must have an oxide content of more than 70% (SiO_2 + Al_2O_3 + Fe_2O_3) and be less than 40% residue on a 0.045-mm sieve [18]. The chemical reactions involved in the transformation of alkali-activated aluminosilicates into geopolymers are influenced mainly by the oxide composition, granularity, and specific surface of the precursor materials [20]: the oxide composition and glasses or minerals of the silicate raw material affect the geopolymerization process [21, 22]. Innovative geopolymers strengthen and improve the durability and self-repairing capacity of concrete; providing the bio-bonding also acts as the eco-third reaction, with low cost, locally sourced and enabled by local climate, using waste, like from thermal power plants, such as fly ash, reducing waste in landfills.

10.2.5.5 Biomineralization

Biomineralization is a biological principle-based mechanism whereby bacteria hibernate in concrete. These bacteria are activated when the concrete is exposed to moisture, and they consume compounds such as calcium lactate to secrete calcium carbonate that fills cracks. Microbial-inspired self-healing technology is a promising process where microorganisms in concrete help repair cracks.

Microbial agents, capable of surviving in various environments like soil and water, play a key role in this process by precipitating beneficial chemicals. The growth of these microorganisms depends on factors like pH, temperature, and moisture content, often using spores rather than active cultures. One innovative method, microbial-induced calcite precipitation (MICP), utilizes bacteria to precipitate calcium carbonate ($CaCO_3$), a form of biomineralization. Ureolytic bacteria hydrolyze urea and produce ammonia that raises pH and facilitates $CaCO_3$ formation in calcium-rich environments (Figure 10.1). This approach significantly enhances cracked concrete's compressive strength and stiffness, leading to bacterial concrete's development. Bacteria such as *Sporosarcina pasteurii* and *Bacillus sphaericus* are widely used due to their ability to hydrolyze urea and produce calcium carbonate. Recent research has focused on optimizing their growth conditions and metabolic activity to enhance mineral precipitation. A comparative study was conducted on concrete enhanced with *Bacillus pasteurii* and *Bacillus sphaericus*, incorporating class F fly ash. The addition of fly ash facilitated the growth of *B. sphaericus* (CCFBS) and *B. pasteurii* (CCFBP) in specific proportions, improving the mechanical properties.

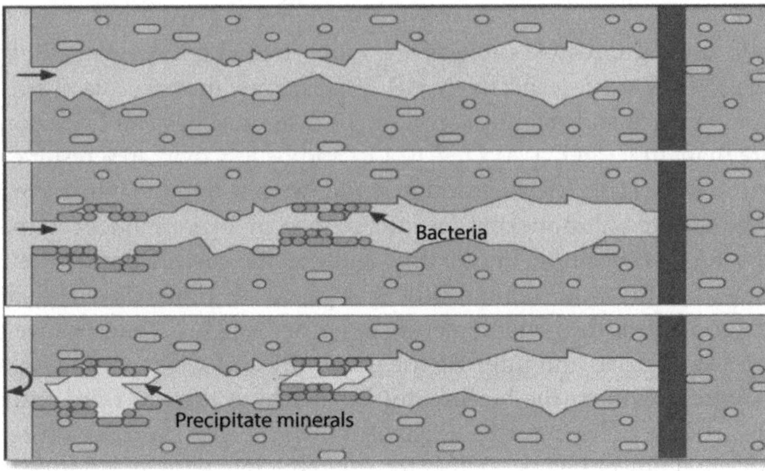

Figure 10.1 Schematic representation of microbial-based self-healing phenomena [21].

Compared to conventional concrete, CCFBS with 10% fly ash exhibited a 10.80% higher compressive strength than the control sample [22].

Additionally, non-ureolytic bacteria like *Bacillus cohnii* can induce mineralization without urea, reducing the risk of ammonia release and making the process more environmentally friendly [23]. The efficiency of biomineralization depends on the availability of nutrients that support bacterial activity and mineral production. Recent research has focused on optimizing nutrient formulations to enhance the process. Calcium sources, such as calcium lactate and calcium nitrate, have been explored as alternatives to traditional calcium chloride, improving the efficiency of biomineralization while reducing costs [23]. Additionally, organic compounds like yeast extract have been shown to enhance bacterial activity and mineral production, further improving the effectiveness of the process [24]. These advancements in nutrient optimization have made biomineralization more practical and cost-effective for large-scale applications. One of the most significant advancements in biomineralization is its integration into self-healing concrete. Autonomous healing is achieved by embedding bacteria and nutrients directly into the concrete mix. The bacteria are activated when cracks form, precipitating calcium carbonate to seal the cracks and prevent water and chloride ingress [11, 25]. This self-healing capability repairs cracks and improves the overall durability and lifespan of concrete structures, reducing the need for frequent maintenance and repairs [26]. Self-healing concrete has been successfully applied in various field trials, demonstrating its potential to revolutionize the construction industry.

A multi-technique experimental study by L.V. Zhang *et al.* [27] investigated *S. pasteurii* immobilized in alkali-activated slag (AAS) composites. As observed in calorimetric curves, adding calcium additives to AAS paste slowed the acceleration stage. Biomineralization reduced pore volume and water sorptivity, as confirmed by MIP analysis. M.P. Smitha *et al.* [28] demonstrated that *Bacillus megaterium* at an optimum concentration of 10^5 cells/mL enhanced compressive (11.3%), split tensile (97.5%), and flexural (8.6%) strength, respectively, at 28 days. Additionally, chloride ion penetration (26.8%) and water permeability (98.7%) significantly decreased. Salehi *et al.* examined four bacterial strains (*S. pasteurii, S. ureae, B. megaterium,* and *B. licheniformis*) in lightweight fiber concrete (LWFC) with varying polypropylene fiber content (0%, 0.5%, and 1%). *B. licheniformis* with 1% fiber showed the highest tensile strength (32.65%) and lowest water absorption (28.39%). Calcium carbonate precipitation and fiber orientation were more pronounced in *B. licheniformis* and *B. megaterium* compared to *S. pasteurii* and *S. ureae* [29].

10.3 Nanomaterials in Concrete

Incorporating graphene, carbon nanotubes (CNTs), and nano-silica into concrete has shown significant potential for enhancing its durability and mechanical properties. These materials improve concrete microstructure, improving performance in various environmental conditions. Carbon nanotubes (CNTs) significantly enhance bond strength when hybridized with nanoclay, achieving improvements of up to 174.9% for specific rebar sizes and cement contents [30]. The optimal CNT concentration ranges from 0.01% to 0.05%, leading to substantial increases in bond strength and overall durability. Similarly, graphene oxide improves hydration processes and microstructural densification, enhancing mechanical properties and durability [31]. It notably boosts resistance to water absorption, chloride penetration, and corrosion, all critical for longevity in aggressive environments. Nano-silica also contributes to concrete's durability by refining its pore structure and improving mechanical properties, with optimal dosages between 1.5% and 3% [32, 33]. It reduces water absorption and gas permeability while increasing compressive strength. However, integrating these nanomaterials must be carefully considered, as even small additions can significantly increase production costs despite their substantial performance improvements.

10.4 Advanced Coatings and Sealants

When protecting and enhancing concrete surfaces, choosing between a sealer and a coating can be crucial. Both options offer unique advantages but serve different purposes and are suited for various applications. A concrete sealer is a protective layer applied to concrete surfaces to shield them from damage caused by water, chemicals, and other contaminants. There are two main types of concrete sealers: penetrating sealers and topical sealers. Penetrating sealers, such as silane, siloxane, and siliconate, absorb into the concrete and create a chemical barrier that repels moisture and substances. Topical sealers, including acrylic, epoxy, and polyurethane, form a protective film on the surface. The benefits of using concrete sealers include water resistance, preventing cracks and spalling due to freeze–thaw cycles, stain protection against oil and grease, enhanced durability by reducing wear and tear, and UV protection to prevent discoloration and degradation. These sealers are commonly used for driveways, walkways, patios, pool decks, and interior floors, such as in basements and garages, to provide a long-lasting barrier against moisture, stains, and environmental damage.

A concrete coating is a thicker protective layer applied to concrete surfaces, offering durability and decorative enhancement. These coatings can be functional, esthetic, or both, with various types available [34]. Epoxy coatings are known for their high-gloss finish and durability, making them ideal for commercial and industrial settings. Polyurea and polyaspartic coatings provide fast curing times, UV stability, and excellent resistance to chemicals and abrasion. Acrylic coatings, primarily used for decorative purposes, come in various colors and finishes [35, 36]. The concrete coating enhances durability by resisting abrasion, chemicals, and heavy traffic while improving aesthetic appeal with numerous color, texture, and pattern options. They allow for customization through decorative elements like flakes and metallic pigments, and some, like polyurea and polyaspartic, cure quickly to minimize downtime. Typical applications include garage floors for stain and abrasion resistance, commercial spaces such as retail stores and warehouses for durability and aesthetics, and outdoor areas like patio, pool decks, and driveways for protection and visual enhancement.

Bio-based coatings show great potential for enhancing durability and minimizing corrosion in concrete structures, making them a promising choice for sustainable construction. New cutting-edge technologies have made an enormous difference in the coatings industry over the last decade, leading to more environmentally conscious solutions. These innovations consist of UV curing processes, low or no-solvent treatments, waterborne

(WB) products, and hyperbranched (HYP) and high-solid (HS) coatings. However, these technologies have been merged in some instances to create high-performance (HP) sustainable coatings. Initially, these techniques were mainly applied to generate organic and/or inorganic treatments for multiple usage areas, some of which were not ultimately sustainable. Nevertheless, these technologies have been extensively incorporated over the last few decades to develop bio-based entities for producing coatings, targeting sustainability in the production process and raw materials.

Bio-based coatings are derived from renewable resources, making them a sustainable alternative that reduces environmental impact and promotes eco-friendly construction practices. These coatings are typically non-toxic and free from harmful solvents, ensuring they are safer for human health and the environment. In addition to their environmental benefits, bio-based coatings also enhance the durability of concrete. They provide excellent water resistance, protect against corrosion, and offer defense against chemical damage, ultimately extending the service life of concrete structures. Bio-based coatings improve the durability of concrete structures by providing practical barriers against water penetration, which helps prevent deterioration caused by freeze-thaw cycles and chemical attacks. They enhance the resistance of concrete to corrosion of reinforcing bars and other environmental factors, thereby prolonging the structure's service life. These coatings are often made from non-toxic materials, contributing to a safer and more sustainable construction approach.

10.5 Chemical Admixtures for Corrosion Protection

10.5.1 Employment of Smart Corrosion Inhibitor

Corrosion inhibitors are a widely acknowledged and cost-effective method to control steel corrosion and enhance the durability of reinforced concrete (RC) structures [37, 38]. Traditional inhibitors are classified into inorganic (e.g., nitrites, molybdates, phosphates) [39–41] and organic types (e.g., amines, alkyl amines, imidazolines) [42, 43] (Figure 10.2). However, inorganic inhibitors often pose risks to human health and the environment, while organic inhibitors can be inefficient at low concentrations and degrade quickly in concrete. Consequently, developing next-generation inhibitors with low toxicity, high efficiency, long-term stability, and intelligent performance is crucial to address these limitations [44].

While traditional corrosion inhibitors enhance steel corrosion resistance, they have notable drawbacks. Nitrite, for instance, is restricted in

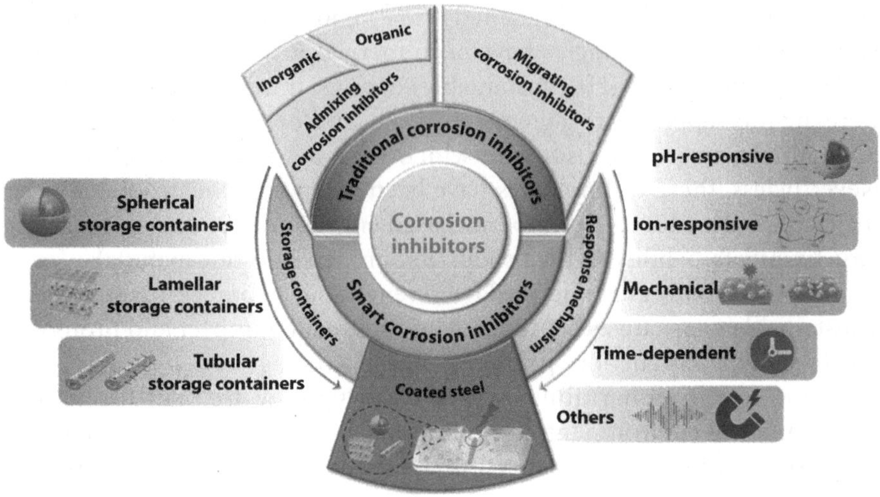

Figure 10.2 Schematic representation of corrosion inhibitors in RC structures [44].

many countries due to toxicity and carcinogenicity. Similarly, high concentrations of amine-based inhibitors can prolong cement paste setting time, potentially compromising concrete durability. To address these issues, researchers are exploring advanced solutions. Recently, smart corrosion inhibitors utilizing micro/nano-containers have shown promise in mitigating these limitations while enabling targeted repair of corroded steel.

Smart corrosion inhibitors typically consist of a core-shell composite structure, where corrosion inhibitors are encapsulated within micro/nano-containers. This encapsulation isolates the inhibitors from the environment, preventing unwanted leakage and minimizing negative impacts on concrete and ecosystems. Containers such as polymer microcapsules, mesoporous silica, layered double hydroxides, halloysite nanotubes, graphene, and carbon nanotubes are categorized into spherical, lamellar, and tubular types, offering innovative response capabilities. These containers release inhibitors in response to stimuli like pH, light, humidity, heat, magnetism, or ions, enabling controlled and targeted action.

Smart corrosion inhibitors based on polymeric microcapsules demonstrate high corrosion inhibition efficiency but raise environmental concerns. Toxic materials like resins and polyurea are commonly used for shell preparation, posing ecosystem risks. Future research should priorities replacing these materials with environmentally friendly alternatives and adopt green approaches from other fields of corrosion protection [45, 46]. Another challenge lies in the limited models available to analyses the corrosion protection behavior of smart inhibitors in cementitious materials. Existing models are often condition-specific, restricting their applicability. Developing predictive, multi-scale evaluation models is essential for assessing the durability of RC structures. Additionally, integrating surface dynamics with molecular dynamics simulations could offer a novel approach to predicting the corrosion protection performance of smart inhibitors [47].

10.5.2 Use of Supplementary Cementitious Materials (SCMs)

Concrete is made of cement, the most demanding material that requires large quantities of raw materials from natural resources and emits a lot of CO_2 [48]. To address this concern, supplementary cementitious materials (SCMs) are utilized to replace cement to some extent, reducing cement utilization and environmental impacts [49]. Furthermore, employing SCMs, which are by-products, also helps reduce industrial waste. This can also reduce concrete costs through reduced cement content in certain instances. Silica fume, fly ash, and ground-granulated blast-furnace slag are the most used SCMs to improve the properties of concrete and reduce its environmental impact [50]. SCMs affect concrete differently at room and high temperatures; silica fume improves room temperature strength but increases strength loss at high temperatures; fly ash and slag improve high-temperature performance. Research [51] has demonstrated that the strength of concrete can diminish as temperatures rise, but the degradation rate varies with the type of aggregate used. Calcareous (e.g., limestone, dolomite) aggregates perform better than siliceous aggregates (e.g., quartzite, sandstone) when subjected to high heat. In contrast, lightweight aggregates have shown superior strength retention compared to heavier types at elevated temperatures [51]. However, existing SCMs are losing dominance due to changes in energy production, creating a demand for alternatives [52].

A wide range of by-products and waste from industrial, agricultural, and municipal processes can be utilized as new SCMs. Several potential alternative supplementary cementitious materials (SCMs) are being explored as substitutes for traditional SCMs. These include the following:

1. Mine tailings: By-products of mining operations can be treated to enhance their pozzolanic properties, although they may initially have low reactivity due to their crystalline nature.
2. Aluminum dross (AD): This is a by-product of aluminum smelting that contains toxic materials requiring treatment before it can be safely used in concrete.

3. Iron and steelmaking slags: Various slags, such as basic oxygen furnace (BOF) and electric arc furnace (EAF) slag, show potential as SCMs, especially with appropriate treatment to enhance their properties.
4. Wood waste ash (WWA): This is a by-product from wood combustion, which has variable properties and can contain heavy metals, necessitating careful handling.
5. Concrete sludge waste (CSW): Leftover material from concrete mixing processes can be processed to improve its properties and reduce environmental impact.
6. Recycled concrete powder (RCP): Produced from recycling concrete structures, it serves as an alternative SCM but requires treatment to enhance its effectiveness.

These materials are being investigated as part of a broader effort to find sustainable and environmentally friendly alternatives to conventional SCMs, which are under threat due to reduced availability. These by-products and waste are waste management problems, and the reuse of their products is limited owing to their diverse chemical and physical characteristics. Alkali-reactive silica (ASR) can be formed in high-alkali environments, but this can be prevented by adding glasses with low alkali content. ASR occurs between amorphous silica in some natural aggregate, reacting with the cement's hydroxyl ions [53, 54]. The process can take years to develop but can cause extensive structural damage over time, making early detection a challenge. Thus, identifying best practices in reducing ASR is critical. Among them, the potential for the beneficial use of new waste by-products and different types of fibers to further reduce ASR holds promise. As laid out in ASTM C1260, accelerated performance testing methods are a fast, practical solution to access the efficacy of SCMs and other materials to help prevent ASR and help ensure the reliability and speed of these solutions [55, 56].

However, these materials offer a partial substitute for cement in cementitious composites, thus reducing costs and pollution. Transportation is the main cost-intensive cost for new SCMs since their scale may exceed that of existing SCMs. Pre-treatment methods like particle size reduction and calcination can improve reactivity, but their environmental impact needs assessment.

10.6 Electrochemical Techniques

Electrochemical treatments for concrete structures have been used as non-destructive forms of treatment against concrete deterioration since the

1960s and are mainly used to mitigate damage resulting from the corrosion of embedded reinforcement. The classification of most of these treatments depends primarily on their purposes, such as whether they are focused on prevention or rehabilitation. One is a cathodic protection system, and the other is electrochemical realkalization and chloride extraction methods. These methods are discussed in detail below.

10.6.1 Cathodic Protection Systems

The first widely adopted electrochemical treatment for corrosion control was cathodic protection (CP), which was introduced in 1959. It involves applying direct current from an external power source to the embedded steel, polarizing it to a more negative potential, thereby decreasing its corrosion susceptibility. During the CP process, the applied current density is commonly restricted to 10–20 mA/m^2 of reinforcement, with a treatment duration that usually extends over a 15–25-year design life [8]. This technique can be used for the internal rehabilitation of deteriorating reinforced concrete structures and is also used preventively to mitigate corrosion in a specific environment when corrosion is expected to occur during a service life based on exposure conditions, the so-called CP method. Over the years, technology has evolved, resulting in the development of several systems, which can be broadly categorized into three types: impressed current cathodic protection (ICCP), hybrid cathodic protection (HCP), and galvanic cathodic protection (GCP) [57]. Asset owners must consider the sustainability of CP systems' sustainability, environment, health and safety, construction, technical performance, operational, and monitoring aspects. Among corrosion engineers, it is generally accepted that ICCP offers the most powerful and reliable solution for managing corrosion in structures with extensive deterioration. However, if future maintenance interventions are anticipated, HCP and GCP are also viable alternatives, even in aggressive environments.

(a) Galvanic Cathodic Protection

Galvanic cathodic protection systems utilize sacrificial anodes made from materials such as zinc, magnesium, or aluminum, which corrode preferentially to protect the steel reinforcement in concrete structures. Recent advancements in this field have significantly improved their effectiveness and versatility. One key development is high-purity zinc and aluminum alloys, which offer longer service life and more consistent current output due to their optimized compositions. Additionally, chemically activated zinc anodes have been introduced to function effectively in low-conductivity environments, such as dry or carbonated concrete. Another innovation is using embedded galvanic anodes, including discrete ones for localized protection or patch repairs and mesh ones for uniform protection over larger areas [58]. Self-regulating anodes have also been developed, which adjust their current output based on environmental conditions like moisture and chloride levels, ensuring optimal protection without the risk of overprotection. Furthermore, hybrid systems that combine galvanic anodes with small power supplies or sensors are employed to enhance performance in challenging environments, making galvanic cathodic protection systems more adaptable and reliable for various applications.

(b) Impressed Current Cathodic Protection (ICCP)

Impressed current cathodic protection systems utilize an external power source to deliver a protective current to the steel reinforcement in concrete structures, effectively preventing corrosion. Recent advancements in ICCP technology have significantly enhanced its efficiency, durability, and applicability. One significant development is using advanced anode materials, such as mixed metal oxide (MMO) anodes, which are titanium-based and highly durable, offering long service life even in aggressive environments. Additionally, conductive ceramic anodes have been introduced, providing excellent resistance to chemical attacks and stable performance over time [59]. Carbon-based anodes, including conductive carbon coatings or fibers, are also gaining popularity due to their lightweight and cost-effective properties [60]. Another key advancement is the implementation of distributed anode systems, such as conductive coatings or overlays applied to the concrete surface, which ensure uniform current distribution across the structure [61]. Furthermore, embedded anodes installed within the concrete structure (e.g., in slots or holes) offer a less intrusive solution while providing targeted protection to specific areas [62]. These innovations have made ICCP systems more reliable and adaptable for various applications, from bridges to marine structures.

10.6.2 Electrochemical Realkalization and Chloride Extraction Methods

Electrochemical realkalization and chloride extraction are critical methods for rehabilitating concrete structures affected by chloride-induced corrosion. These techniques leverage electrochemical processes to enhance the durability and longevity of reinforced concrete. The following sections outline the key aspects of these methods.

10.6.2.1 Electrochemical Chloride Extraction (ECE)

Mechanism: ECE involves applying a direct current to a concrete structure, where the reinforcing steel acts as the cathode and a temporary anode is placed on the surface. This process extracts chloride ions from the concrete matrix [63].

Effectiveness: The efficiency of chloride removal is influenced by the total charge passed, with optimal results achieved at 1,500 to 2,000 Ah/m^2. Water is a more effective electrolyte than $Ca(OH)_2$ [64].

Innovations: Recent advancements include using anodic nickel electrodes and prefabricated high conductivity magnesium phosphate cement mortar rods, which enhance extraction efficiency and simplify the process.

10.6.2.2 Electrochemical Realkalization

Purpose: This method aims to restore the alkaline environment of concrete, which is essential for the passivation of steel reinforcement. It counteracts the effects of carbonation and chloride ingress.

Application: Realkalization typically involves injecting alkaline solutions or using electrochemical methods to increase the pH around the steel reinforcement, thereby preventing corrosion. While electrochemical methods promise to extend the service life of concrete structures, concerns remain regarding potential adverse effects on the concrete matrix, such as reduced binding capacity and durability. Further research is needed to optimize these techniques and mitigate any negative impacts.

10.7 Role of AI and ML in Designing Corrosion-Resistant Concrete

Integrating AI and ML in designing corrosion-resistant concrete is transforming materials science and structural engineering. These technologies enhance the assessment, prediction, and optimization of concrete materials, leading to improved durability and service life of structures. The following sections outline the key roles of AI and ML in this domain.

10.7.1 Corrosion Assessment and Prediction

Machine learning algorithms have been employed to evaluate the corrosion of reinforced concrete structures, significantly improving the accuracy of assessments compared to traditional methods [65]. Various AI techniques [66], such as adaptive neuro-fuzzy inference systems (ANFIS) and genetic programming and multi expression programming (MEP), have demonstrated high predictive accuracy for corrosion in different concrete types. The MEP model showed the highest performance among the tested models. It achieved a coefficient of determination (R^2) of 92.1 for training and 87.9 for testing in predicting corrosion in cement concrete. For sulfur concrete, the MEP model reached an impressive R^2 of 99.5 for training and 97.7 for testing. The findings suggest that machine learning-based models can assist engineers in estimating the corrosion of concrete pipes in sewer systems. This can lead to better maintenance strategies and improved infrastructure longevity, ultimately benefiting public health and safety.

10.7.2 Material Discovery and Optimization

The MOSS platform exemplifies the use of AI in discovering corrosion-resistant materials, facilitating data collection and analysis for material scientists [67, 68]. Hybrid ML techniques, like the multi-objective slime mold and optimized support vector regression, have been developed to assess the corrosion behavior of eco-friendly concrete, yielding superior predictive performance [69].

10.7.3 Structural Design Enhancement

ML models have been utilized to optimize the design of corroded reinforced concrete beams, achieving high predictive accuracy for shear strength, which is critical in structural integrity [70]. While AI and ML offer significant advancements in corrosion-resistant concrete design, challenges remain in integrating these technologies into practical applications and ensuring their widespread adoption in the industry.

10.8 Future Considerations

Sustainability means providing for current society's needs in an economically viable manner while ensuring environmental protection for the future. However, this must be done with those involved in the design and construction of new structures and with the proactive, preventive maintenance of existing ones. Finally, when evaluating the lifetime environmental impact, both the embodied carbon and global warming potential of the construction materials and the buildings' anticipated service life must be considered. In recent years, there has been much focus on reducing the environmental impact of new concrete structures. This encompasses activities like adopting alternative cement, including Portland-limestone cement, maximizing the use of supplementary waste-products in concrete, developing an FRP (fiberglass) reinforcement guide, and establishing sustainable concrete and building standards and durability codes that extend the service life of new buildings.

Safety: This evolving subject of concern with the safety of existing structures in these environments, where the structures are subjected to de-icing chemicals and other corrosion-inducing conditions, has been growing over the past few decades. These settings promote the corrosion of materials, especially in reinforced concrete, where the steel of a structure is susceptible to the degradation caused by saltwater, humidity, and de-icing products, which can cause steel reinforcement deterioration and consumption. Increased corrosion weakens the structural integrity and lowers the strength and durability of the impacted structures over time, leading to the risk of failure. This is where all future initiatives are expected to converge with interests in the increasing resilience of materials and the increased prevention of their corrosion. This could include creating advanced corrosion-resistant materials, like new coating systems, corrosion inhibitors, and more durable reinforcement options such as fiber-reinforced polymers (FRPs) or corrosion-resistant alloys. More effective monitoring systems, such as sensor technologies and real-time diagnostic tools, will also play a vital role in identifying corrosion early to enable maintenance before severe damage results. Another key focus will be on predictive models relying on data analytics and artificial intelligence to evaluate the remaining life of structures and inform maintenance strategies. These models can predict corrosion-susceptible areas, which helps to improve resource allocation and target maintenance activities accordingly.

Technology: Concrete repair and corrosion control processes represent high potential pathways for sustainability. New technologies that allow the assessment of what is possible today and whether it is worth the investment in terms of cost or time to restore the previous quality of a building should provide a better process for evaluating existing structures. Nonetheless, several structures are not sufficiently assessed or kept for a complete repair until rust and damage have taken root, which hinders the possibilities of most techniques and assessments to prolong their life span. For structures with extensive deterioration, a more conservative approach, which makes repairs more complex, is frequently required. Consequently, the fact that an evaluation can be timely is (in most cases) fundamental to the fact that corrosion control methods can be adopted, and their effects can be durable. By identifying issues early, design professionals can make more informed decisions and take proactive steps to extend the structure's life, ultimately leading to more sustainable and cost-efficient outcomes.

This interest has extended into machine learning (ML) to improve predictive accuracy of concrete properties and corrosion performance [71, 72]. These techniques could help determine when a critical structure evaluation will be necessary. ML can enable the prediction of corrosion initiation (or other distress mechanisms) in new and existing structures more accurately by building a comprehensive database with structure types and deterioration mechanisms and relying on reliable data generation methods. This data could help create maintenance plans in the design phase or guide evaluation strategies. Indeed, as per Pacheco and Tape [72], there are best practices for corrosion monitoring, so systems for structural health monitoring and corrosion control will only evolve further.

10.9 Conclusion

This chapter explores a wide array of innovative approaches to concrete corrosion control, each contributing significantly to enhancing the durability and longevity of reinforced concrete structures. The use of corrosion-resistant reinforcements, including advancements in stainless steel innovations and fiber-reinforced polymers, provides substantial improvements in combating the degradation of concrete. Additionally, incorporating coated rebars offers another method of protection, extending the lifespan of concrete structures exposed to harsh environments. Self-healing concrete, a groundbreaking concept, demonstrates remarkable potential in mitigating corrosion-induced damage. Through mechanisms such as autogenic repair, utilizing hollow fibers and microencapsulation, and developing alkaline-activated geopolymer binders, self-healing systems offer an exciting avenue for autonomous restoration of concrete integrity. Biomineralization further augments these efforts, using biological processes to facilitate the repair and maintenance of concrete in corrosive environments. Advanced coatings and sealants also play a critical role in safeguarding concrete from corrosion, offering effective barriers to aggressive environmental factors. Chemical admixtures for corrosion protection, such as smart corrosion inhibitors and supplementary cementitious materials (SCMs), are key contributors to mitigating the effects of chloride-induced corrosion. Innovations in chloride binding further strengthen the resilience of concrete structures exposed to deicing salts and seawater. Electrochemical techniques have emerged as essential tools in corrosion monitoring and prevention, providing real-time assessments of corrosion activity.

Additionally, the role of artificial intelligence (AI) and machine learning in designing corrosion-resistant concrete represents a new frontier, enabling predictive modelling, material optimization, and efficient design processes. Looking ahead, the future of concrete corrosion control lies in further integrating these innovative technologies, particularly in developing sustainable materials and systems. However, challenges remain in the widespread implementation and standardization of these methods. Addressing cost-effectiveness, material compatibility, and environmental impact will be crucial to successfully adopting these innovations globally. As research continues to evolve, there is hope for even more advanced solutions to ensure the long-term durability of concrete structures, thus mitigating the economic and safety risks of corrosion.

AI disclosure statement: The author acknowledges the use of AI-based tools for minor language editing or grammatical correction. However, all intellectual contributions, analyses, and conclusions presented in this chapter are entirely the author's own.

References

1. Wang, J., Wang, Q., Zhao, Y., Li, P., Ji, T., Zou, G., Qiao, Y., Zhou, Z., Wang, G., Song, D., Research Progress of Macrocell Corrosion of Steel Rebar in Concrete. *Coatings*, 13, 5, 853, 2023.
2. Mundra, S. and Provis, J.L., Mechanisms of passivation and chloride-induced corrosion of mild steel in sulfide-containing alkaline solutions. *J. Mater. Sci.*, 56, 14783–14802, 2021.
3. Corrosion Costs and Preventive Strategies in the United States. *NACE Int.*, 57, 2002. ISBN: 1-57590-089-3.
4. Koch, G., Cost of corrosion, in: *Trends in oil and gas corrosion research and technologies*, pp. 3–30, 2017.
5. Surana, V. and Deshpande, A.V., Stainless Steel Bars as Reinforcement. *Challenges for Existing and Oncoming Structures, IABSE Symposium, Prague*, 2022.
6. Zhang, J., Wu, H., He, W., Fiber reinforced polymer (FRP) material and stainless steel tube composite and preparation technology thereof, 2013.

7. Islam, S.Z. and Young, B., Use of Stainless Steel as Structural Members in Bridge Construction and Fibre Reinforced Polymer Strengthening, in: *IABSE-JSCE Joint Conference on Advances in Bridge Engineering-III*, pp. 164–173, 2015.
8. Yan, L., Deng, W., Wang, N.-W., Xue, X., Hua, J., Chen, Z., Anti-Corrosion Reinforcements Using Coating Technologies—A Review. *Polymers*, 14, 21, 4782, 2022.
9. Kumar, R., Review on the mechanism and mitigation of cracks in concrete. *Appl. Eng. Science Appl. Engg. Sci.*, 16, 100154, 2023.
10. Barbu, A.M. and Stoian, M.M., Innovative technologies in constructions. Self-repairing concrete used in road infrastructure, in: *IOP Conference Series: Earth and Environm. Sci*, vol. 664, IOP Publishing, p. 012082, 2021.
11. Seifan, M., Samani, A.K., Berenjian, A., Bioconcrete: next generation of self-healing concrete. *Appl. Microbio. Biotech.*, 100, 6, 2591–2602, 2016.
12. Wu, M., Johannesson, B., Geiker, M., A review: Self-healing in cementitious materials and engineered cementitious composite as a self-healing material. *Const. Build. Mater.*, 28, 1, 571–583, 2012.
13. Ter Heide, N., Crack healing in hydrating concrete Thesis Delft University of Technology, 2005.
14. Dry, C.M., Smart multiphase composite materials that repair themselves by a release of liquids that become solids, Smart structures and materials, in: *Smart materials*, pp. 62–70, Proceedings of the SPIE, 1994.
15. Dry, C.M., Three designs for the internal release of sealants, adhesives, and waterproofing chemicals into concrete to reduce permeability. *Cement. Conc. Res.*, 30, 12, 1969–77, 2000.
16. Joseph, C., Lark, R., Jefferson, T., Gardner, D., *Potential application of self-healing materials in the construction industry*, Cardiff University, Cardiff Wales, 2009.
17. Al Bakri Abdullah, A.M.M., Hussin, K., Bnuhussain, M., Ismail, K.N., Ahmad, M., II, Chemical Reactions in the Geopolymerisation Process Using Fly Ash-Based Geopolymer: A review. *Aust. J. Basic. Appl. Sci.*, 5, 7, 1199–1203, 2011.
18. Project PS Innovative solutions for transport infrastructure protection by the use of building materials with special self-maintenance and self-repair properties, p. 6, 2019.
19. NIRD URBAN-INCER, Scientific documentation on materials with self-repair properties and self-maintenance technologies currently developed nationally and internationally for transport infrastructure components made of concrete, 2019.
20. Mehdi, B., Geopolymer technology, from fundamentals to advanced applications: a review. *Mater. Tech.*, 24, 2, 79–87, 2009.
21. Nair, P.S., Gupta, R., Agrawal, V., Self-healing concrete: A promising innovation for sustainability-a review. *Mat. Today Proceed*, 65, 1410–1417, 2022.
22. Jagannathan, P., Studies on the mechanical properties of bacterial concrete with two bacterial species. *Mater. Today Proc.*, 5, 2, 8875–8879, 2018.
23. Wiktor, V. and Jonkers, H.M., Quantification of crack-healing in novel bacteria-based self-healing concrete. *Cem. Conc. Comp.*, 33, 7, 763–770, 2011.

24. Achal, V., Pan, X., Fu, Q., Zhang, D., Biomineralization based remediation of As(III) contaminated soil by Sporosarcina ginsengisoli. *J. Hazard. Mater.*, 201, 178–184, 2012.
25. Jonkers, H.M., Thijssen, A., Muyzer, G., Copuroglu, O., Schlangen, E., Application of bacteria as self-healing agent for the development of sustainable concrete. *Ecolog. Eng.*, 36, 2, 230–235, 2010.
26. De Belie, N., Gruyaert, E., Al-Tabbaa, A., Antonaci, P., Baera, C., Bajare, D., Darquennes, A., Davies, R., Ferrara, L., Jefferson, T., Litina, C., A Review of Self-Healing Concrete for Damage Management of Structures. *Adv. Mat. Interface*, 5, 17, 1800074, 2018.
27. Zhang, L.V., Suleiman, A.R., Allaf, M.M., Marani, A., Tuyan, M., Nehdi, M.L., Crack self-healing in alkali-activated slag composites incorporating immobilized bacteria. *Constr. Build. Mater.*, 326, 126842, 2022.
28. Smitha, M.P., Suji, D., Shanthi, M., Adesina, A., Application of bacterial biomass in bio cementation process to enhance the mechanical and durability properties of concrete. *Cleaner Mat.*, 3, 100050, 2022.
29. Salehi, P., Dabbagh, H., Ashengroph, M., Effects of microbial strains on the mechanical and durability properties of lightweight concrete reinforced with polypropylene fibre. *Constr. Build. Mater.*, 7, 126519, 2022.
30. El-hadad, A.S., Al Hariri, M.O.R., Gamal, A.A., Elgammal, M., Badawy, A.H., El-Feky, M.S., Unlocking the potential of carbon nanotubes and nano clay in concrete: assessing the bond strength and failure modes. *Eng. Res. J. (Shoubra)*, 53, 4, 89–97, 2024.
31. Djenaoucine, L., Microstructural Analysis and Performance Evaluation of Graphene Oxide-Enhanced Cement Mortar and Concrete: Implications for Mechanical Strength and Durability, PhD diss., Universidad Politécnica de Madrid, 2024.
32. Labaran, Y.H., Atmaca, N., Tan, M., Atmaca, K., Aram, S.A., Talib, A., Nano-Enhanced Concrete: Unveiling the Impact of Nano-Silica on Durability, Strength, and Cost Efficiency. https://doi.org/10.21203/rs.3.rs-4411190/v1, 2024.
33. Bhandari, S., Rajan, J., Bhoir, D., Avhad, S., Bagul, S., Experimental Study on Performance of Fibre Reinforced Nanosilica Concrete. *Indian Sci. J. Res. Engg. Manag.*, 8, 6, 1–5, 2024. https://doi.org/10.55041/ijsrem35628.
34. Elnaggar, E.M., Elsokkary, T.M., Shohide, M.A., El-Sabbagh, B.A., Abdel-Gawwad, H.A., Surface protection of concrete by new protective coating. *Const. Build. Mat.*, 220, 245–252, 2019.
35. Diamanti, M.V., Brenna, A.N.D.R.E.A., Bolzoni, F.A.B.I.O., Berra, M.A.R.I.O., Pastore, T., Ormellese, M.A.R.C.O., Effect of polymer modified cementitious coatings on water and chloride permeability in concrete. *Constr. Build. Mater.*, 49, 720–728, 2013.
36. Nguyen, T.A., Pham, T.L., Dinh, T.M.T., Thai, H., Shi, X., Application of nano-SiO2 and nano-Fe2O3 for protection of steel rebar in chloride contaminated concrete. Epoxy nanocomposite coatings and nanomodified mortars. *J. Nanosci. Nanotechnol.*, 17, 427–436, 2017.

37. James, A., Bazarchi, E., Chiniforush, A.A., Aghdam, P.P., Hosseini, M.R., Akbarnezhad, A., Martek, I., Ghodoosi, F., Rebar corrosion detection, protection, and rehabilitation of reinforced concrete structures in coastal environments: a review. *Constr. Build. Mater.*, 224, 1026–1039, 2019.
38. Raja, P.B., Ghoreishiamiri, S., Ismail, M., Natural corrosion inibitors for steel concrete—a review. *Surf. Rev. Let.*, 22, 1550040, 2015.
39. Frontini, M.A., Schreiner, W., Vázquez, M., Valcarce, M.B., Nitrite corrosion inhibition in chloride-rich electrolytes correlated to the electrical properties of surface films on carbon steel. *Constr. Build. Mater.*, 227, 116650, 2019.
40. Shi, J., Wu, M., Ming, J., In-depth insight into the role of molybdate in corrosion resistance of reinforcing steel in chloride-contaminated mortars. *Cem. Concr. Compos.*, 132, 10462, 2022.
41. Mandal, S., Singh, J.K., Lee, D.E., Park, T., Effect of phosphate-based inhibitor on corrosion kinetics and mechanism for formation of passive film onto the steel rebar in chloride-containing pore solution. *Materials*, 13, 3642, 2020.
42. Huang, Z., Song, Z., Zhang, Y., Donkor, S., Jiang, L., Chu, H., A state-of-the-art review of self-healing stimuli-responsive microcapsules in cementitious materials. *J. Ind. Eng. Chem.*, 119, 153–180, 2022.
43. Verma, C., Olasunkanmi, L.O., Ebenso, E.E., Quraishi, M.A., Substituents effect on corrosion inhibition performance of organic compounds in aggressive ionic solutions: A review. *J. Mol. Liq.*, 251, 100–118, 2018.
44. Liu, Y. and Shi, J., Recent progress and challenges of using smart corrosion inhibitors in reinforced concrete structures. *Constr. Build. Mat.*, 411, 134595, 2024.
45. Barbaz-Isfahani, R., Saber-Samandari, S., Salehi, M., Novel electrosprayed enhanced microcapsules with different nanoparticles containing healing agents in a single multicore microcapsule. *Int. J. Biol. Macromol.*, 200, 532–542, 2022.
46. Sun, J., Li, W., Li, N., Zhan, Y., Tian, L., Wang, Y., Effect of surface modified nano-SiO2 particles on properties of TO@CA/SR self-healing anti-corrosion composite coating. *Prog. Org. Coat.*, 164, 106689, 2022.
47. Zhang, Y., Qiang, Y., Ren, H., Cao, J., Cui, L., Zong, Z., Chen, D., Xiang, T., Inhibitor loaded functional HNTs modified coatings towards corrosion protection in reinforced concrete environment. *Prog. Org. Coat.*, 170, 106971, 2022.
48. Chandra Paul, S., Mbewe, P.B., Kong, S.Y., Šavija, B., Agricultural solid waste as source of supplementary cementitious materials in developing countries. *Materials*, 12, 1112, 2019.
49. Snellings, R., Li, X., Avet, F., Scrivener, K., Rapid, Robust, and Relevant (R3) reactivity test for supplementary cementitious materials. *ACI Mater. J.*, 116, 155–162, 2019.
50. Toutanji, H., Delatte, N., Aggoun, S., Duval, R., Danson, A., Effect of supplementary cementitious materials on the compressive strength and durability of short-term cured concrete. *Cem. Concr. Res.*, 34, 311–319, 2004.

51. Ramzi, S. and Hajiloo, H., The Effects of Supplementary Cementitious Materials (SCMs) on the Residual Mechanical Properties of Concrete after Exposure to High Temperatures—Review. *Buildings*, 13, 103, 2023.
52. Jhatial, A.A., Nováková, I., Gjerløw, E., A Review on Emerging Cementitious Materials, Reactivity Evaluation and Treatment Methods. *Buildings*, 13, 526, 2023.
53. Zapała-Sławeta, J., Alkali silica reaction in the presence of metakaolin-the significant role of calcium hydroxide. *IOP Conf. Ser. Mater. Sci. Eng.*, 245, 022020, 2017.
54. Elshazli, M.T., Ibrahim, A., Eidelpes, E., Ilevbare, G.O., Degradation mechanisms in overpack concrete of spent nuclear fuel dry storage systems: A review. *Nucl. Eng. Des.*, 414, 112632, 2023.
55. Tang, S.W., Cai, X.H., He, Z., Zhou, W., Shao, H.Y., Li, Z.J., Wu, T., Chen, E., The review of pore structure evaluation in cementitious materials by electrical methods. *Constr. Build. Mater.*, 117, 273–284, 2016.
56. Dŏgruyol, M., Determination of ASR in Concrete Using Characterization Methods. *Buildings*, 14, 657, 2024.
57. Brueckner, R., Cobbs, R., Atkins, C., A review of developments in cathodic protection systems for reinforced concrete structures, in: *MATEC Web of Conferences, EDP Sciences*, vol. 361, p. 02001, 2022.
58. Yahaya, M.S., Nahar, K., Kumar, D., Usman, H.M., Gambo, A.S., Umar, T.A., Sulaiman, M., Advancements in Anode Materials for Cathodic Protection: Nanostructured Alloys, Surface Modifications, and Smart Monitoring. *Buletin Ilmiah Sarjana Teknik Elektro*, 6, 3, 281–307, 2024.
59. SP0290 and NACE, *Impressed Current Cathodic Protection of Reinforcing Steel in Atmospherically Exposed Concrete Structures*, NACE International, Houston, TX, USA, 2007.
60. Chess, P. (Ed.), *Cathodic protection of steel in concrete*, pp. 124–126, E & FN Spon, London, UK, 1998.
61. Chess, P.M. and Broomfield, J.P. (Eds.), *Cathodic protection of steel in concrete and masonry*, CRC Press, 2013.
62. Advances in Anode Materials for Impressed Current Cathodic Protection Systems. *Corros. Sci.*, 6, 281–307, 2020.
63. Nguyen, T.H.Y., Verification of practical effectiveness of various applications of electrochemical chloride extractions for cement concrete. *Doctoral Dissertation*, 1997.
64. Lima, T.T., A State-of-the-Art Review of Electrochemical Chloride Extraction for Concrete Structures. *Open J. Civ. Engg.*, 12, 1–24, 2023.
65. Jia, H., Qiao, G.F., Han, P., Machine learning algorithms in the environmental corrosion evaluation of reinforced concrete structures - A review. *Cem. Conc. Comp.*, 133, 104725, 2022.
66. Sabour, M.R., Dezvareh, G., Pourrostami Niavol, K., Application of Artificial Intelligence Methods in Modeling Corrosion of Cement and Sulfur Concrete in Sewer Systems. *Environ. Proc.*, 8, 4, 1601–1618, 2021.

67. Yin, B., Josselyn, N., Zhang, Z., Rundensteiner, E.A., Considine, T.A., Kelley, J.V., Rinderspacher, B.C., Jensen, R., Snyder, J.F., MOSS: AI Platform for Discovery of Corrosion-Resistant Materials, in: *Proceedings of the 32nd ACM International Conference on Information and Knowledge Management*, pp. 5128–5132, 2023.
68. Sun, Z., Li, Y., Bei, Y., Han, T., Liu, R., Wang, L., Su, L., Compressive strength resistance coefficient of sustainable concrete in sulfate environments: Hybrid machine learning model and experimental verification. *Mat. Today Commun.*, 39, 108667, 2024.
69. Nguyen, T.A. and Ly, H.B., Hybrid Machine learning Techniques-Aided design of corroded reinforced concrete beams. *Comput. Stru.*, *298*, 107388, 2024.
70. Angst, U.M., Isgor, O.B., Hansson, C.M., Sagüés, A., Geiker, M.R., Beyond the chloride threshold concept for predicting corrosion of steel in concrete. *Appl. Phys. Rev.*, 9, 1, 63, 2022.
71. Li, Z., Pei, T., Ying, W., Srubar III, W.V., Zhang, R., Yoon, J., Ye, H., Dabo, I., Radlińska, A., Can domain knowledge benefit machine learning for concrete property prediction? *J. Am.Ceramic Soc.*, 107, 3, 1582–1602, 2024.
72. Pacheco, J. and Tepke, D., Best Practices for Measurement, Sensing, and Quantifying Corrosion in existing Reinforced Concrete Structures. *Sustain. Resilient Infrastruct.*, 8, 2, 284–291, 2023.

11

Role of Nanotechnology in Concrete Corrosion Prevention

Nabil Lahrache[1], Khalid Bouiti[1], Ichraq Bouhouche[1], Najoua Labjar[1*], Hamid Nasrellah[2], Said Laasri[3], Ayoub Cherrat[1,4] and Souad El Hajjaji[1]

[1]*Laboratory of Spectroscopy, Molecular Modeling, Materials, Nanomaterials, Water and Environment, CERNE2D, ENSAM, Mohammed V University in Rabat, Rabat, Morocco*

[2]*Higher School of Education and Training, El Jadida, Morocco*

[3]*Laboratory of Engineering Sciences for Energy, ENSA- El Jaddida, Morocco*

[4]*Forensic Institute of Gendarmerie Royale, Rabat, Morocco*

Abstract

Nanotechnologies are being integrated into intelligent concrete, representing a significant advance in the construction industry, with benefits ranging from improved structural performance to improved air quality. The construction sector is calling for intelligent concrete, as traditional concrete suffers from a lack of durability and concrete efficiency. To counter this, intelligent concrete combined with advanced nanotechnologies, possessing self-detecting and self-healing properties, will enable it to recognize damage at an early stage and seal cracks. Researchers are investigating ways of reducing the carbon footprint of concrete by replacing clinker and using alternative binders. This chapter will present smart technologies in the concrete construction sector based on the maintenance of structural conditions and construction methods with a focus on autonomous crack repair properties. The influence of nanomaterials on mechanics, long-lasting autodetection function, self-repair, and self-adapting reinforced cement highlights the potential of smart concrete to transform building methods. The use of nanomaterials such as phase-change materials and thermal insulation nanoparticles can improve energy efficiency, leading to a reduction in heating and cooling energy requirements. In addition, these materials extend the lifespan and reduce the maintenance requirements of nanotechnology-based intelligent concrete structures, helping to reduce

**Corresponding author*: najoua.labjar@ensam.um5.ac.ma

Ruby Aslam, Qihui Wang, Zhitao Yan and Afroz Aslam (eds.) Corrosion in Concrete Structures: Integrating Advanced Technologies and Sustainable Practices, (317–342) © 2026 Scrivener Publishing LLC

the number of materials used, the generation of waste over time, and the emission of greenhouse gas. The challenges and perspectives are also discussed in this chapter.

Keywords: Corrosion by concentration pile, spike corrosion, fretting corrosion, bio-corrosion, hot corrosion, nanohybrid compound

11.1 Introduction

Corrosion means attack, the destruction of metals when they come into contact with corrosive media, which may be environmental chemical reagents. According to ISO, corrosion is the set of physical–chemical steps using a metal and its environment as reagents and producing changes in the metal's properties, such as the degradation of metal functions [1].

As one of the world's most widespread construction materials, concrete makes a significant contribution to the creation of building structures. Unfortunately, concrete compositions are limited in terms of durability, strength, and adaptability to environmental constraints. Recently, nanotechnology has revolutionized concrete, offering new ways of improving performance and functionality [2].

These products are used in many fields, including life sciences [3, 4], nature [5, 6], energy [7], electronics industry [8, 9], chemicals [3], and polymers [10]. They can be employed in several contexts, such as bioinspired materials, medicine, biomimetics, renewable energy, metal protection, and sanitation [11–15]. Research into these molecules has enabled the synthesis of heterostructured nanomaterials such as layered nanohybrids. These products have been used in a number of areas, such as corrosion inhibition (due to their specific physicochemical and mechanical properties), which have not been achieved by other similar nanohybrids [16–18]. These products are prepared by intercalation reactions [19, 20] by reversibly inserting inorganic, organic, or ion molecules in a product while retaining its structural characteristics [21–25].

This class of materials has dimensions that fall within the nanometric scale, generally between 1 and 100 nm. At this tiny scale, materials have unique properties that distinguish them from atoms. The properties of materials change dramatically at the nanometric scale. Nanoparticles can have a large surface area and high reactivity. For example, we find these products in sun creams and stain-resistant clothing as well as drug delivery systems. Nanomaterials are employed for the development of new corrosion-resistant coatings or improve existing ones [26].

By taking advantage of the properties of nanomaterials, researchers have succeeded in overcoming previous constraints and developing concrete composites with enhanced mechanical properties, improved durability, and intelligent functionality. This chapter looks at advances in this field, examining the impact of nanotechnologies on improving concrete's mechanical properties, durability, and intelligent functions [27]. The chapter will discuss the deterioration behavior of concrete and the use of intelligent nano-hybrid compounds to inhibit this corrosion.

11.2 Role of Nanotechnology in Corrosion Prevention

The nanometric scale offers excellent potential for a range of fields other than corrosion, from medicine (making it easier to administer drugs) to electronics, where they facilitate the development of ultra-miniaturized devices [28].

11.2.1 Inhibition of Steel Corrosion by Nanohybrids

Because of its physical and chemical properties, steel is used in industries. Zinc, as a sacrificial coating protects steel, but recently nanographene sheets have been used and have given a good reduction in the corrosion of this metal [29]. Birbilis and colleagues [30] were able to achieve good corrosion reduction in steel using graphene-coated nickel and copper. Other researchers [31] found this result with the use of graphene nano-oxide [32]. Researchers demonstrated that zinc coating with a high concentra-tion of GO increased the corrosion resistance of steel [33]. Others [34] used GO sheets as sediment and phosphates, which captured metal ions and helped inhibit the corrosion of steel. Researchers [35] have reported that GO-modified ZnO species have antibacterial properties, so these pur-ists reduce metal deterioration. Other researchers have used a nanohybrid ZnO-graphene film and achieved high steel inhibition [36]. In another study, three nanohybrid silver nanoparticle products (HPDS, HPDO, and HPDE) were used and showed high inhibition of steel corrosion in acidic media [37].

The stages of corrosion inhibition are represented in Figure 11.1(a–d).

11.2.2 Inhibition of Copper Corrosion by Nanohybrids

Transition metals (MXenes) are widely used, but they oxidize in the presence of oxygen to form metal oxides (such as TiO_2), which limits their practical applications [38]. Figure 11.2 illustrates the layer formed on the

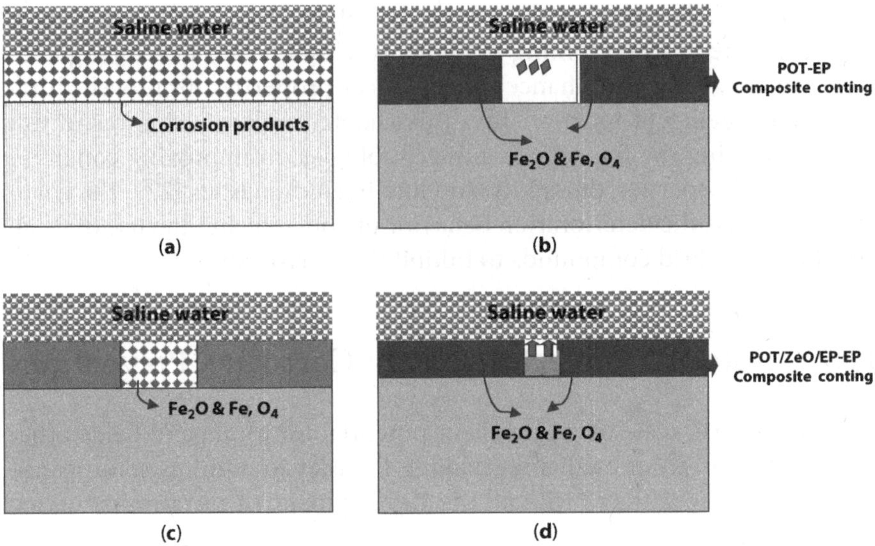

Figure 11.1 Nanohybrid inhibition of iron corrosion [39].

copper surface by electrostatic interaction. Surface modification in water, which stabilizes these metals by covering oxidation-sensitive defects on the nanofilm, alters their intrinsic conductivity and dispersity [39]. Copper corrodes in alkaline environments, which affects the efficiency of devices made of copper. Adding corrosion inhibitors to electrolytes prevents copper from corroding while maintaining its high conductivity. The addition of anti-corrosion additives to electronic devices has become a necessity to guarantee their long-term operation. To remedy this problem, researchers have used the most extensively studied transition metal Ti_3C_2Tx (T indicates terminations) as a conductive substrate. The optimal Ti_3C_2Tx@PEI (M@nP, n being the PEI/MXene mass ratio) obtained was used as a nanohybrid inhibitor for copper and gave high copper protection [40].

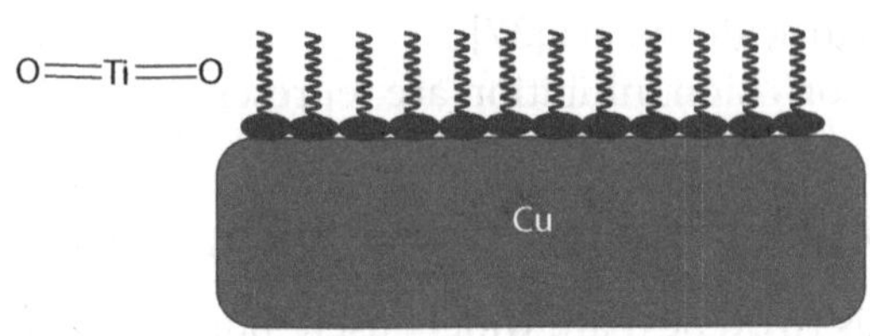

Figure 11.2 Fixation of nanohybrid TiO_2 on the copper surface by electrostatic interaction [40].

11.2.3 Examples of Nanohybrids Used to Prevent Metal Corrosion

Cited below are some examples of nanohybrids used to inhibit the corrosion of metals against aggressive elements, such as:

a- Metal oxide-based polymer nanocomposite coating
In this section, metal oxide-based compounds are added to polymers such as:

a-1- Nano-silica (SiO_2)
Anti-corrosion SiO_2 nanofillers added to coating polymers form a corrosion-resistant nanohybrid [40]. They also serve to improve the host's mechanical strength and hydrophobicity [41] as shown in Figure 11.3. This molecule blocks the entry of corrosive ions by occupying the gaps in the polymer [42]. Researchers have shown that the use of an epoxy resin with SiO_2 fillers inhibited the corrosion of DSS 2205 stainless steel [43].
Other researchers used poly (vinylpyrrolidone)/acrylate/silane + SiO_2 and obtained a corrosion resistance of 83.4% [44].
a-2- Nano-zinc oxide (ZnO)
ZnO nanoparticles added to polymers increase their hardness, decrease their refractive index, increase their hydrophobicity, and enhance their corrosion inhibition of metals [45], while one study of zinc monoxide to metal increased their corrosion protection [46]. Another study interspersed ZnO with epoxy polyamide polymer, resulting in a good metal-corrosion-inhibiting nanohybrid [47].
a-3- Cerium oxides (CeO_2)
(CeO_2) is used as a catalyst [48], UV absorber [49], antioxidant [50], and coating additive for corrosion protection [51].

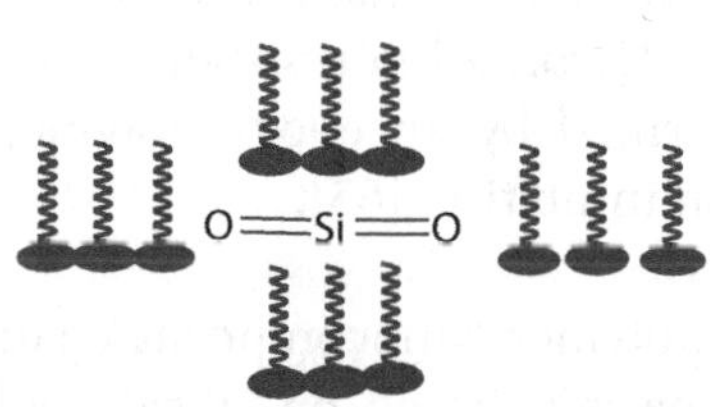

Figure 11.3 Product obtained by the addition of a nanohybrid to a polymer [51].

One study added CeO_2 to a polymer to form an anticorrosive nanohybrid on various metal substrates [52, 53]. Another study used the polyurethane nanohybrid WCNT-CeO_2 and achieved good corrosion reduction [54].

Another team studied nanohybrid epoxy resin with CeO_2 and obtained a corrosion inhibition rate of 75% [55].

a-4- Nano-zirconia (ZrO_2)

Researchers have added ZrO_2 to hydroxyapatite (HA) and TiO_2 to form a nanohybrid with coating performance. The mixture of ZrO_2 and polymer (PMC) forms a nanohybrid resistant to metal corrosion, with high resistance to cracking, wear, and high hardness. A study employing a coating formed from polymeric epoxy resin with ZrO_2 formed a nanohybrid resistant to metal corrosion [56].

a-5- Nano-iron oxide (FeO, Fe_2O_3, Fe_3O_4)

Iron nano-oxides FeO, Fe_2O_3, and Fe_3O_4 have exceptional physical and thermal properties [57]. One study incorporated Fe_3O_4 into a polyaniline polymer to form a good metal corrosion inhibitor [57]. Another study developed a nanohybrid using an epoxy-acrylate resin and Fe_3O_4, which inhibits corrosion by 98.21% [58].

a-6- Nano-alumina (Al_2O_3)

Nano-alumina (Al_2O_3) has excellent strength, large electrical insulation, and resistance to wear and deterioration [59]. The addition of Al_2O_3 to polymers forms nanocomposites with good properties [60]. Researchers have used the coating obtained from an epoxy resin polymer with an Al_2O_3 filler and achieved good corrosion reduction [61].

a-7- Nano-titania (TiO_2)

Nano-titanium (TiO_2) has good electrical conductivity and UV adsorption, so it serves as an additive to a coating, increasing its UV resistance and serving as an anti-corrosion, anti-microbial, self-cleaning, and self-smoothing agent [62]. Studies have shown that the addition of TiO_2 to polymers increases its corrosion resistance [66]. A study on the nanohybrid formed by an epoxy polymer and TiO_2 gave good corrosion inhibition [63].

b- Silicate-based polymer nanocomposite coating

Nanohybrid coatings based on silicate polymers to subtitling of a polymer with silicates reduce water absorption and

therefore protect metals against corrosion, increase nanohybrid metal and ensure the stacking pattern of the coating, improves the alignment of polyaniline chain, and ensure biocompatibility and adhesion.
One study used a silicate-based polymeric nanohybrid coating as an anticorrosion coating and achieved a high inhibition rate. Another study worked with a modified epoxy-resin-based coating and silicates and demonstrated increased metal protection [54, 64, 65].

11.3 Types of Nanomaterials Used in Concrete Corrosion Prevention

The use of nanomaterials makes it possible to decrease the concentration of cement in concrete by replacing it. Nanomaterials can reduce micropore formation by acting as fillers to produce dense concrete. They also reduce the growth of micropores in these structures. Nanomaterials have excellent filling power, refinement of the cement, and manufacturing inter-sector region concrete. They manipulate or modify the cementitious matrix system to obtain a new product. They eliminate microvoids and porosity with new binders smaller than cement particles [66].

Several methods have been proposed by researchers to inhibit corrosion in reinforced concrete, such as the use of nano-admixtures, employment of nanoparticles, and application of nanocoatings. Nanotechnology is one of the most widely used fields of research in many sectors, such as construction materials. The application of nanomaterials has improved characteristics such as strength, elasticity, and wear resistance [67].

11.3.1 Use of Nano-Admixtures

Nano-admixtures are complementary nano-elements to concrete, other than H_2O, cement, and aggregates. Admixtures are chemical compounds designed to improve the performance of fresh or hardened concrete. This may include changes in setting time, corrosion resistance, permeability, durability, and workability [68]. New-generation products such as nanoparticles, additives, and admixtures are needed to create new-generation materials. Research using nanomaterials in cement and concrete is of great interest to achieve exceptional properties in cementitious materials so that cement hydration is more efficient, and resistance to high temperatures and corrosion becomes higher than in its initial form [69]. The mode of action of nano-additives in cements is to help improve the mechanical and physical properties of hardened cement. Research has shown that nano-additives can act as crystallization centers, accelerating hydration or retarding it. They can make the cement microstructure denser, more homogeneous, and highly resistant to corrosion. In the case of nanoparticles, moisturizers must be thin to prevent their growth [70]. A research team used nano-ground granulated blast furnace slag as a substitute product to extend the service life of buildings, incorporating it in proportions of 2%, 4%, 6%, 8%, 10%, and 12% of cement mass. They showed that this product increases the acid resistance, chloride ingress, and corrosion resistance of cement using NGGBFS [71].

Ceramic nanoparticles are mainly composed of pure silica in non-crystalline form (Al_2O_3, SiO_2, and Fe_2O_3). The key factors in the effectiveness of nanomaterial addition in cement are their ability to penetrate the hardened matrix and the pozzolanic reaction with calcium hydroxide. Nano-SiO_2 is obtained from the sol-gel process or the vaporization of silica. It has been used in several studies, in the first of which [72] it was found that this nano-product increased the impermeability of hardened cement by reducing the thickness of the clay layer caused by the decrease in the volume of the upper pores. Other studies have added nanoalumina [73], nanotitanium oxide [74], nanoiron oxide [75], calcium nano-carbonate [76], nano-carbon [77], nanoclay [78], self-healing [79] and layered hydroxides [80].

Some exploit the synergetic effect of nanoparticles' exceptional properties by mixing two or more nanoparticles to achieve a high level of protection for cement-based structures. Researchers have used both CuO and Fe_2O_3 nanoparticles to reduce the cement's permeability to Cl-. This mixture combines the properties of both nanoparticles, resulting in a synergistic effect that creates a cement with high mechanical properties and toughness [81]. Other researchers have added a mixture of $CuFeO_2$ oxide nanoparticles and CuO/Fe_2O_3 nanoparticles to cement, resulting in a more corrosion-resistant structure having a resistance of 86.87% [82].

Concrete reprocessing is a method of protecting concrete structures and extending their service life. These substances help to develop strength and enhance durability by increasing the surface area of nanoparticles, which reduce the concentration of calcium in water and air and produce calcium silicate hydrate (CeSeH). They strengthen the aggregate structure, enabling the formation of a strong bond between the aggregates and the cement paste. They use free lime (calcium hydroxide) during cement hydration to form the gel [81].

A team has studied the influence of nano-TiO_2 (NT) and nano-ZnO (NZ) at different nanoparticle levels on improving the characteristics of reinforced concrete, reducing the rate of its corosion. It includes a study of their effect on workability. They found a compressive strength of 7.28 after 90 days. They also obtained 6% to 36% improvement in compressive strength and 8% improvement in tensile strength. The disadvantage of using NZ is that it takes a long time to demold the elements and reduce the degree of hydration because the reaction between ZnO and CH increases the percentage of NZ and the mechanical properties of the concrete, but they obtained opposite results when using NZ. They formed a poor concrete for NZ use. In contrast, NT accelerates hydration and generates a large quantity of C-S-H, as demonstrated by the SEM test, thus improving the concrete's properties [82].

11.3.2 Employment of Nanoparticle

To prevent corrosion of reinforced steel in concrete, research has been carried out to guarantee the durability of these structures and extend their service life [83].

Another research team have shown that steel-reinforced concrete reinforced with EG/AgNPs inhibitor offers better corrosion resistance (inhibition efficiency is 95%) than that of the inhibitor without NPs [84]. Research has been carried out to guarantee the durability of these structures and extend their service life. Hegazy *et al.* investigated the effects of colloidal copper nanoparticles as a steel anticorrosion paint [83]. They prepared this product using the reduction method of Cu (or copper oxide) and iron oxide (or iron oxide) or using copper (II) chloride ($CuCI_2$) by using a 0.5% by-weight solution of copper nanoparticles to prevent these reinforcements [85].

Corrosion of this steel is a key factor in the deterioration of concrete structures. The factor behind this corrosion is the proliferation of Cl- and carbon. Once corrosion has started, corrosion products are deposited on the steel surface. This formation in this limited space causes expansive stresses that lead to cracks and bursts in the concrete. Liquids and corrosive gases can penetrate these cracks, causing the concrete to degrade. These cracks allow the aggressive substances that cause cracking to pass through, reaching the steel afterwards and inducing corrosion of the steel, leading to concrete damage and system failure. Concrete ensures the corrosion resistance of steel reinforcement, serving as a barrier to corrosive environmental agents. Concrete that is not exposed to the influence of corrosive factors has a pH value between 12.5 and 13.5 [86].

The use of conventional steel results in surface irregularities, causing stress build-up and fatigue crack formation. On the other hand, with

Table 11.1 Examples of recent studies of nanoparticle used in concrete.

Authors	System details	Key findings
Wang *et al.* [83]	Eight nanofillers (nano-SiO_2, nano-TiO_2, nanoZrO_2, untreated multiwall carbon nanotubes (MWCNT), hydroxyl-functionalized MWCNT, nickel-coated MWCNT, multilayer graphenes (MLG), and nanoboron nitride (nano-BN) were studied.	All nanofillers improved the bond strength between these nanoproducts and cement by, respectively, 2.1 MPa, 35.1%; 2.32 MPa, 38.8%; and 2.56 MPa, 42.8% and 2%, 0.3%, and 0.3% by weight. Scanning electron microscope images demonstrate the presence of nanofillers, which enhance the products and cause their interfacial hydrophobicity. Those from the energy dispersive spectrometer show that the local concentration of nanofillers in the ITZ is greater. These results confirm that nanofillers transfer water migration to the aggregates.
Sumesh *et al.* [84]	Calcium hydroxide	The addition of NPs avoids the formation of various gel crystals produced by the high reactivity and large surface area of the NPs present in the concrete. The presence of various NPs in the OPC matrix has been shown to slow the crystallization rate of calcium hydroxide in the transition regions, also forming an excess C-S-H gel that acts as a barrier against corrosion.

(*Continued*)

Table 11.1 Examples of recent studies of nanoparticle used in concrete. (*Continued*)

Authors	System details	Key findings
Mustakim *et al.* [85]	Nanosilica	The results showed that concretes without this product had better characteristics than those containing nanosilica. It was noted that the material synthesized by incorporating 1.5% nanosilica had the highest strength. This improvement is due to the filling of the nanopores inside the GP concrete with nanosilica particles. The chemical composition of nanosilica, with its great characteristics, improved the strength of GP concrete to an optimum level of 1.5%. For quantities above the optimum level, a decrease in strength is obtained due to the large number of unreacted nanoproducts producing nanosilica agglomerations which prevent the dissolution of these materials in concrete.

the employment of nanoparticles as modifiers, surface irregularities are decreased, thus preventing cracking [87]. Using corrosion inhibitors is an effective way of protecting steel against corrosion. Although these products are toxic, they are widely used for their exceptional effectiveness against corrosion [88], such as benzotriazole (BTA) and its derivatives [89]. These protective effects can be extended by adding nanomaterials. A range of carriers can be employed, such as mixed-oxide nanoparticles, layer-by-layer assembled nanocontainers, and 2D layered materials [90]. Tubular nanoclays such as halloysites (HNTs) can be used to reinforce resins against corrosion, but they depend on the chemical nature of their constituent substances. These HNTs are aluminosilicate clay minerals ($Al_2Si_2O_5(OH)_4$) found in abundance in many parts of the world; they are inexpensive and non-toxic and offer mechanical strength [91]. HNTs are used in many sectors for the delivery of active molecules. These products are Portland cement (PC) constituents with good mechanical properties [92]. Maness *et al.* [93] showed that TiO_2 increased the production of malondialdehyde, indicating that this product underwent a reduction reaction upon exposure to light, enabling *Escherichia coli*, resulting in bed death of *Escherichia coli* membrane, and leading to the bedridden death of Escherichia coli. Kim *et al.* [94] affirmed that NT eliminated *Bacillus cloacae*. Venieri *et al.* claimed that NT eliminated *Pneumoniae* destruction with a particle size of 21 nm [95].

Other examples of the use of nanomaterials are described in Table 11.1.

11.3.3 Nanocoatings

Coatings made from polymer nanocomposites offer more impressive properties than those of virgin polymers, as well as high thermal stability and good corrosion resistance, which has prompted researchers to use these products in their research. The corrosion monitoring techniques linear polarization resistance, potentiodynamic polarization, and electrical resistivity were used to determine corrosion resistance in seawater. After 365 days, the results of the morphology and thermal tests showed that the addition of 5% of EG/AgNPs to reinforced concrete enhanced corrosion resistance, with the formation of a thin barrier layer on the concrete. This effectiveness is due to the creation of a hydrated calcium silicate gel (C-S-H) in the concrete that closes the concrete pores. Maximum efficiency was found to be 94.74% [99]. To prevent the corrosion of rebar steel, researchers used a surface pre-treatment/modification technique employing nanosilica on the zinc phosphate conversion coating. Resistance was determined using electrochemical techniques and accelerated corrosion tests [92]. A polymer coating was based on nanophase-modified fly ash on steel reinforcement. In another work, five films were produced using ordinary Portland cement (CC), CPO replaced by 40% by weight fly ash (CF), and CF mixed with 2% by weight nano-$CaCO_3$ (CFC), nano-SiO_2 (CFS), and nanoZrO_2 (CFZ). These samples were analyzed using electrochemical methods after exposure to Cl-, and performance was determined using voltage tests. The microstructure of these coatings and their chemical compositions were determined using SEM, XRD, and LRS. The corrosion rates

Table 11.2 Examples of recent studies of nanocoatings used in concrete.

Authors	System details	Key findings
Lee *et al.* [96]	Nanosilica (NS) nanomaterials	The use of nanomaterial nanosilica (NS) was produced from rice husk ash and used with the replacement material carbon nanotubes (0.01%–0.07%) instead of cement and rock grains (RD) instead of river sand. Five different mixes were prepared: among them a polycarboxylate-based water-reducing admixture (to facilitate the workability of the mix for embedding). Compression tests, minimum load at fracture, durability (sorption), and H_2O absorption and Cl- infiltration test, followed by a potentiodynamic polarization test, were carried out to demonstrate the corrosion resistance of the composite cement mortar. The results were 12% resistance when using NGS and 76% resistance when adding CNT.
Chuah *et al.* [97]	0D nanoparticles, 1D nanofibers, and 2D nanosheets (nanosilica and carbon nanotubes), carbon nanotubes (CNT), and nanosheets such as graphene oxide (GO)	This study carried out on the effect of incorporating these nanomaterials has demonstrated an improvement in the workability, hydration, microstructure, and mechanical properties of cement-based composite materials.

(*Continued*)

Table 11.2 Examples of recent studies of nanocoatings used in concrete. (*Continued*)

Authors	System details	Key findings
Nguyen *et al.* [98]	Evaluated the microstructure and corrosion protection of Zn electrodeposited layers and Zn-graphene oxide nanocomposites using brightening additives (surfactants) and various amounts of graphene oxide (colloid)	Zn and Zn-GO nanocomposite films were electrodeposited on steel. The incorporation of GO into the Zn structure and the presence of brightening additives enabled the formation of a high-strength film. The adhesion strength of the film was twice that of the pure Zn coating. Potentiodynamic polarization and electrochemical impedance spectroscopy (EIS) confirm this observation. Good corrosion resistance is obtained.

of CFC-, CFS-, and CFZ-coated bars are lower than those of CF-coated bars. The accelerated post-study was carried out after compression test, which showed that the initiation time of cracking in nanomaterial-containing coatings was greater than that of conventional coatings. Open-circuit potential, corrosion current, and polarization resistance were used to determine the electrochemical parameters of reinforcement cracking, confirming that corrosion is negligible [100].

Compared with virgin polymers, nanocomposite polymer coatings have excellent properties: high thermal stability and corrosion resistance, impermeability to gases, and barrier properties [101]. The diffusion circuit increases with torsion, leading to an increase in concrete with nanocomposite content and an increase in nanocomposite properties (appearance and volume) [102]. Nanocomposites make polymer nanocomposite coatings highly gas-impermeable (50 to 500 times less) at low concentrations [103]. Few studies have used these coatings or polymer/clay in concrete structures [104]. Researchers have demonstrated the performance of polymer/SiO_2 coatings [105].

11.4 Challenges and Limitations

In summary, the incorporation of nanomaterials has been shown to have a significant impact on the durability and corrosion resistance of cement-based concretes. These products render these concretes non-hydratable, seal out chlorides, and enhance their mechanical properties.

Nano-silica behaves like a pozzolanic material and fills the voids in cement paste, resulting in a more compact, corrosion-resistant texture. Nano-alumina, on the other hand, strengthens particle cohesion and reduces the amount of water in the concrete, thus improving the product's mechanical properties. Carbon nanotubes, with their attractive appearance and excellent mechanical properties, can inhibit concrete corrosion. However, the quantity and dispersion of these nanomaterials must always be controlled to prevent any adverse effect on the concrete's long-lasting performance.

11.5 Future Perspectives

Nanomaterials have a significant influence on concrete, opening up new opportunities for the research and development of new nanocomposites, exploiting the properties of mixtures, and synthesizing new coatings based on nanoparticles. Encouraging results indicate that the incorporation of nanomaterials can significantly improve the performance and properties of concrete and can be used as a concrete corrosion inhibitor. Further studies are required to fully understand all of the effects caused by these nanomaterials, such as the impact on long-term durability and the ecological consequences of making this concrete. In addition, we need to optimize methods for dosing and dispersing these products in concrete as well as strategies to assess their environmental impact.

The potential application of nanomaterials in construction products other than concrete is also worth exploring. This will make it possible to manufacture nanomaterials that can be used in construction.

11.6 Conclusion

In this chapter, we presented the role of nanotechnology in corrosion prevention and we also defined the types of these nanomaterials. The use of nanohybrids ensures uniform coating distribution. The integration of nanotechnologies in intelligent concrete is a tool that promises to revolutionize the construction sector. This study demonstrates the environmental benefits and potential uses of nanotechnology-based intelligent concrete. With a diminution in the consumption of old cement aggregates, a reduction in energy consumption for heating and cooling, and a decrease in gas emissions, nanotechnology-based intelligent concrete offers excellent advantages in terms of energy efficiency and the environment. Despite these encouraging results, more research is needed into the role of nanotechnology in the inhibition of corrosion of intelligent concrete. Long-term performance, durability of nanotechnology-enhanced concrete structures, and the scalability and cost-effectiveness of nanotechnology applications in the construction sector are areas requiring further research and exploration. However, there are disadvantages in their use that need to be overcome, such as production cost, which is higher than that of other inhibitors. There are also environmental problems caused by their use, as they are non-biodegradable materials. In this chapter, we have defined the corrosion of concrete reinforcing steel and the different types of nanomaterials used in concrete corrosion prevention.

Future research should therefore focus on synthesizing new nanoproducts with the performance, durability, and environmental impact of nanotechnology-based concrete structures.

AI disclosure statement: The author confirms that no artificial intelligence (AI)-generated content or AI-assisted writing tools were used in the preparation of this chapter (including summaries and the abstract). All text, analysis, and conclusions are the result of the author's original work.

References

1. ISO 8044:2019- Corrosion of metals and alloys - Basic terms and definitions.
2. Samaei, S.R. and Hassanabad, M.G., Investigating and providing solutions to reduce air pollution in large cities. *2th International Conference on Creative Achievements of Architecture, Urban Planning, Civil Engineering and Environment in the Sustainable Development of the Middle East*, 2023.
3. Mehdi, A., Reye, C., Corriu, R., From molecular chemistry to hybrid nanomaterials. Design and functionalization. *Chem. Soc. Rev.*, 563–574, 40, 2011, Doi: 10.1039/b920516k.
4. Darder, M., Aranda, P., Ruiz-Hitzky, E., Bionanocomposites:. A new concept of ecological, bioinspired, and functional hybrid materials. *Adv. Mater.*, 19, 1309–1319, 2007, Doi: 10.1002/adma.200602328.

5. Ruiz-Hitzky, E., Darder, M., Wicklein, B., Castro-Smirnov, F.A., Aranda, P., Clay-Based Biohybrid Materials for Biomedical and Pharmaceutical Applications. *Clays Clay Miner.*, 67, 2019, 44–58, 2019, Doi: 10.1007/s42860-019-0005-0.
6. Wen, B., Li, Y., Liang, C., Chen, Y., Zhao, Y., Wang, Q., Recent Progress on Porous Carbons for Carbon Capture. *Langmuir*, 40, 8327–8351, 2024. Doi; 10.1021/acs.langmuir.3c03876.
7. Tiwari, T., Kaur, G.A., Singh, P.K., Balayan, S., Mishra, A., Tiwari, A., Review Emerging bio-capture strategies for greenhouse gas reduction: Navigating challenges towards carbon neutrality. *Sci. Total Environ.*, 929, 1–27, 2024,Doi: 10.1016/j.scitotenv.2024.172433.
8. Mendrela, P., Stanek, W., Simla, T., Thermo-ecological cost - System evaluation of energy-ecological efficiency of hydrogen production from renewable and non-renewable energy resources. *Int. J. Hydrogen Energy*, 50, 1–14, 2024, Doi: 10.1016/j.ijhydene.2023.06.150.
9. Kuillaa, T., Bhadra, S., Yao, D., Kim, N.H., Bose, S., Lee, J.H., Recent advances in graphene based polymer composites. *Prog. Polym. Sci.*, 35, 1350–13755, 2010, Doi: 10.1016/j.progpolymsci.2010.07.005.
10. Zhang, W., Zheng, K., Ren, J., Fan, Y., Strong- ductile and lightweight bionanocomposites constructed by bioinspired hierarchical assembly. *Compos. Commun.*, 17, 97–103, 2020, Doi: 10.1016/j.coco.2019.11.017.
11. Joseph, B., Krishnan, S., Sagarika, V.K., Tharayi, A., Kalarikkal, N., Thomas, S.,Bionanocomposites as industrial materials, current and future perspectives:a review. *Emergent Mater.*, 3, 1–13, 2020, Doi: 10.1007/s42247-020-00133-x.
12. Kabeb, S.M. and Abu Bakar, N.H., Polymer blend nanocomposites for polymer electrolyte membrane fuel cell (PEMFC) applications. *Micro Nano Technol.*, 17, 479–493, 2023, Doi: 10.1016/B978-0-323-99549-8.00002-9..
13. Mishra, D., Singh, P., Pandey, V., Yadav, A., Khare, P., The emerging role ofnanotechnology in agri-food sector: recent trends and opportunities. *Smart Nanoarchitectures Technol. Challenges Appl. Micro Nano Technol.*, 1–19, 2024,Doi: 10.1016/B978-0-323-99682-2.00014-1.
14. Kumar, N.S., Smart and innovative nanotechnology applications for water-purification. *Hybrid Adv.*, 3, 1–23, 2023, Doi: 10.1016/j.hybadv.2023.100044.
15. Sanchis-Gual, R., Coronado-Puchau, M., Mallah, T., Coronado, E., Review Hybrid nanostructures based on gold nanoparticles and functional coordination polymers: Chemistry, physics and applications in biomedicine, catalysis and magnetism. *Coord. Chem. Rev.*, 480, 1–23, 2023, Doi: 10.1016/j.ccr.2023.215025.
16. Constantino, V.R.L., Figueiredo, M.P., Magri, V.R., Denise, E., Rodrigues Cunha, V.R., Alcântara, A.C.S., Perotti, G.F., Review Biomaterials Based on Organic Polymers and Layered Double Hydroxides Nanocomposites: Drug Delivery and Tissue Engineering. *Pharmaceutics*, 15, 1–54, 2023, Doi: 10.3390/pharmaceutics1502041.

17. Gomez-Romero, P., Pokhriyal, A., Rueda-García, D., Bengoa, L.N., González-Gi, R.M., Hybrid Materials: A Metareview, Chemistry of. *Materials*, 36, 8–27, 2024, Doi: 10.1021/acs.chemmater.3c01878.
18. Odetallah, M. and Kuss, C., A Review of Chemically Induced Intercalation and Deintercalation in Battery Materials. *Energy Technol.*, 11, 1–23, 2023, Doi: 10.1002/ente.202201060.
19. Laipana, M., Xiange, L., Yua, J., Martine, B.R., Zhu, R., Zhu, J., He, H., Clearfield, A., Sun, L., Layered intercalation compounds: Mechanisms, new methodologies, and advanced applications. *Prog. Mater Sci.*, 109, 1–37, 2020, Doi: 10.1016/j.pmatsci.2019.100631.
20. Choy, J.-H., Kwon, S.-J., Gyeong-Su, High-T_c Parsuperconductors in the two dimensional limit: [(Py-CnH2n+1)2HgI4]-Bi2Sr2Cam−1CumOy (m = 1 and 2). *Science*, 280, 1589–1592, 1998, Doi: 10.1126/science.280.5369.1589.
21. Osada, M., Emerging ceramic nanosheets: From tailored synthesis to electronic applications. *J. Ceram. Soc. Jpn.*, 131, 851–862, 2023, Doi: 10.2109/jcersj2.2312.
22. Tawalbeh, M., Khan, H.A., Al-Othman, A., Review: Insights on the applications of metal oxide nanosheets in energy storage systems. *J. Energy Storage*, 60, 1–19, 2023, Doi: 10.1016/j.est.2023.106656.
23. Bhattacharjee, J. and Subhasis, R., Exploration of Molybdenum Oxide Compounds-A Review. *Catal. Res.*, 4, 1–28, 2024. Doi:10.21926/cr.2403011.
24. Singh, A.P., Saxena, R., Saxena, S., Maurya, N.K., The Future of Protection: Unleashing the Power of Nanotech against Corrosion. *Asian J. Curr. Res.*, 9, 23–44, 2024, Doi: 10.56557/ajocr/2024/v9i38727.
25. L.C., Yule, Bentley, C.L., West, G., Shollock, B.A., Unwinn, P.R., Scanning electrochemical cell microscopy: A versatile method for highly localised corrosion related measurements on metal surfacesn. *Electrochim. Acta*, 298, 80–88, 2019, Doi: 10.1016/j.electacta.2018.12.054.
26. Venkatraman, A., Katona, R.M., Maestas, D., Roop, M., Noell, P., de Oca Zapiain, D.M., An active learning framework for the rapid assessment of galvanic corrosion. *Mater. Degrad.*, 8, 1–11, 2024, Doi: 10.1038/s41529-024-00476-4.
27. Tian, Q., Huang, J.Z.J., Hou, J., Zhou, Z., Yongzhou, Z.L., Sun, M., Decker, J.HJ., Rollins, K.M., Ellsworth, J.C., Building the future: Smart concrete as a key element in next-generation construction. *Constr. Build. Mater.*, 429, 1–22, 2024, Doi: 10.3390/app13031507.
28. Khosravi, I., Shahryari, Z., Moghadas, S.M.J., Sarraf, H.T., Yeganeh, M., Theapplication of graphene oxide as corrosion barrier. *Corros. Prot. Nanoscale Micro Nano Technol.*, Chapter : 8, 127–140, 2020, Doi: 10.1016/B978-0-12-819359-4.00008-8.
29. Cui, G., Bi, Z., Zhang, R., Liu, J., Yu, X., Li, Z., Review A comprehensive review on graphene-based anti-corrosive coatings. *Chem. Eng. J.*, 373, 104–121, 2019, Doi: 10.1016/j.cej.2019.05.034.
30. Lil, X., Long, Y., Ma, L., Li, J., Yin, J., Guo, W., Coating performance of hexagonal boron nitride and graphene layers. *2D Materials*, 8, 1–18, 2021, Doi: 10.1088/2053-1583/abe777.

31. Karimi Azar, M.M., Gugtapeh, H.S., Rezaei, M., Evaluation of corrosion protection performance of electroplated zinc and zinc-graphene oxide nanocomposite coatings in air saturated 3.5 wt. % NaCl solution. *Colloids Surf., A*, 2020601, 1–11, 2020, Doi: 10.1016/j.colsurfa.2020.125051.
32. Xie, Y., Chen, M., Xie, D., Zhong, L., Zhang, X., A fast, low temperature zinc phosphate coating on steel accelerated by graphene oxide. *Corros. Sci.*, 128, 1–8, 2017, doi: 10.1016/j.corsci.2017.08.033.
33. Hu, C., Yang, Y., Lin, Y., Wang, L., Ma, R., Zhang, Y., Feng, X., Wu, J., Chen, L., Shao, L., GO-based antibacterial composites: Application and design strategies. *Adv. Drug Deliv. Rev.*, 178, 1–31, 2021, Doi: 10.1016/j.addr.2021.113967.
34. Razavizadeha, O. and Ghorbani, M., Surface modification of carbon steel by ZnO-graphene nano-hybrid thin film. *Surf. Coat. Technol.*, 363, 1–11, 2019, Doi: 10.1016/j.surfcoat.2019.01.094.
35. A.M., Elsharif, Abdulazeez, I., Almarzooq, M.A., Haladu, S.A., Synthesis, and experimental evaluation of novel 4-(-3-(2-hydroxyethoxy)-3-oxopropenyl)-1,2-phenylene nanohybrid derivatives as potential corrosion inhibitors for mild steel in 1 M HCl. *J. Ind. Eng. Chem.*, 116, 474–488, 2022, Doi: 10.1016/j.jiec.2022.09.037.
36. Lakhan, M.N., Hanan, A., Wang, Y., Leed, H.K., Arandiyan, H., Integrated MXenes and Metal Oxide Electrocatalysts for Oxygen Evolution Reaction: Synthesis, Mechanisms, and Advances. *Chem. Sci.*, 15, 15540–15564, 2024, Doi: 10.1039/D4SC04141K.
37. Cui, M., Ren, S., Chen, J., Liu, S., Zhang, G., Zhao, H., Wang, L., Xue, Q., Anticorrosive performance of waterborne epoxy coatings containing water-dispersible hexagonal boron nitride (h-BN) nanosheets. *Appl. Surf. Sci.*, 397, 77–86, 2017, Doi: 10.1016/j.apsusc.2016.11.141.
38. Liu, Y., Fan, B., Xu, B., Yang, B., Ambient-stable polyethyleneimine functionalized Ti3C2Tx nanohybrid corrosion inhibitor for copper in alkaline electrolyte. *Mater. Lett.*, 337, 1–4, 2023, Doi: 10. 1016 / j.matlet. 2023. 133979.
39. Liu, L., Zhao, M., Pei, X., Liu, S., Luo, S., Yan, M., Shao, R., Sun, Y., Xu, W., Xu, Z., Improving corrosion resistance of epoxy coating by optimizing the stress distribution and dispersion of SiO2 filler. *Prog. Org. Coat.*, 179, 1–11, 2023, Doi: 10.1016/j.porgcoat.2023.107522.
40. Wonnie Ma, I.A., Ong, G., Shafaamri, A., Evaluation of natural oil polyol hydrophobic acrylic-based coating incorporated with SiO2 nanoparticles for enhanced corrosion protection. *Pigm. Resin Technol.*, 53, 1–10, 2023, Doi: 10.1108/PRT-07-2022-0092.
41. Liu, Y., Chen, L., Tang, Y., Zhang, X., Qiu, Z., Synthesis and characterization of nano SiO2@octadecylbisimidazoline quaternary ammonium salt used as acidizing corrosion inhibitor. *Rev. Adv. Mater. Sci.*, 61, 186–194, 2022, Doi: 10.1515/rams-2022-0006.
42. Xiao, T., Du, C., Liu, Y., Electrochemical Evaluation on Corrosion Behavior of SAF 2507 Duplex Stainless Steels in Blended Concrete with Metakaolin

and ultrafine Slag Admixtures. *Int. J. Electrochem. Sci.*, 16, 1–11, 2021, doi: 10.20964/2021.06.15.

43. Randis, D.B., Darmadi, F.G., Sonief, A.A., Akpan, E.D., Ebenso, E.E., The potential of nanocomposite-based coatings for corrosion protection of metals: A review. *J. Mol. Liq.*, 390, 1–17, 2023, Doi: 10.1016/j.molliq.2023.123067.
44. Al Beladi, M., Geesi, M.H., Riadi, Y., Alahiane, M., Aljohani, T.A., Berisha, A., Reka, A., Kaiba, A., Ouerghi, O., Synergistic inhibitive effect of a hybrid zinc oxide-benzalkonium chloride composite on the corrosion of carbon steel in a sulfuric acidic solution. *Green Process. Synth*, 12, 1–19, 2023, Doi: 10.1515/gps-2023-0139.
45. Slosarczyk, A., Klapiszewska, I., Skowronska, D., Janczarek, M., Jesionowski, T., Klapiszewski, L., Review A comprehensive review of building materials modified with metal and metal oxide nanoparticles against microbial multiplication and growth. *Chem. Eng. J.*, 466, 1–20, 2023, Doi: 10.1016/j.cej.2023.143276.
46. Mostafaei, A. and Nasirpouri, F., Epoxy/polyaniline–ZnO nanorods hybrid nanocomposite coatings: Synthesis, characterization and corrosion protection performance of conducting paints. *Prog. Org. Coat.*, 77, 146–159, 2014, Doi: 10.1016/j.porgcoat.2013.08.015.
47. Yang, C., Lu, Y., Zhang, L., Kong, Z., Yang, T., Tao, L., Zou, Y., Wang, S.,Defect Engineering on CeO2-Based Catalysts for Heterogeneous Catalytic Applications. *Small Struct.*, 2, 1–25, 2021, Doi: 10.1002/sstr.202100058.
48. Bao, B., Sun, Y., Li, X., Li, L., Yu, Y., Tuning the UV absorbing ability of CeO2 nanoparticles with F– doping. *Flat Chem*, 39, 1–9, 2023, Doi: 10.1016/j.flatc.2023.100494.
49. Matussin, S.N., Harunsani, M.H., Khan, M.M., CeO_2 and CeO_2-based nanomaterials for photocatalytic, antioxidant and antimicrobial activities. *J. Rare Earths*, 41, 167–181, 2023, Doi: 10. 1016/j.jre.2022.09.003.
50. Chen, M., Ma, Y., Chen, L., Zhang, D., Zhou, H., Chen, J., Lubrication, wear resistance and corrosion protection on aluminum alloy toward multifunctional coatings with nano-CeO_2 inhibition effect. *Prog. Org. Coat.*, 186, 1–13, 2024, Doi: 10.1016/j.porgcoat.2023.108066.
51. Le, T.T., Nguyen, T.V., Nguyen, T.A., Nguyen, T.T.H., Thai, H., Tran, D.L., Dinh, D.A., Nguyen, T.M., Lu, L.T., Thermal, mechanical and antibacterial properties of water based acrylic Polymer/SiO2–Ag nanocomposite coating. *Mater. Chem. Phys.*, 232, 362–366, 2019, Doi: 10.1016/j.matchemphys.2019.05.001.
52. Eduok, U., Faye, O., Tiamiyu, A., Szpunar, J., Fabricating protective epoxy-silica/CeO_2 films for steel: Correlating physical barrier properties with material content. *Mater. Des.*, 124, 58–68, 2017, Doi: 10.1016/j.matdes.2017.03.062.
53. Ramezanzadeha, B., Bahlakehb, G., Ramezanzadeh, M., Ramezanzadeha, B., Bahlakehb, G., Ramezanzadeh, M., Polyaniline-cerium oxide (PAni-CeO_2) coated graphene oxide for enhancement of epoxy coating corrosion

protection performance on mild steel. *Corros. Sci.*, 137, 111–126, 2018. Doi:10.1016/j.corsci.2018.03.038.

54. Wenjie, X., Yang, C., Su, W., Zhong, N., Xia, X., Effective corrosion protection by PDA-BN@CeO_2 nanocomposite epoxy coatings. *Colloids Surf., A*, 657, 1–10, 2023, Doi: 10.1016/j.colsurfa.2022.130448.
55. Safavi, M.S. and Etminanfar, M., A review on the prevalent fabrication methods, microstructural, mechanical properties, and corrosion resistance of nanostructured hydroxyapatite containing bilayer and multilayer coatings used in biomedical applications. *J. Ultrafine Grained Nanostruct. Mater.*, 52, 1–17, 2019, Doi: 10.22059/JUFGNSM.2019.01.01.
56. Pourhashem, S., Saba, F., Duana, J., Rashidie, A., Guan, F., Nezhad, E.G., Hou, B., Polymer/Inorganic nanocomposite coatings with superior corrosion protection performance: A review. *J. Ind. Eng. Chem.*, 88, 29–57, 2020, Doi: 10.1016/j.jiec.2020.04.029.
57. Zhang, S., Zhang, Y., Wu, S., Effects of ZnO, FeO and Fe_2O_3 on the spinel formation, microstructure and physicochemical properties of augite-based glass ceramics. *Nternational J. Miner. Metall. Mater.*, 30, 1207–1217, 2023, Doi: 10.1007/s12613-022-2489-1.
58. Li, A. and Zhu, A., Preparation of Fe_3O_4/PANI nanocomposite and its metal anticorrosive activity. *Prog. Org. Coat.*, 161, 1–9, 2021, Doi: 10.1016/j.porgcoat.2021.106477.
59. Liu, Y., Yu, Q., Cai, M., Yu, B., Zho, F., Effects of Nanoadditives on the Anticorrosion Performance of Nanocomposite Coatings: A Review. *Nano Mater.*, 7, 12249–12272, 2024, Doi: 10.1021/acsanm.4c01347.
60. Eraslan, F.S. and R.G., Chemical composition optimization of Al2O3-TiO_2 composite coatings for enhanced wear and corrosion resistance. *Surf. Coat. Technol.*, 474, 1–12, 2023, Doi: 10.1016/j.surfcoat.2023.130053.
61. Mahmoud, M.E., Khalifa, M.A., El-Sharkawy, R.M., Youssef, M.R., Effects of Al2O3 and BaO nano-additives on mechanical characteristics of high-density polyethylene. *Mater. Chem. Phys.*, 262, 1–11, 2021, Doi: 10.1016/j.matchemphys.2021.124251.
62. Liang, X., Hua, C., Zhang, M., Zheng, Y., Song, S., Cai, M., Huang, Y., He, C., Fan, X., Preparation of Anticorrosive Epoxy Nanocomposite Coating Modified by Polyethyleneimine Nano-Alumina. *Coatings*, 13, 1–11, 2023, Doi: 10.3390/coatings13030561.
63. Wang, B., Wei, S., Huang, W., Wang, Y., Liang, Y., Li, J., Shi, R., Guo, L., Xue, J., Preparation, anticorrosion mechanism, and application of nano-Ti polymer-containing coating. *Eur. Polym. J.*, 195, 1–17, 2023. Doi:10.1016/j.eurpolymj.2023.112200.
64. Peng, T., Xiao, R., Rong, Z., Liu, H., Hu, Q., Wang, S., Li, X., Zhang, J., Polymer Nanocomposite-based Coatings for Corrosion Protection. *Chem. Asian J.*, 15, 3915–3941, 2020, Doi: 10.1002/asia.202000943.
65. Bouiti, K., Alsharabi, H., Bouhlal, F., Abidi, B., Labjar, N., Bensemlali, M., El Hajjaji, S., Response surface methodology for optimizing corrosion

inhibition: investigating the synergistic efect of Eriobotrya japonica extract and potassium iodid. *Euro-Mediterr. J. Environ. Integr.*, 9, 469–481, 2024, Doi: 10.1007/s41207-023-00457-0.

66. Nguyen, T.A., Nguyen, T.H., Pham, T.L., Dinh, T.M.T., Thai, H., Pplication of nano-SiO_2 and nano-Fe_2O_3 for protection of steel rebar in chloride contaminated concrete: epoxy nanocomposite coatings and nano-modified mortars. *J. Nanosci. Nanotechnol.*, 17, 427–436, 2017, Doi: 10.1166/jnn.2016.12396.
67. Cubides, Y., Castaneda, H., August, Corrosion protection mechanisms of carbon nanotube and zinc-rich epoxy primers on carbon steel in simulated concrete pore solutions in the presence of chloride ions. *Corros. Sci.*, 109, 145–161, 2016, Doi: 10.1016/j.corsci.2016.03.023.
68. Boris, R., Wilińska, I., Pacewska, B., Antonovič, V., Investigations of the Influence of Nano-Admixtures on Early Hydration and Selected Properties of Calcium Aluminate Cement Paste. *Materials*, 15, 1–20, 2022, Doi: 10.3390/ma15144958.
69. Abd El-Hamid, H.K. and Radwan, M.M., Influence of nano-silica additions on hydration characteristics and cytotoxicity of calcium, aluminate as biomaterial. *Heliyon*, 5, 21–35, 2019. Doi:10.1088/1742-6596/2857/1/012019.
70. Guo, C., Wang, E., Hou, X., Chen, J., Zhang, W., Ye, J., Qin, S., Characterization and mechanism of early hydration of calcium, aluminate cement with anatase-TiO2 nanospheres additive, Constr. *Build. Mater.*, 20, 119922, 2020.
71. Johnpaul, V., Santhiya, K., Balasundaram, N., Enhancing the Durability and Morphology Characteristic of Concrete Using Nano-based GGBFS. Iran. *J. Sci. Technol. Trans. Civ. Eng.*, 3, 1–22, 2022, Doi: 10. 1007/s40996-022-00997.
72. Hou, P., Cheng, X., Qian, J., Zhang, R., Cao, W., Shah, S.P., Characteristics of surface treatment of nano-SiO2 on the transport properties of hardened cement pastes with different water-to-cement ratios. *Cem. Concr. Compos.*, 55, 26–33, 2015.
73. Behfarnia, K. and Salemi, N., The effects of nano-silica and nano-alumina on frost resistance of normal concrete. *Constr. Build. Mater.*, 48, 580–584, November2013.
74. Tao Meng, Y.Y., Qian, X., Zhan, S., Qian, K., Effect of nano-TiO2 on themechanical properties of cement mortar. *Constr. Build. Mater.*, 29, 241–245, 2012. Doi:10.1016/j.conbuildmat.2011.10.047.
75. Soto-Pérez, L., Vazquez, N., Molina, O., Jo, M., Hwang, S., Effect of iron-oxide nano particles on the durability of fly ash cement paste, in: *Nanotechnology in Construction*. Konstantin Sobolev, S.S.P., (Ed.), Part XII Durability, pp. 473–478, Springer, 2015. Doi: 10.1007/978-3-319-17088-6.
76. Shaikh, F.U.A. and Supit, S.W.M., Mechanical and durability properties of high volume fly ash (HVFA) concrete containing calcium carbonate ($CaCO_3$) nano particles. *Constr. Build. Mater.*, 70, 309–321, November2014.
77. Kyoung Lee, H., Nam, I.-W., Kim, H.-K., Carbon nanotube/cement composites for crack monitoring of concrete structures. *Compos. Struct.*, 180, 741–750, November 152017.

78. Ali Mmohammad, M., Langaroudi, M., Mohammadi, Y., Effect of nano-clay on workability, mechanical, and durability properties of self-consolidating concrete con taining mineral admixtures. *Constr. Build. Mater.*, 191, 619–634, December 102018.
79. Koleva, D,.A,., An innovative approach to control steel reinforcement corrosion by self-healing. *Materials*, 11, 309–334, 2018.
80. Cao, Y., Dong, S., Zheng, D., Wang, J., Zhang, X., Du, R., Song, G., Lin, C., Multifunctional inhibition based on layered double hydroxides to comprehensively control corrosion of carbon steel in concrete. *Corros. Sci.*, 126, 166–179, September2017.
81. Tiwari, A. and Chowdhury, S., An overview of application of nanotechnology in construction materials, in: *Proceedings of the International Symposiumon Engineeringunder Uncertainty: Safety Assessment and Management (ISEUSAM-)*, Springer, pp. 483–492, 2012.
82. Sobhy, C.S., Tawfik, T.A., Abd El Hafez, G.M., Serag Faried, A., Insights on the influence of nano-Titanium dioxide and nano-Zinc oxide on mechanical properties and inhibiting of steel reinforcement. *Case Stud. Constr. Mater.*, 16, 1–18, 2022, Doi: 10.1016/j.cscm.2022.e01017.
83. Wang, X., Dong, S., Ashour, A., Zhang, W., Han, B., Effect and mechanisms of nanomaterials on interface between aggregates and cement mortars. *Constr. Build. Mater.*, 240, 1–17, 2020. Doi; 10. 1016/ j. conbuildmat. 2019. 117942.
84. Sumesh, M., Alengaram, U.J., Jumaat, M.Z., Mo, K.H., Alnahhal, M.F., Incorporation of nano-materials in cement composite and geopolymer based paste and mortar—A review. *Constr. Build. Mater.*, 148, 62–84, 2017.
85. Mustakim, S.M., Das, S.K., Mishra, J., Aftab, A., Alomayri, T.S., Assaedi, H.S., Kaze, C.R., Improvement in fresh, mechanical and microstructural properties of fly ash-blast furnace slag based geopolymer concrete by addition of nano and micro silica. *Silicon*, 13, 2415–2428, 2021.
86. Kelestemur, O. and Yıldız, S., Effect of various dual-phase heat treatments on the corrosion behavior of reinforcing steel used in the reinforced concrete structures. *Constr. Build. Mater.*, 23, 78–84, 2009.
87. Hegazy, M., Badawi, A., El Rehim, S.A., Kamel, W., Influence of copper nanoparticles capped by cationic surfactant as modifier for steel anti-corrosion paints. *Egypt. J. Pet.*, 22, 549–556, 2013.
88. Abd El Haleem, S.M., Abd El Wanees, S., Bahgat, A., Environmental factors a ecting the corrosion behaviour of reinforcing steel. VI. Benzotriazole and its derivatives as corrosion inhibitors of steel. *Corros. Sci.*, 87, 321–333, 2014.
89. Sheban, M., Abu-Dalo, M., Ababneh, A., Andreescu, S., E ect of benzotriazole derivatives on the corrosion of steel in simulated concrete pore solutions. *Anti-Corros. Methods Mater.*, 54, 135–147, 2007.
90. Tian, H., Li, W., Liu, A., Gao, X., Han, P., Ding, S.-J., Yang, C., Wang, D., Controlled delivery of multi-substituted triazole by metal-organic framework for inhibition of mild steel corrosioninneutral chloride solution. *Corros. Sci.*, 131, 1–6, 2018.

91. Rawtani, D. and Agrawal, Y.K., Multifarious applications of halloysite nanotubes: A review. *Rev. Adv. Mater. Sci.*, 30, 282–295, 2012.
92. Nguyen, T.A., Nguyen, T.H., Pham, T.L., Dinh, T.M.T., Thai, H., Shi, X., Application of nano-SiO_2 and nano-Fe_2O_3 for protection of steel rebar in chloride contaminated concrete: epoxy nanocomposite coatings and nano-modified mortars. *J. Nanosci. Nanotechnol.*, 17, 1, 427–436, 2017, Doi: 10.1166/jnn.2017.12396.
93. Maness, P.-C., Smolinski, S., Blake, D.M., Huang, Z., Wolfrum, E.J., Jacoby, W., Bactericidal Activity of Photocatalytic TiO2 Reaction: toward an Understanding of Its Killing Mechanism. *Appl. Environ. Microbiol.*, 65, 4094–4098, 1999.
94. Kim, H.-K., Kim, H.-R., Yoon, S.-J., Lee, K.-B., Kim, J., Kim, B.-C., Colorimetric Aptasensor for Detecting Bacillus carboniphilus Using Aptamer Isolated with a Non-SELEX-Based Method. *Chemosensors*, 9, 121, 2021. Doi:10.3390/chemosensors9060121.
95. Venieri, D., Fraggedaki, A., Kostadima, M., Chatzisymeona, E., Binas, V., Zachopoulos, A., Kiriakidis, G., Mantzavinos, D., Solar light and metal-doped TiO2to eliminate water-transmitted bacterial pathogens: Photocatalyst characterization and disinfection performance. *Appl. Catal., B*, 154–155, 93–101, 2014, Doi: 10.1016/j.apcatb.2014.02.007.
96. Lee, H.S., Balasubramanian, B., Gopalakrishna, G.V.T., Kwond, S.-J., Karthicka, S.P., Saraswathy, V., Durability performance of CNT and nanosilica admixed cement mortar. *Constr. Build. Mater.*, 159, 463–472, 2018, Doi: 10.1016/j.conbuildmat.2017.11.003.
97. Chuah, S., Pan, Z., Sanjayan, J.G., Wang, C.M., Wen, H.D., Nano reinforced cement and concrete composites and new perspective from graphene oxide. *Constr. Build. Mater.*, 73, 113–124, 2014, Doi: 10.1016/j.conbuildmat.2014.09.040.
98. Divya Bharathi, K., Udaya Bhat, K., Devadas Bhat, P., Arun Kumar, D.S., Rahman, M.R., The cohesion strength of electrodeposited Zn/GO nanocomposite coating on stainless steel. *Diamond Relat. Mater.*, 152, 1–9, 2025. Doi; 10.1016/j.diamond.2024.111922.
99. Asaada, M.A., Ismaila, M., Tahirb, M.M., Huseiena, G.F., Rajac, P.B., Asmara, Y.P., Enhanced corrosion resistance of reinforced concrete: Role of emerging eco-friendly Elaeis guineensis/silver nanoparticles inhibitor. *Constr. Build. Mater.*, 188, 555–568, 2018, Doi: 10.1016/j.conbuildmat.2018.08.140.
100. Tong, T., Fan, Z., Liu, Q., Wang, S., Tanc, S., Yu, Q., Investigation of the effects of graphene and graphene oxide nanoplatelets on the micro- and macro-properties of cementitious materials. *Constr. Build. Mater.*, 106, 102–114, 2016. Doi; 10.1016/j.conbuildmat.2015.12.092.
101. Woo, R.SC., Zhu, H., Chow, M.MK., Leung, C.KY., Kim, J.-K., Barrier performance of silane–clay nanocomposite coatings on concrete structure. *Compos. Sci. Technol.*, 68, 2828–2836, 2008, Doi: 10. 1016/ j.compscitech.2007.10.028.

102. Cai, Y., Hou, P., Cheng, X., Du, P., Ye, Z., The effects of nanoSiO2 on the properties of fresh and hardened cement-based materials through its dispersion with silica fume. *Constr. Build. Mater.*, 148, 770–780, 2017, Doi: 10.1016/j.conbuildmat.2017.05.091.
103. Oroujenia, M., Kaboudina, B., Xiab, W., Jönssonc, P., Ossipov, D.A., Conjugation of cyclodextrin to magnetic Fe3O4 nanoparticles *via* polydopamine coating for drug delivery. *Prog. Org. Coat.*, 114, 154–161, 2018, Doi: 10.1016/j.porgcoat.2017.10.007.
104. Xie, T., Ozbakkaloglu, T., Xie, T., Ozbakkaloglu, T., Behavior of low-calcium fly and bottom ash-based geopolymer concretecured at ambient temperature. *Ceram. Int.*, 41, 5945–5958, 2015, Doi: 10.1016/j.ceramint. 2015.01.031.
105. Sujoy, B. and Das, C., *Introduction to Ceramics: Fabrication, Characterizations and Applications*, pp. 1–456, CRC Press, Book 1st Edition, 2024. Doi: 10.1201/9781003470571.

Index

Printed in the USA/Agawam, MA
June 30, 2026

907143.075